PALEOCEANOGRAPHY

Paleoceanography

Thomas J. M. Schopf

HARVARD UNIVERSITY PRESS
*Cambridge, Massachusetts
and London, England   1980*

Copyright © 1980 by the President and Fellows of Harvard College
All rights reserved
Printed in the United States of America

Library of Congress Cataloging in Publication Data

Schopf, Thomas J M
    Paleoceanography.

    Bibliography: p.
    Includes indexes.
    1. Submarine geology. 2. Marine sediments.
3. Oceanography. I. Title.
QE39.S34   551.4'6   79-12546
ISBN 0-674-65215-0

Dedication

Firstly, to the freedom to follow one's intellectual curiosity as encouraged by personnel of marine laboratories where I have been lucky enough to do work: Marine Biological Laboratory, Woods Hole, and the nearby Woods Hole Oceanographic Institution; Bermuda Biological Station; Friday Harbor Laboratories of the University of Washington, San Juan Island; the West Indies Laboratory of Fairleigh Dickinson University, St. Croix; and Scripps Institution of Oceanography. Secondly, to five children with the hope that they may always enjoy the seas, past and present: Kenneth, Jennifer, Carl, Miles, and Whitney. Thirdly, to the memory of my father, James Morton Schopf (1911–1978), for his unbridled enthusiasm, unrepentant encouragement, and continued perseverance and optimism. And lastly, to Ros.

CONTENTS

ABOUT THIS BOOK

Paleontologists traditionally have two scientific interests: knowledge of fossils combined with an intimate appreciation of historical geology. For example, when the famous paleontologist Carl Dunbar was cited for his contributions, on the zoological side was his work on brachiopods, and on the geological side was his book (coauthored with John Rodgers) *Principles of Stratigraphy*. That book of the late 1950s, and others of its genre, focused on interpretations of rocks per se. However, information on the ancient oceans in which the rocks formed has grown enormously during the last two decades. Even paleontologists specializing in Precambrian times are now making paleoceanographic interpretations. One can hardly pick up any general geological journal these days and not find reference to ancient ocean circulation, climate, bathymetry, chemistry, temperature, etc.

At the University of Chicago, I have taught for the past ten years a course called Paleoceanography. Owing to the lack of *any* book in the field, assignments have been from a variety of readings in the primary literature. In this book, my aim has been to treat succinctly each of the major topics of paleoceanography (but with enough depth to equip the reader to understand the current literature). In a 30-lecture quarter, my general attack has been to divide the material as follows: ocean volume (4 lectures), bathymetry (5), water studies (4), temperature (5), chemistry (6), climatology (3), and biology (3).

Each of the seven chapters is organized in this way: each begins with a brief statement of what the present pattern of the particular oceanographic factor (such as tides, temperature, etc.) is like; this is followed by a discussion of the methods whereby ancient conditions can be estimated; the chapter then concludes with a summary of that particular aspect over geologic time. Something of the nature of the book is revealed by the fact that of the 114 figures, approximately 10 percent are world maps, another 10 percent have "latitude" as one axis, an additional 10 percent plot something against "depth," and nearly 20 percent have as an axis "m.y. before present," "b.y. before present," "b.y. after earth's origin," or "thousands of years before present." Thus about 50 percent of the illustrations are an attempt to convey broad summary knowledge.

This book began with my training in stratigraphy and my need for paleoceanographic guidelines in order to figure out early Paleozoic problems (Schopf, 1966:22). Berger (1978c) was absolutely correct that paleoceanography is the latest descendant of this most geological of all geological subjects. I have kept the needs of paleontologists specializing in Paleozoic rocks particularly in mind, and therefore have focused on what may be of use over *all* of geologic time rather than concentrating on the far more exhaustively investigated, but much shorter Pleistocene and later Tertiary. This book is primarily intended for paleontologists, stratigraphers, and geologists, although I believe that others needing this information will find the breadth of coverage of interest.

There is one basic philosophical question which must be considered because it permeates consciously—and unconsciously—the work of all those who write on paleoceanography. That is the resolution of how one treats the role of history per se (what is technically called historicity) versus the role of the steady state. This theme is repeated over and over again in discussions of ocean volume, bathymetry, water studies, temperature, chemistry, climatology, and biology. If one consciously supports a steady-state model (as I generally do), then the importance of negative feedback mechanisms becomes of critical importance. However, these mechanisms are often not understood, and if "understood" are the subject of considerable dispute. And even though the steady-state configuration may seem the most reasonable, perturbations about a "mean" value are sometimes of such enormity that non-steady-state historical factors must be considered. I find it instructive that the contrast of historical versus equilibrium views is not limited by any means to science. In a summary of Norbert Elias's magnum opus on the history of manners in Western civilization, the reviewer wrote that this book had been largely ignored by those who were "uninterested in history, regarding equilibrium rather than development as the normal state" (Thomas, 1978). Undoubtedly, many examples of these contrasting viewpoints abound in other fields of human endeavor.

In matters orthographic, "paleoceanographic" rather than "paleooceanographic" or "paleooceanographic" is preferred (contrary to previous urgings [Schopf, 1975]). The correct combining form is *pale,* not *paleo*—the extra *o* being added merely for euphonious connection, and not being required in this word. Stratigraphers may question why "early" and "late" are not capitalized, as in "early Ordovician." My feeling is that where these terms are defined and used precisely they should be capitalized, but that for general reading they need not. Finally, several other matters of expression have been discussed (often heatedly!) in the literature (including eucaryote vs eukaryote, benthic vs benthonic, and Precambrian vs pre-Phanerozoic); it is hoped readers will have forbearance and accept choices made here for the sake of discussion.

I should say something about the nature of the literature. The purpose of citations is threefold: to document the basis for arguments with reasonable fidelity to those who have done the most important work; to cite reasonable examples of the more general points; and to provide a lead into the extensive literature. Many confirming papers are not cited because coverage cannot be encyclopedic. The problem of choice is exacerbated in a book which draws upon geological, geochemical, oceanographic, atmospheric, climatological, and paleontological literature. My aim has been to try to present a balanced view, and I have included references to approximately 900 of the 3,500 papers I have used. Approximately 45 percent of the references cited are to papers published in 1974 and later; this suggests a half-life of about 5 years for the literature of paleoceanography, a rate which is characteristic of rapidly changing fields.

What this book does not have—and could not have—is an accurate set of paleogeographic maps. That field is simply too complex and changing too rapidly to include. Instead, my Chicago colleagues Alfred M. Ziegler and Christopher Scotese have been working for the past several years on an atlas of paleogeography, the first base maps for which are now prepared and published in 1979 in the *Journal of Geology* and in *Annual Reviews of Earth and Planetary Sciences*. The best procedure would be to use this book together with whatever set of paleogeographic maps is most amenable and most accurate for local use.

Any author who seeks to take several themes finds them in different states of preparedness. Some topics, like paleotides, had a burst of energy and then stabilized within the last couple of years. Others, like ocean chemistry, underwent their major conceptual development somewhat longer ago, and are generally simmering, but with new additions now and then. Yet other fields are just now in the exponential phase of growth, such as paleobathymetry applied to the deep sea with the promise of revealing ocean circulation and its sedimentological consequences over all of geologic time. The results from such rapidly growing fields are hardest to capture. Paleoceanography as a whole is like a foraminiferan with pseudopodia extended in various degrees, some moving out rapidly, others more slowly, and some being resorbed into the main body.

Several persons helped me to avoid more mistakes in fact and in judgment than I have unwittingly made. I owe a special debt of gratitude to Wolfgang Berger, who reviewed for me the whole of an earlier draft of the manuscript, and to Jan Veizer, who thoroughly examined Chapters 1, 4, and 5. I also thank for critical review Peter J. Wyllie (Chapter 1), George Platzman and Carl Wunsch (Chapter 3), and Julian Goldsmith (Chapter 5). Special assistance was provided on more limited topics by R. M. Garrels, H. C. Jenkyns, E. D. McCoy, W. S. McKerrow, W. Michaelis, J. J. McCarthy, S. Moorbath, P. J. Müller, M. Mullin, A. T. S. Ramsay, J. W. Schopf, I. B. Singh, and U. von Rad. I wish it had been possible to take all of their excellent advice. The chairmen of my department (Geophysical Sciences) at the University of Chicago have materially aided the work by allowing funds for Xeroxing relevant literature. They have in fact made the book possible by saving me an immense amount of time in these short-sighted days of shortened library hours (curiously imposed to "save money," as though the reputation of an academic institution was based on how much money it saved). In recognition of their blind faith in the efforts of a colleague, royalties have been assigned to the long-standing Gurley Fund of the Department, the purpose of which is to aid paleontological research.

Thomas J. M. Schopf

Friday Harbor
August 26, 1979

PALEOCEANOGRAPHY

1. OCEAN VOLUME

Plus ça change, plus c'est la même chose.

The earth is approximately 4.6 b.y. old, and, as will be developed, the oceans as we know them may have been in existence for half of that time. The general view is that from 4.6 to 2.5 b.y. the earth's crust and oceans exhibited unidirectional trends. From 2.5 b.y. to the present, the pattern largely has been one of recycling materials added to the surface earlier in the earth's history. Thus the earth is viewed as having an earlier period in which most of the major changes occurred and a later period in which stasis has prevailed. It is important to recall, however, that all that is meant by "stasis" is that secular change has been slower than on the order of 10^9 years, and therefore is not clearly visible in only 2.5×10^9 years.

The purpose of this chapter is to present ideas on (1) the origin of the world ocean, (2) how the present ocean volume is distributed as a function of depth (the hypsometric curve), (3) how knowledge of the thickness of the continental crust can be used to determine ocean volume in the past, and (4) what controls changes in sea level. I will also explore how the hypsometric curve and sea level may have changed over geologic time. Thus this chapter sets the stage for subsequent studies in paleoceanography.

Origin of the Ocean

The ocean came from the earth's interior; the question is, when and at what rate did this occur? The starting point for modern discussions of degassing of the earth's interior is the view presented by W. W. Rubey a quarter of a century ago: Rubey concluded that there has been a slow, steady degassing of the earth's interior over geologic time and that this process is continuing today.

More generally, three patterns have been suggested for the rate of accumulation of water in the ocean basins, as shown in Figure 1-1. In curves A, C, and D, respectively, the oceans formed rapidly

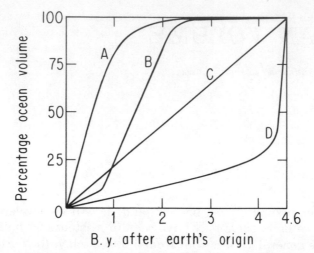

Figure 1-1 Proposed time courses for accumulation of water on the surface of the earth. Curve *A* represents a very early accumulation of ocean water. Curve *B* passes through a point based on isostatically balancing the thickness of the Precambrian crust during and after the Archean with the required volume of water (see text). Curve *C* shows a constant rate of accumulation of water (calculated to be 1 m/m.y. by Dietz, 1964a). Curve *D* represents a very late accumulation of ocean water. Curve *B* is preferred in this book.

and early (Kuenen, 1950), at a constant rate (Rubey, 1951), or rapidly and late in geologic history (Revelle, 1955). Much recent evidence has been published which allows us to evaluate these patterns, and to account for the 1.37×10^9 km³ of water in the ocean.

Water added to the earth's surface via degassing from the earth's interior is from two sources: terrestrial basalt and ocean basalt. (1) The amount of water derived from basalt which has been extruded on land can be calculated as follows. The annual total volume of new basalt is 1 km³ (both Kuenen, 1950, and Hess, 1962, accepted this estimate of Sapper [1927], which was based on the contribution of volcanics over the past 400 years). If these basalts could contribute 0.5 percent of their volume in water, and if this water is derived from the mantle, then this is an addition of 0.005×10^9 km³ of water per billion years, or 0.0175×10^9 km³ of water in the past 3.5 b.y. (2) The amount of water derived from basalt which has been extruded on the ocean bottom at ridge crests, etc., can be calculated as follows. The oldest rocks found in the oceans are from the Jurassic period, which began about 0.19 b.y. ago. Apparently, approximately every 0.2 b.y. the oceans obtain a new basaltic floor by sea-floor spreading (Hess, 1962; Dietz, 1965). This covers an area of 287.3×10^6 km^2 (Menard and Smith, 1966: Table 5). Oceanic crust is on the order of 10 km thick. Therefore, every billion years, there are 5 renewals of the ocean floor, yielding a total of 14.365×10^9 km^3 of basalt. If 0.5 percent of this volume is water, and if all of this water is derived from the mantle, then as much as 0.071×10^9 km^3 of water would be added each billion years, or 0.25×10^9 km^3 during the past 3.5 b.y.

These calculations are of *maximum* volumes of water added to the earth's surface from terrestrial and oceanic basalts. If all of the water being added from basalt to the modern ocean is merely recycled ocean water, the necessity for a large initial degassing is increased. Nevertheless, the calculation of the preceding paragraph shows that the contribution of water from oceanic basalt *could* have been approximately 10 times that from terrestrial basalt. In addition, *if* the 0.5 percent of water currently in basalt *both* is "new" water and is expelled completely from basalt (total 0.25×10^9 km^3), then the amount of water in the oceans (1.4×10^9 km^3) could *not* be accounted for by the gradual expulsion of water from basalt over the past 3.5 b.y. A more definitive statement on this topic does not seem possible at the present time.

Several lines of evidence indicate that slight degassing of the earth's interior is or could be still going on today. The basic argument is that patterns of abundance of noble and other gasses, and some other elements, correspond more closely with solar or meteorite abundances than with geologically altered terrestrial abundances, and that there must be some rejuvenating process which maintains these primitive patterns. By far the most important line of evidence is from ^{3}He/^{4}He. There is *no* clear evidence that mantle derived H$_2$O is being added to the oceans. The lines of evidence for degassing are as follows.

(1) An enrichment of ^{3}He in ocean water relative to the atmospheric ^{3}He/^{4}He ratio first was reported in 1969 by Clarke, Beg, and Craig. Since then, ^{3}He has been found to be enriched significantly relative to ^{4}He both in deep ocean waters and in glassy rims of submarine pillow basalts (summarized by Craig and Lupton, 1976). This excess ^{3}He "is due to leakage into ocean water of a remnant of the earth's primordial He^3—there appears to be no other possible mechanism" (Clarke, Beg, and Craig, 1969). The leak rate was initially estimated in 1969 at "about 2 atoms cm^{-2}sec^{-1}," and later shown to be closer to 4 ± 1 atoms/cm^2/sec (Jenkins, Edmond, and Corliss, 1978). In the North Atlantic, at 3,200 m, the ^{3}He peak is at

the same level as a "sharp local maximum in dissolved iron, copper and zinc" (Jenkins et al., 1972), possibly also of mantle origin.

More recently, ^{3}He derived from the mantle has been detected in the Galapagos rift zone and the Red Sea brines (Lupton, Weiss, and Craig, 1977a, 1977b; Jenkins, Edmond, and Corliss, 1978). ^{3}He is the best indicator that mantle degassing is still going on. It is being released at the margins of plates both in the ocean and on land (Wakita et al., 1978). The constant ratio of ^{3}He/^{4}He which is being released "implies that a remarkably constant proportion of radiogenic and primordial He concentration exists in the upper mantle" (Dymond and Hogan, 1978). Whether or not this mantle constituent also can be used as a tracer to estimate the amount of water, CO_2, or other mantle volatiles is yet to be seen.

As for the remaining points, 2 through 7, concerning present degassing, either they have been investigated to a much lesser extent than the ^{3}He/^{4}He approach, or the evidence is more seriously contested. Many of these points are still being debated, and their general acceptance or rejection is yet to come.

(2) ^{20}Ne appears to lack radioactive source, and therefore it is a tracer for primordial neon. It accompanies primordial ^{3}He (Craig and Lupton, 1976). A 2 to 5 percent enrichment in ^{20}Ne has been found in exhalants both from the Kilauea volcano and from the East Pacific Rise volcanics.

(3) ^{129}Xe is thought to have been produced from now extinct 17 m.y. ^{129}I very early in the history of the earth. ^{129}Xe apparently is still being added to the atmosphere (Butler et al., 1963; Boulos and Manuel, 1971). Indeed, Wasserburg (1964) asserted that radiogenic ^{129}Xe constituted at that time "the only evidence for the contribution of primordial gas" (The source of the Xe is disputed, however, by Fanale and Cannon, 1971.)

(4) ^{40}Ar is much more common than ^{36}Ar or ^{38}Ar in the present atmosphere. The source of ^{40}Ar is radioactive ^{40}K, and the sources of ^{36}Ar and ^{38}Ar are primordial. The present excess of ^{40}Ar may represent a historical remnant of Precambrian ultramafic rocks, and thus indicate a catastrophic early degassing (Schwartzman, 1973; Ozima, 1975). The ^{40}Ar/^{36}Ar ratio has been determined in Devonian Rhynie Chert, 380 m.y. old. The ratio is so close to modern values that it eliminates the possibility of a single-stage degassing model characterized by *continuous* significant degassing over geologic time (Cadogan, 1977). Instead, it is evidence for a two-stage model, with catastrophic early degassing followed by very gradual later degassing—the same model used later to account for ^{87}Sr/^{86}Sr ratios.

(5) Boron is enriched in basalts which have been serpentinized

in the crust. Fresh water contains very little boron (0.01 ppm) relative to sea water (4 ppm), and yet boron constantly is being removed from sea water by clays. In order to replenish boron in sea water, an additional source of boron seems to be required (Thompson and Melson, 1970). Such a source may be boron of juvenile water transferred by means of serpentinized basalts, which, on exposure, lose boron to sea water. In opposition to this view, Harriss (1969) provided a mass balance for boron which went so far as to omit submarine vulcanism as a source, asserting that "it can be neglected."

(6) A mass balance of H_2O, Cl, and Br based on their occurrence in liquid inclusions in lavas suggests that each of these has a residence time close to the age of the earth (Anderson, 1974, 1975). This may indicate that continuous gradual degassing is occurring, a conclusion supported by other mass balance calculations on Cl (Schilling, Unni, and Bender, 1978). However, the possible error in the measured values from inclusions is so large that the values do not exclude the possibility that 80 to 90 percent of the ocean water degassed early, with only 10 to 20 percent of the present volume added by gradual degassing during the subsequent 2.5 b.y. to the present.

(7) The customary view of atmospheric history is to make a comparison with the solar abundances of elements and then to note that the earth's atmosphere is depleted significantly in the nonradiogenic rare gases (Ne, Ar, Kr, and Xe), as well as H and He. If the solar abundances are the proper choice for comparison, the conclusion follows that the earth's atmosphere must be secondary (Brown, 1949), possibly having been added slowly over geologic time. However, the proper comparison for initial element abundances on the earth may not be with solar abundances, but rather with primitive earthlike materials, namely, chondritic meteorites. Such a comparison shows that nonradiogenic rare gases in fact have been conserved, not swept away by solar wind, etc., and that the earth "exhibits a normal undifferentiated meteoritic inventory in its atmosphere" (Fanale, 1971:89). Consequently, the present atmosphere (with the exception of O_2) can be thought of simply as a historical remnant of initial conditions (further discussed in Chapter 5).

Let me now examine several other lines of evidence which are consistent with the view that although *some* degassing now may be going on, the early catastrophic degassing provided *most* of the water in the present ocean.

(1) In a comprehensive simulation of the origin of present ocean chemistry, the modern composition was obtained by simultaneous solution (using a high-speed computer) of the thermodynamic reac-

tions governing chemical erosion of the "average igneous rock" (Lafon and Mackenzie, 1974). The conclusion was that chemical equilibrium can be reached rapidly. This is consistent with what is known about early ocean chemistry. The results of the simulation are described further in discussions of changes in ocean chemistry over geologic time in Chapter 5.

(2) Unique to the mineral beryl is the property of accumulating rare gas in channels parallel to the c-axis, at the time of beryl crystallization. The comparison of substantial amounts of ^{40}Ar and ^{4}He in ancient (> 2.5 b.y.) beryl with an order of magnitude lower values in more recent (< 1.0 b.y.) beryl indicates a substantial rate of degassing prior to 2.5 b.y. (Fanale, 1971).

(3) Remnant magnetism in Precambrian rocks 2.5 to 2.7 b.y. old indicates that core formation had occurred even earlier (Hanks and Anderson, 1969). In fact, geochemical differentiation of Archean rocks is advanced sufficiently to indicate that "core-mantle differentiation was already essentially complete" in the Archean (Cloud, 1971b:851). And "once core formation begins, enough gravitational energy is released to carry it quickly to conclusion, and an almost completely molten upper mantle must result" (Fanale, 1971:92). This suggests a catastrophic early degassing.

(4) $^{87}Sr/^{86}Sr$ values of Archean sedimentary rocks (3.8 to 2.5 b.y.) are like those of the mantle ($\sim.703$; see also p. 187). Later strontium isotope values are intermediate between mantle values and (Rubidium rich) continental rocks ($\sim.720$). The interpretation given to this (Veizer, 1976b) is that values for Archean sedimentary rocks represent weathering from Archean *mafic* "continents" and ocean basins and that the later values are owing to a mixture of strontium isotopes of both low (basaltic) and high (granitic) values (Davies and Allsopp, 1976).

A crust of constant volume during the past 2.5 b.y. (but one that is being generated constantly and then being eroded and recycled again) can account for the substantial differences in the strontium and in the lead isotopic compositions of oceanic and continental crust (Armstrong and Hein, 1973; and Armstrong earlier). The rate of mixing of continent, ocean basin, and mantle inherent in plate tectonics is sufficient to account for observed isotopic patterns, given stability of the system for some 2.5 b.y. If one considers only the lead isotopes of North America, this leads to the result that "the bulk of the North American continent was formed during the interval 3500 − 2500 m.y. ago" (Patterson and Tatsumoto, 1964). If the volume of the world ocean has been close to its present value for about 2.5 b.y. (see below), then this, together with the results of the isotopic mixing models, leads logically to the conclusion that most of the earth's degassing occurred early in its history.

(5) Hydrogen and oxygen isotopes have been measured in igneous rocks which were subject to meteoric-hydrothermal influences. For samples from the Precambrian at 0.65, 1.3 to 1.5, and 2.5 b.y., the very oldest gneisses have anomalously low values (of deuterium). This led Taylor (1977) to support the interpretation that the water which was involved in their metamorphism was "juvenile water." He wrote, "The presence of such H_2O could be accounted for if these ancient granitic magmas formed from an upper mantle not yet heavily contaminated with hydroxyl minerals from subducted oceanic lithosphere." All of the younger rocks have the more usual isotopic values. Therefore, Taylor favored a major isotopic fractionation during "early high-temperature out-gassing."

In summary of the recent arguments concerning degassing of the earth's interior, some evidence indicates that continual but very slight, and slow, addition of juvenile materials to the earth's surface probably is going on today. This certainly seems reasonable, for, as Cloud (1976b) wrote, "the primitive Earth cannot have been completely outgassed, for a dry planet would be incapable of the subsequent tectonic history observed on ours." The main evidence that degassing of the mantle is *still* going on comes from the finding that gases from basalts which have been erupted onto the ocean floor have compositions unlike those of the present ocean or atmosphere. This is seen most strikingly in (1) the $^3He/^4He$ ratio, which near plate boundaries is enriched considerably in ^3He relative to atmospheric background. Other but weaker lines of evidence of degassing have focused on the addition of (2) ^{20}Ne, (3) ^{129}Xe, (4) ^{40}Ar, and (5) B. In addition, there are (6) arguments from mass balance of H_2O, Cl, and Br from liquid inclusions, and (7) notions on the origin of the atmosphere which are consistent with "remnant degassing."

Rapid degassing of the earth's interior would have resulted in (1) relatively rapid equilibrium with an ocean chemistry like that of the modern world. Rapid degassing is consistent with all of the lines of evidence cited above, and is supported (2) by the finding of excess volatiles in samples of the mineral beryl which were formed at or more than 2.5 b.y. ago, (3) by the formation of the core (and hence thorough mantle melting) before 2.7 b.y. ago, (4) by a mass balance of lead and strontium isotopes, and (5) by deuterium values in ancient gneisses.

The general conclusion supported here is that most of the mantle degassing occurred early in the earth's history, between 4.6 and 2.5 b.y. This is the chief period of rapid continental and oceanic evolution. Subsequently, the earth's crust and oceans have been slowly recycled.

Hypsometric Curve

The cumulative curve which shows the percentage of land at different elevations is known as the hypsometric curve (*hypso* = height) (Figures 1-2, 1-3; Table 1-1). The hypsometric curve for the earth as a whole has two steps, as was realized by John Murray from soundings of the Challenger Expedition more than a century ago. During the earlier part of the twentieth century, Alfred Wegener emphasized in *The Origin of Continents and Oceans* (4th edition, 1929) that the major implication of the hypsometric curve is that the two levels represent rocks of different density. The plateau in elevation near sea level reflects a light block composed of granite (density ≃2.8). The second plateau at a depth of about 4 km represents a heavy block which is basalt (density 3.3). These masses of rock of

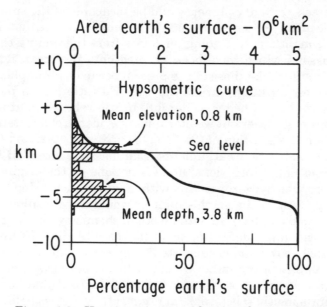

Figure 1-2 Hypsometric curve showing the cumulative percentage of the earth's solid surface at any given level of elevation or depth. At the left is the frequency distribution of elevations and depths for 1 km intervals. Note the concentration of land at the two levels of 0 to +1 km, and −4 to −6 km. Data from Sverdrup, Johnson, and Fleming, 1942:19.

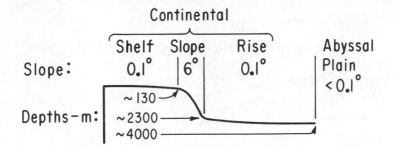

Figure 1-3 Physiographic regions of the continental margin. Note (in the text) the considerable variation in both the slopes and the depths of the different regions. Data from Emery and Uchupi, 1972; Drake and Burk, 1974

different density and chemical composition are in isostatic equilibrium relative to the outer rim of the earth in which they "float."

In contrast to the hypsometric curve of the earth, that of the moon does not show any "steps" (Figure 1-4). This implies that the lunar surface is not differentiated into discrete blocks of different chemical composition and density. Without clearly established differentiation of the crust into blocks of different chemistry, the lunar

Table 1-1 Depth distribution of world ocean. After Menard and Smith, 1966:4314.

Depth, km	Area		Cumulative area (shallower than deeper limit of depth interval)	
	10^6 km²	Percent	10^6 km²	Percent
0–0.2	27.123	7.49	27.123	7.49
0.2–1	16.012	4.42	43.135	11.91
1–2	15.844	4.38	58.978	16.29
2–3	30.762	8.50	89.740	24.79
3–4	75.824	20.94	165.565	45.73
4–5	114.725	31.69	280.289	77.42
5–6	76.753	21.20	357.042	98.62
6–7	4.461	1.23	361.503	99.85
7–8	0.380	0.10	361.883	99.96
8–9	0.115	0.03	361.998	99.99
9–10	0.032	0.01	362.031	100.00
10–11	0.002	0.00	362.033	100.00

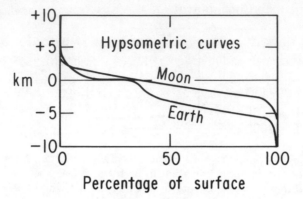

Figure 1-4 Hypsometric curve of the moon compared with that of
the earth. The moon's curve shows no preferential levels, whereas
the earth's reflects two: ocean basins and continents. The moon has
lacked a process which differentiates its crust into various levels.
After Dietz and Holden 1965:637.

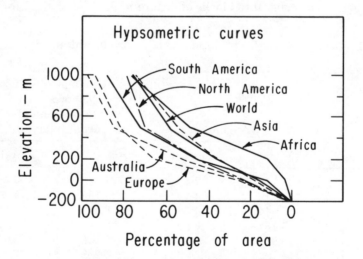

Figure 1-5 Hypsometric curves of the 6 individual continents and
the earth total between −200 m and +1 km. For each continent, the
area between the edge of the continental shelf (taken as −200 m) and
the highest land elevation has been normalized to equal 100 percent.
Note that Africa stands high, but that other continents are very similar to each other. After Forney, 1975.

surface probably is not differentiated into discrete "plates," such as exist on earth.

The shape of the hypsometric curve has several additional implications for geology. The portion of the hypsometric curve of greatest interest for shallow inland seas is between − 200 and + 200 m. Africa may have undergone a late Tertiary uplift (Bond, 1978a) relative to the Americas, Australia, and Europe. Except in Africa, a eustatic change in sea level would transgress a similar percentage of land area, up to a coverage of 30 to 40 percent (Figure 1-5). Above this point, the hypsometric curves for different continents strongly diverge. Thus if 75 percent of one continent were covered (as is claimed for North America in the Ordovician or Cretaceous), that might be equivalent to less than 40 percent coverage on another continent.

During the Permian period, the area of continental regions covered by shallow marine seas was reduced from approximately 45 percent in the early Permian to less than 15 percent in the latest Permian (Schopf, 1974). A change in sea level of approximately 310 m is required to effect this change in area on the present hypsometric curve (Forney, 1975; see Table 1-2). Bond (1976, 1978b) used the hypsometric curve to estimate that Cretaceous seas rose about 200 m in order to account for the areal extent of the flooding. Owing to the fact that erosion always will reduce the continental block to base level (sea level), the hypsometric curve for the Permian or Cretaceous world certainly was approximately like that of the present time. In general, the area covered by shallow marine seas can be estimated from changes in sea level by using the hypsometric curve.

Table 1-2 Percent coverage of continental crust by shallow marine seas for stages of the Permian. For comparison, the present coverage by shallow seas of continental crust is approximately 28.3 × 10⁶ km², or 16 percent. Data from Forney, 1975:775.

	Area of shallow marine seas × 10⁶ km²	Percent coverage of continental crust	Permian sea level relative to present sea level (m)	M.y. after beginning of Permian
Permian A	72	41	+250	14
B	65	37	+200	24
C	59	34	+170	38
D	23	13	−40	46
Triassic	61	35	+180	56

The width of any particular continental shelf may vary according to the tectonic state (documented by Hayes, 1964). Leading
edges of continents have narrow coasts (less than 6.4 km wide) and
are associated with young mountain ranges. Rocky bottoms are
most common on shelves less than 4.5 km wide. The tectonically
stable coasts of trailing edges of continents are on average approximately 18 km wide and have the highest frequency of shell debris
(but climatic influences can be as or more significant than the tectonic state, as discussed on p. 193). Regions with large amounts of
sediment are on average 73 km wide and are characterized by sand
and mud. Thus tectonic setting is important for predicting whether
a shelf is a site of erosion or of deposition (summarized by Emery,
1970), and if depositional, the type of the sediment which is deposited.

The depth of the shelf break varies considerably (from less than
10 m to more than 500 m), with a world mean of 132 m (Shepard,
1963:257). Off the east coast of the United States the shelf break
now changes from nearly 300 m in Labrador, where the shelf margin appears to be going down in response to the rising of the adjacent land area (because of glacial rebound), to less than 10 m where

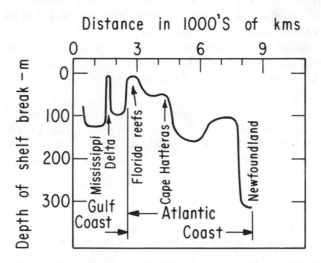

Figure 1-6 Depth of shelf break between Labrador shelf and the
western part of the Gulf of Mexico. Greater depths of shelf break in
the north are believed to result from sinking of the shelf as the adjacent land surface rises because of isostatic rebound; shallower depths
in the south are caused by rapid deposition of sediment. After Emery
and Uchupi, 1972:22.

the Mississippi delta has built out (Figure 1-6). There is enormous variation in shelf width along a single continental margin.

Seaward of the shelf break is the continental slope (Figure 1-3), and it too varies considerably in width and steepness (average of 100 km wide, ≃6° dip, with common range 3° to 12°, for the Western North Atlantic; Emery and Uchupi, 1972:53). Slopes on the order of only 1° to 4° appear sufficient to permit extensive slumping (Figure 1-7; also see Lewis, 1971). Shear stresses may be generated by an external source such as earthquakes, but strong currents at times of highest river runoff also are effective, as off the Magdalena River and the Congo River (Heezen, 1963). The effect of such a shear stress on mass movement of deep-sea carbonate may be enhanced greatly by dissolution processes both for large-scale slumping and sliding and for creep of near-shore sediment (Johnson, Hamilton, and Berger, 1977).

Slumping emplaces large volumes of sediments on the lower continental slope and the continental rise and may be the chief

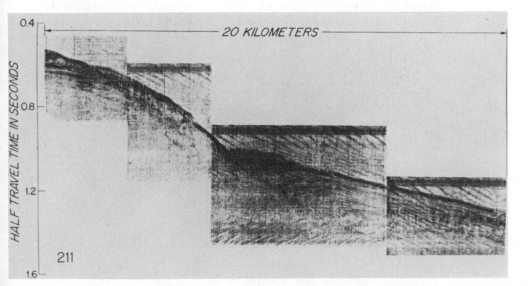

Figure 1-7 Photograph of seismic profile from the continental margin southeast of New England to illustrate sediment slumping at a low angle of repose. 1 sec half travel time = 750 m in water, or about 1 km in sediment. Travel time is used because sound velocity changes from layer to layer and is not precisely known. From Uchupi, 1967:facing p. 636. Reproduced by permission of Pergamon Press and E. Uchupi.

mechanism for the deposition of sediments at the base of the continental slope. Paleoslopes of deposition can be determined from information on upper and lower levels of deltas. The Bordon siltstone (Mississippian) of Illinois has foreset beds with slope of 6 to 19 m/km ($\sim 1°$) which extend for several kilometers (Swann, Lineback, and Frund, 1965). In shelf sediments, a slope of 1 to 2 m/km ($\sim 0.1°$) is estimated for western United States upper Cambrian deposits (Lohmann, 1976).

The isostatic balance which is responsible for the hypsometric curve has smaller-scale implications. Vogt (1974) suggested that the height of a volcano was limited by the thickness of the plate from which it was drawing its isostatic support. As the oceanic crust thickens away from the ridge crests, the height of volcanoes increases.

The general two-step *shape* of the hypsometric curve will remain the same as long as there are blocks of different density which "float" in the mantle. The absolute relief between "continent" and "ocean basin" will depend upon their absolute difference in density, as shown in Figure 1-8. If blocks are differentiated chemically only slightly, little difference in density will exist, and little relief will be

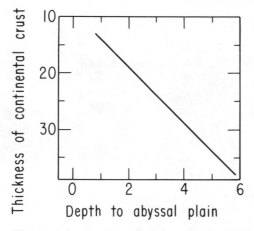

Figure 1-8 Graph portraying depth of the continental crust (i.e., depth to the M discontinuity under continents with 0.3 km above sea level as the reference elevation) versus depth to abyssal plain in oceans, computed from balance of crustal columns. The thicker the continental crust, the greater the difference in density between continents and ocean basins, and the greater the depth to the abyssal plain. After Hess, 1962:616.

found. In such a situation, with very little relief, the ocean "basin" will be very shallow. At present, the continental and oceanic blocks are differentiated chemically to a large extent, and the ocean basin is relatively deep. The degree of differentiation is reflected in the thickness of the continental and oceanic crust. On the average, as continental regions become concentrated in lighter granite, their "root" increases in thickness. At the same time as light material is transferred to continents, oceanic crust is enriched in heavier minerals and the thickness of oceanic crust is reduced. Thus if one knows the thickness of the continental crust, one can use isostatic relationships to estimate the relief between the average continental elevation and the average oceanic depth.

If one knows the average depth of the ocean basins, then on a globe of fixed size, this determines the volume of the ocean. Consequently, if we can chart the history of the thickness of continental crust over geologic time, then we have a gauge on both the average depth of the ocean and the average volume of the ocean.

Thickness and Extent of the Continental Crust

The earth is 4.6 b.y. old, but the oldest rocks are approximately 800 m.y. younger, at 3.8 b.y. old (Moorbath, 1976). Because of the increasing chance for older rocks to have been eroded away, or to have become deeply buried, one may not expect to find many of the rocks in the 4.6 to 3.8 b.y. interval. However, rocks 3.8 to 2.5 b.y. old (the "Archean" interval) occur in North and South America, Greenland, Scandinavia, Africa, India, the Soviet Union, Antarctica, and Australia, and one would expect to find older rocks if they existed.

One possible explanation for the absence of these older rocks is that they exist, but are unrecognized owing to an intense metamorphic "thermal event" at about 3.6 b.y. which "reset" radiometric clocks. Some support for this is provided by examination of the oldest rocks in North America (3.55 b.y. old granite gneisses in Minnesota), which intrude basalt lavas. The rocks in this area "are similar to more recent crustal rocks and do not represent a protocrust" (Goldich, Hedge, and Stern, 1970). In addition, "even the oldest rocks from Greenland indicate heterogeneous U and Pb distribution prior to 3800 m.y. ago" (Oversby, 1978:237).

A second possible explanation for the absence of 4.6 to 3.8 b.y. old rocks is that this is the interval of time which was required for formation of a protocrust. Shaw (1976) proposed that following core formation and mantle solidification the whole earth was covered by a thin (14 km) layer of continental crust overlying basalt. Calcula-

tions of the length of time required for the formation of a protoarchean crust were 0.5 to 1.0 b.y.

Archean rocks include gneisses and a unique series of green-
stones (= metabasalts and andesite volcanics) intruded by granites.
These rocks are estimated to cover at least 20 percent of the area of
present-day continents (Engel, 1969:265), and North America had
at least 50 percent its present areal size 2.5 b.y. ago (Muehlberger,
in Cloud, 1971a). A few of the lava flows have pillow structures and
quench features that indicate extrusion below the sea surface. Results of studies of phase relations for a 3.3 b.y. old peridotite indicated a temperature of extrusion of 1,650°, about 400° higher than
that for comparable types of rock forming today (Green et al., 1975).

If this high temperature at a shallow depth is a true indication
of a steeper temperature gradient of Archean time (see p. 141), then
granite may not have been able to become completely differentiated
from rocks of basaltic composition. Accordingly, this crustal material of intermediate composition would not have been dense
enough to be subducted into the mantle (Green, 1975). This crustal
material could have accumulated at shallower depths, at least until
this protocrust reached a thickness of about 25 km (Veizer, 1976a).
At that thickness, crustal melting would have released granitic liquids and the differentiation process would continue.

As the crustal temperature gradient lessened during the Archean, the density difference between basaltic and granitic rocks became more important, and reprocessing protocrust would have led to
modern continent and ocean basin differentiates. Veizer (personal
communication, 1978) reported that his model calculations (1976b)
indicated that an equivalent of about two-thirds of the present-day
area of granitic crust appeared in the Archean. The basic physical
property which drives the machinery forward is the density difference which develops as the geothermal gradient lessens and the
rocks become chemically differentiated.

For the *early* part of the Archean, several authors suggested
that such continental crust as was then in existence was thinner
($\simeq$20 to 25 km) than the present-day crust (38 km), and that thickening took place during the Archean (Condie and Potts, 1969;
Condie, 1973; Green, 1975; Naqvi, Rao, and Narain, 1978; Tarney
and Windley, 1977).

An example of proposed crustal thickening during the middle to
late Archean is the South Pass area, a greenstone belt of southwestern Wyoming. In this region, the crust thickened from "about 15 to
at least 35 kms in 300–500 $\times$ 10^6 years" (Condie, 1972:111). Similar inferences indicating successive thickening of shield areas were

described for the Archean Swaziland System in South Africa (summarized by Engel, 1969) and for the Archean gneisses in West Greenland (Myers, 1976). Alternatively, Archean continental crust 2,800 m.y. old in West Greenland has been estimated to have been "ca. 35 km" thick (Wells, 1976), an estimate based on coexisting mineral phases in this metamorphic terrain. If so, then Archean continental crust may not have been significantly thinner than modern crust—there was just much less of it until the Archean was over.

One indication of massive early formation of continents is seen in the $^{87}Sr/^{86}Sr$ ratios. The ~ 2700 m.y. old Archean granite-gneisses and greenstone volcanics have values typical of the mantle rather than of the older sialic basement rocks. Moorbath et al. (1977) concluded that their study of Rhodesian Archean rocks "demonstrates again the widespread, major production of continental crust from mafic lithosphere or upper mantle at discrete, widely separated times during the Archean, when continental accretion greatly predominated over continental reworking."

A second indication of massive early formations of continents is based on the observation that present-day spacing of volcanoes is correlated with the thickness of crustal thickness (Vogt, 1974; Mohr and Wood, 1976). Volcanoes of the major North American greenstone belt (the Abitibi of southern Canada) are spaced such that the crustal thickness was probably 34 to 45 km (Windley and Davies, 1978). These values, however, are 5 to 10 km greater than most estimates made from geochemical criteria.

In general, the Archean interval appears to mark the widespread development of continental crust 35 to 40 km thick. The origination of widespread typical granites may have begun about 3.0 b.y. ago in South Africa (see Veizer, 1976a:21); the rate of origination of granites then increased, and approximately "50 to 60 percent of all granitoids of the earth's crust have an age of 2.7 to 2.6 billion years" (Ronov, 1968). The rate of production of granites then rapidly decreased. The curve for the rate of production of granites is therefore sigmoidal—a typical shape for a growth curve with saturation (e.g., bacteria growing on a limited nutrient). In the case of granite production, the limiting factor may be the amount of K in the crust. Once the crust is sufficiently thick to form granites ($\simeq 3.0$ b.y.), the process goes rapidly to completion ($\simeq 2.7$ to 2.5 b.y.) as the required "nutrients" are utilized. Subsequent granites are then largely dependent upon recycling of these initial "nutrients."

Let us now consider the consequences of crustal thickening for the volume of water in the ocean. As shown in Figure 1-8, an increase in crustal thickness from 15 or 20 km in the early Archean to approximately 35 km at the end of the Archean corresponds to an increase in depth to abyssal plain from 2 to 5 km. This change in ocean volume is depicted in the sigmoidal shape of curve B of Figure 1-1. Shallow-water deposits in the Archean are indicated by detrital sediments, algal beds, and shrinkage cracks (von Brunn and Hobday, 1976). Therefore, Archean continents, like their modern descendants, were at base level.

If this historical summary generally is correct, it negates strongly the suggestion that the volume of the oceans increased some 25 percent (or even more) during the past 100 m.y. (curve D, Figure 1-1). In addition, little if any support is given to the idea that the earth has been expanding (Carey, 1976; Steiner, 1977). Instead, it appears that the continental crust and ocean volume increased rapidly during the Archean, and that the volume of the ocean has been in a steady state for the past 2.5 b.y., as depicted in curve B, Figure 1-1.

Although much if not most of the continental crust was formed by 2.4 b.y. ago, some has been added subsequently. The accretion is said to occur by addition to continents of lavas extruded in island arcs and derived from subducting oceanic crust, which in turn was derived from subridge mantle (Jakeš and White, 1971). Condie (1976:231) calculated that North America has grown by about 20 percent over the last billion years, and Rogers (1977) found that "new sial" of late Paleozoic-early Mesozoic time represents approximately 5 percent of the western part of North America. One cannot doubt that some continental accretion occurs. For example, the Kodiak Islands south of Alaska have been added since Mesozoic time (Carden et al., 1977). And andesitic volcanism "apparently involves both addition of new material to the crust and fractionation of preexisting crust" (Eichelberger, 1978).

The question is, how much continental growth has been caused by continental accretion over the past 2 b.y.? Whether it is 2 percent, 20 percent, or 40 percent is unresolved. In addition, whether crustal formation has been accompanied by continental destruction and continental recycling is also open to question, although the bulk of present evidence favors the case for continental permanence (Moorbath, 1977).

To summarize: Archean was a time with a high geothermal gradient, worldwide magmatism, and few continental platforms (Strong and Stevens, 1974). The world then passed from the Archean into the Proterozoic and its lower geothermal gradient, linearly arranged magmatic zones, widespread continental platforms and modern sized ocean basins.

Changes in Sea Level

Changes in ocean depth can be both apparent and real. Oceanwide average changes in mean sea level are referred to as eustatic changes. They cannot be of the same magnitude everywhere owing to the irregular distribution of gravitational effects on the ocean surface (Mörner, 1976). In addition, continental shelves are themselves not fixed and exhibit much variation in degree of uplift and subsidence, most recently in regard to glacial loading and subsequent rebound. Climate also affects ocean volume through changes in evaporation-precipitation regimes.

At least six factors can lead to the fact (or appearance) of changes in sea level which are significant geologically. Major changes in sea level on the order of as much as 100 to 500 m may have four possible causes (1 to 4, as follows).

(1) The transfer of water from the ocean to the formation of continental glaciers, and back again at the time of melting, may cause 100 to 200 m of sea-level change, and requires approximately 10,000 to 20,000 years. At the time of maximum extent of Pleistocene glaciers, sea level was lower, but the apparent lowering may be 60 to 150 m depending upon the part of the coast which is examined. From region to region, the record appears to be so variable that no single world-wide eustatic sea-level curve seems to apply, given present data (J. D. Milliman, personal communication, 1979). If all of the ice in Greenland and Antarctica were to melt, sea level would rise approximately 60 m. Periods of extensive glaciation have occurred four times in the last billion years (late Precambrian, late Ordovician–early Silurian, late Carboniferous–early Permian, and the Miocene–Pleistocene interval).

(2) The second major cause of a change in sea level is related to the rate of sea-floor spreading, a process which has continued for at least 1 b.y., and possibly for 2.0 to 2.4 b.y. As material is added at ridges, it is hot and relatively buoyant. As time passes, the lithospheric plate moves laterally, cools, becomes denser, contracts, and sinks. If the rate of sea-floor spreading is very rapid, the ocean bottom at any given distance from the ridge crest does not have as much time to cool as it would have had under "normal" conditions, and the ocean bottom stands relatively higher. Since there is a fixed volume of sea water, the ocean surface also is raised, and ocean water spills onto the craton. Conversely, during times of a very slow rate of sea-floor spreading, ocean bottom has more than a "normal" time to cool, with the result that the ocean bottom subsides to a depth greater than "normal." During such periods the oceans drain off the continent back into the ocean basins.

The critical factor in this method of controlling sea level is the changing rate of subcrustal heating (Russell, 1968). Simply to produce oceanic ridge (and thereby displace water) has no effect on sea level because for every km^3 of ridge produced, a km^3 of ridge must be consumed in a steady-state world. Since we have had neither complete flooding nor complete drying up, we can conclude that ridge production and ridge consumption must be in steady state. The explicit coupling of changes in rate of sea-floor spreading to changes in sea level over geologic time was independently developed by several persons in the early 1970s (summarized by Schopf, 1974:137).

Variation in the rate of sea-floor spreading may be the major long-term control over changes in sea level of 300 to 500 m, over time periods of a few to several million years. The change in sea level during the Permian could have resulted from an average reduction in the rate of sea-floor spreading of 5 cm/year (Forney, 1975). More generally, Berger and Winterer (1974:26) estimated that "a 10% change in spreading rate, persisting for 10 million years, would produce a 20 m change in sea level."

It would be useful to have a way to distinguish changes in sea level caused by glaciation from those caused by changes in rate of sea-floor spreading. Changes in sea level associated with glaciation alter the density of the surface of the ocean. During the maximum extent of glaciation in the Pleistocene, ocean salinity was increased by approximately 1‰ (see p. 184). On melting, sea surface water should have a reduced salinity owing to the melt water. Accordingly, during times of maximum density, water is added to the deep ocean, and during times of minimum density, the deep ocean is not renewed to the same extent. This difference might have a sedimentological signal of alternating oxidized shales and unoxidized, organic-rich shales. Conceivably, the black shales of northern Great Britain of the latest Ordovician represent times of glacial retreat during that period of glaciation. If so, the apparent correlation of these black shales with transgressions (Leggett, personal communication, 1979) is merely the by-product of a reduced sea-surface density as the glaciers melted and sea level rose.

(3) A third possible cause of a major change in sea level is simply the displacement of ocean water by eroding continents. To raise sea level by 1 m, the area of the ocean (362×10^{12} m^2) needs to be displaced 1 m, or, we need 362×10^{12} m^3 rock. One m^3 rock (density 2.7 gm/cc) is 2.7×10^6 gms, so we need a total of 9.8×10^{20} gms of rock to raise sea level 1 m. The annual rate of addition of sediment to the ocean (preman) is approximately 2×10^{16} gms (1.8 from mechanical load, 0.4 from chemical load [Garrels and Mackenzie,

1971a:120–121]). Thus displacement of 1 m of sea water at this denudation rate would require approximately 50,000 years. A 100 m rise in sea level by this mechanism would take 5 m.y., which is rapid on a geologic time scale. Counterbalancing the displacement of water by the added sediment is the isostatic sinking of sedimentary basins under the increased weight of sediment. This sinking may reduce greatly the effectiveness of this method of changing sea level, but further analysis of the topic seems to be particularly speculative and is left to the reader to pursue.

(4) The drying up—or filling up—of a large basin like the Mediterranean transfers water to and from the world ocean. Complete drying up of the Mediterranean (as may have occurred during the late Miocene) has been estimated to be able to cause a raising of sea level of 10 m (Berger and Winterer, 1974) and a lowering of sea level by that same amount when the Mediterranean again filled. An isolated North Atlantic in the middle to late Jurassic (Sclater, Hellinger, and Tapscott, 1977) was larger than the present Mediterranean with correspondingly greater potential for altering sea level. Possibly the worldwide regression at the end of the Callovian or in the latest Jurassic was related to the filling of the North Atlantic at the time of continental breakup.

(5) Local or regional (but not worldwide) changes in sea level have been attributed to two causes (here given as 5 and 6). The first of these is that differences in the oceanic age-depth curve (p. 48) may be caused by elevation changes in "bumps" in the underlying asthenosphere (Menard, 1973). These bumps have wavelengths of 1,000 to 2,000 km, and a relief of a few hundred meters, and may account for rapid changes in some age-depth curves. Plates that were moving up or down such bumps would have deposits that appeared to reflect a raising or a lowering of sea level. This may account for anomalous regions (such as Tethys during the latest Permian) which continued to receive sediment while all other regions of the world's continental margin became terrestrial.

(6) A phase change at the M discontinuity corresponding to the transitions between basalt and eclogite "offers a promising explanation for the occurrence of great thicknesses of shallow-water sedimentary deposits" (Joyner, 1967). A series of computer simulations showed that basins 500 m deep may accumulate 4 to 6 km of shallow-water sediments, and basins 1.5 km deep may accumulate 7 to 10 km of clastic sediments, before uplift occurs. The model is appropriate for a regional basin, but not adaptable to a broader scale (summary in Wyllie, 1971:238).

A change in sea level through geologic time has been one of the most interesting topics to geologists for a century. Many "correlations" (but none with any statistical support) have related periods of mountain building to times of rising sea level. The causal mechanism now postulated for this is that times of rapid sea-floor spreading are times of transgression (for reasons reviewed above) as well as times of rapid subduction (which in turn leads to more extensive mountain building along leading margins of continents).

Several authors have presented data on the percentage flooding of continental regions over geologic time (Table 1-3). Most authors

Table 1-3 Estimated areas of Phanerozoic continental seas, in 10^6 km². Data are given with present sea level as the zero base line; present sea level already covers 28.3×10^6 km² of continental crust. The total area of continental crust is approximately 176×10^6 km². Data from Sepkoski, 1976:299; see also Ronov et al., 1977.

Stratigraphic interval	Estimated areas $\times$ 10^6 km²		
	Ronov	Strahov	Termier
Q	5	2	0
Neog.	—	—	10
Paleog.	45	33	20
UK	62	55	35
LK	42	24	28
UJ	33	39	27
MJ	28	—	26
LJ	24	—	24
UT	22	27	16
MT	20	—	22
LT	17	—	18
UP	21	19	21
LP	37	—	38
UC	38	50	31
LC	48	53	42
UD	52	59	33
MD	54	—	42
LD	22	—	28
US	—	—	55
LS	—	—	52
Uθ	—	—	34
Mθ	[60]	72	55
Lθ	—	—	40
Uϵ	[56]	65	44
Mϵ	—	—	42
Lϵ	—	—	45

agree that times of maximum transgression included the middle to late Ordovician, middle to late Devonian, middle Mississippian, middle Pennsylvanian and middle to late Cretaceous. Intervening times, especially those associated with the close of geologic periods, usually are associated with times of maximum regression. The general pattern of greater transgression with increasing geologic age chiefly may reflect the fact that older geologic periods tend to be longer (Veizer, 1971). Certainly the shape of the sea-level curve is strongly influenced by inclusion of data for the past 10 million (or so) years (a very short geologic interval). Exclusion of those data and the use of consistently long geologic intervals reveal no striking secular trend in aerial coverage of shallow seas over geologic time.

Periods of major marine regression also have been linked with episodes of volcanism (Moore, 1976). The mechanism envisaged is that *regression* of sea water off continental margins permits isostatic rebound and marginal unwarping. This in turn may trigger stresses in the lower lithosphere which initiate volcanism. Alternatively, major marine *transgressions* can be related to periods of rapid sea-floor spreading, which in turn should result in more subduction and in more extensive orogeny (Turcotte and Burke, 1978). There is no clear-cut answer to the question of a causal relationship between sea-level changes and vulcanism.

On a worldwide basis, there is a very simple but effective way to obtain data on absolute changes in sea level. Given worldwide continental reconstructions plotted on an equal-area projection, one then (1) marks the region of marine deposits, (2) cuts the whole map into marine and nonmarine portions, and (3) weighs them on an analytical balance. Relative to the total weight, one can then calculate the percentage of land area occupied by marine seas. Then, using the hypsometric curve, one can easily read off the sea-level position necessary to cover that percentage of the land. By doing this through successive intervals of approximately the same geologic duration, one can plot successive changes in sea level. The longer the time interval chosen, the greater will appear the extent of transgressions (see p. 69). But the shorter the time interval, the more "local" factors of preservation, etc., are emphasized. Possibly the level of geologic stage, an interval often of approximately 10 m.y., is the optimum choice given present knowledge of the geologic record.

This method has been effectively used for successive stages of the Permian and the Cretaceous (see pp. 10–11) and as more accurate continental reconstructions become available for other intervals of geologic time, this should be the method of choice for determining sea-level changes on a global basis. The accuracy is probably no better than ± 10 to 20 percent, but no method is better and by utilizing worldwide data this method reduces conventional reliance on a few well-studied sections. Even McKerrow's (1979) detailed and excellent analysis of Ordovician and Silurian changes in sea level (based as it is on correlation of sections with synchronous changes in depth-related brachiopod communities) must sort out local, regional, and global changes for each section. Correlation of sections is sometimes a matter of dispute, and, in any event, no quantitative data have been obtained on the extent of areal coverage or on absolute changes in sea level.

Vail, Mitchum, and Thompson (1978) used geologic interpretations of seismic profiles (i.e., seismic stratigraphy) as evidence for

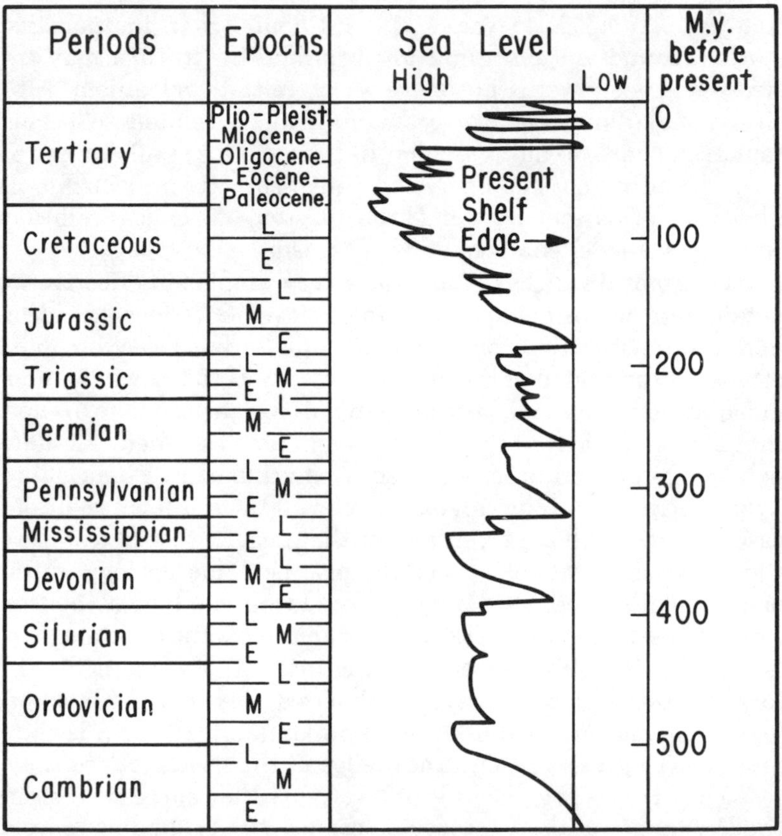

Figure 1-9 Inferred sea-level curve for Phanerozoic time. Note that transgressions were mapped as very slow and regressions as very rapid. After Vail, Mitchum, and Thompson, 1978.

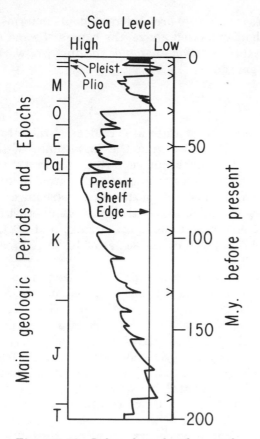

Figure 1-10 Inferred sea-level curve for past 200 m.y. Arrows along the vertical axis mark major changes. Note that transgressions were mapped as being very slow and regressions as very rapid. After Vail, Mitchum, and Thompson, 1978.

worldwide synchrony of changing sea levels. Their interpretation of relative (not absolute) sea level is shown in Figure 1-9 for the Phanerozoic and in greater detail in Figure 1-10 for the Jurassic to Recent. In their article (and companion chapters in the same volume), these authors provide the most detailed predictions of changes in sea level yet attempted for the Mesozoic and Cenozoic. In every instance, the lowering of sea level is considered to have been fairly rapid, and the rise of sea level to have been comparatively slow. However, this may in part be due to the fact that regressions are not recorded until sea level is so low as suddenly to have eroded coarse sediments beyond the shelf break into deeper water. Although many if not most of the changes in sea level are understandable in terms of one or more of the mechanisms cited above, the causes of some regressions remained unexplained (for example, the intense, worldwide early Oligocene regression).

Summary

The most important topics in this chapter are (1) ideas on the rate of addition of sea water to the oceans and the importance of the ^{3}He/^{4}He evidence to show that degassing is still going on; (2) the hypsometric curve and the way this simple plot can be used to infer the ocean's volume and basic properties of the earth's structure; (3) the significance of finding the thickness of the continental crust for determining the history of the volume of the oceans; and (4) the major causes of sea-level changes for time scales of 10^3 to 10^6 years and for magnitudes of 100 to 500 m.

Twenty-five years ago, W.W. Rubey provided the conceptual framework of mantle degassing, and this has since become the central theme for the origin and evolution of sea water. The current question is, what is the *rate* of degassing? At present, a very small amount of degassing is still going on, as shown most conclusively for ^{3}He. However, seemingly nearly all of the ocean water resulted from degassing during the Archean (prior to 2.5 b.y. ago). During that interval of time, primordial continental crust more than doubled in amount to achieve its present extent. Since continental crust and oceanic crust are in isostatic equilibrium, the ocean basins then achieved the same average depth as occurs in modern times. Both the growth of continental crust and the growth of water in the ocean basin may follow a sigmoidal curve, typical of growth in systems with a limiting "nutrient."

Although the volume of ocean water may have been essentially uniform for the past 2.5 b.y., sea level has varied over periods of several thousand to a few million years by as much as a few hundred meters, particularly depending upon glaciation and the rate of seafloor spreading. Yet to be clearly established is the relationship (if any) between vulcanism and change in sea level, as well as detailed curves of sea-level change for the pre-Mesozoic.

2. BATHYMETRY

Factors controlling element distributions in sediments are difficult to induce because the distributions are frequently the result of nonequilibrium processes.

D. W. Spencer, E. T. Degens and G. Kulbicki, 1968

Numerous assertions have been made about the depth of water in which sedimentary rocks were initially deposited, and a presumed greatly increased oceanic depth even has been invoked as the key feature leading to the evolutionary diversification of metazoan life (LaBarbera, 1978). Most of these assertions are only assertions and are unsubstantiated and unquantitative remarks which allude to "shallow" or "deep"—as the case may be. In fact, bathymetry is enormously difficult to determine since criteria for absolute depth seem to exist in only three general cases: the shoreline, the photic zone, and oceanic ridge crests. Deposits can be determined relative to one of these three possibilities, but that is often the best one can do.

In common parlance, depth of water and "energy" are related. Deposits are ascribed to "low-energy" (deeper) or "high-energy" (shallower) conditions. Thus the first topic of this chapter is "energy." That section is followed by discussions of various other ways which have been used to gauge bathymetry. Treated in sequence are sedimentological criteria (grain-size distributions, ripple marks, lithology, and sedimentary sequences), geochemical and mineralogical criteria, and biological criteria. Discussions of bathymetry lead naturally into the topic of general sedimentological regimes. This chapter therefore closes with a comparison of continental platforms versus marginal seas as a model for ancient epeiric seas.

Energy

When geologists talk about a shallow "high-energy" regime, or a deep "low-energy" regime, they usually merely mean that one deposit is coarse grained and another deposit is fine grained. This is a reasonable, although very qualitative, distinction. What one really is talking about is the current strength which must have been present to allow particles of a given size to be deposited.

In actual quantitative terms approximately half of the mechanical energy which is dissipated in the coastal zone comes from wind-generated waves, and half from tidal currents in shallow seas (Table 2-1). *Wave base* is the depth at which surface waves begin to "feel bottom," a depth approximately half of their wavelength. This depth may be as deep as 200 m during major storms, but generally is 20 m or less, as, for example, off the east coast of the United States (Emery, 1966:A9). Less than 5 percent of wave energy is dissipated at wave base (Dietz, 1964b:1279).

Much more important energetically than wave base is *surf base,* which is the depth at which waves peak up and become surf. Waves break when the depth is approximately four-thirds the deep water height of the wave. Approximately 95 percent of the energy of a wave is expended in this process. Only in exceptional circumstances would surf base be deeper than 5 m.

One precise measure of "energy" is breaker height, and this can be measured along the coastline. The zone of high energy is defined (Tanner, 1960) as that in which breaker height is more than 50 cm; the zone of intermediate energy, 10 to 50 cm; and the zone of low energy, less than 10 cm. In terms of action on sediments, the "zero-energy" limit is a breaker about 3 cm high. For the coastal region from Cape Hatteras to the tip of Florida, breaker height is highest (75 to 100 cm) where the shelf is narrowest. Breaker height could also be interpretable in terms of wave force (kg/m²). Kukal (1971) suggested that values of about 100 kg/m² are characteristic of in-

Table 2-1 Estimates of the natural rates of dissipation of mechanical energy in the shallow waters of the world in units of 10^9 kw (1 kw = 10^{10} erg sec^{-1}). Data from Inman and Brush, 1973.

Source	Rate
Wind-generated waves breaking against the shoreline	2.5
Tidal currents in shallow seas	2.2
Large-scale ocean currents in shallow seas (Guiana Current off the northeastern coast of South America, 0.13; Falkland Current over the Argentine Shelf, 0.03)	0.2
All other sources (wind stress on the beach, 0.01; internal waves, 0.01; edge waves; shelf seiche; tsunamis; rivers entering the oceans)	0.1
Total	5.0

land seas, with a maximum of 8,450 in the open ocean, and a maximum of 10 for large rivers. In the geological literature "energy" almost never is used in the quantitative sense discussed in this paragraph.

From a bathymetric point of view, "energy" is defined best in terms of the velocity of water (in cm/sec) which is necessary to transport particles of a certain size. The multiplication of that velocity by the percentage of a stratigraphic section of a given sediment size, summed over for all size classes in the sample, yields an "energy value" for that sample. Visher (1961) suggested that by contouring these values over a basin, an "energy map" could be produced. Such a map generally would follow a bathymetric chart.

The energy imparted to sediments influences the size of particles that remain, as well as their sorting and roundness, surface texture, silt/clay ratios, and the types of fauna and flora which are able to exist in a given region. All of these factors have been used to classify "energy" regimes. As the energy level increases, sediment size increases, the silt/clay ratio increases, sorting may improve, roundness increases, and filter feeders increase in abundance (and detritus feeders decrease in abundance), yielding "stages of maturity" (Folk, 1951:128) or an "energy index." This scheme appears to distinguish clearly lagoonal, barrier, and basinal deposits—i.e., general bathymetry—of the Jurassic Smackover and Haynesville limestones (Plumley et al., 1962).

Relative "energy" has also been estimated from examination of surface textures of sand grains. The V-shaped patterns on the sand grains observable with the scanning electron microscope are oriented irregularly if formed in surf conditions (Krinsley and Donahue, 1968). However, where wave action is less intense, an *en echelon* appearance commonly occurs. The extent of weathering is also a function of wave height ("energy"), as shown in Table 2-2. Chemical features are noticeable on virtually 100 percent of sand grains under "low-energy" conditions in which mean wave height is less than 10 cm. Mechanical features dominate on sand grains under "high-energy" conditions of mean wave height more than 50 cm. Unfortunately, the surfaces of sand grains are altered by diagenesis. Thus although eolian sands are inferred from rocks of Cambrian age, the oldest known case of eolian surface textures is a sandstone of early Triassic age (Krinsley, Friend, and Klimentidis, 1976). This means that a measure of "energy" from surface features of sand grains may be chiefly a tool for the Mesozoic and Cenozoic.

All of the discussion about "energy" is based on an assumption of equilibrium conditions between the supply of "energy" and the

Table 2-2 Summary of surface texture characteristics of sand grains of high-energy and low-energy conditions. Data from Krinsley and Donahue, 1968:744.

High energy (surf)	Medium- and low-energy beach

I. V-shaped patterns, irregular orientation
  a. 0.1 mm average depth
  b. 2 Vs per square mm density
II. Straight or slightly curved grooves
III. Blocky conchoidal breakage patterns
	I. En echelon V-shaped indentations at low energy; as energy increases, randomly oriented Vs replace the en echelon features
	Continuous gradation between high- and low-energy features

supply of sediment. One way to summarize the relationship between sediment supply and depositional energy is shown in Figure 2-1. Wave energy in terms of average breaker height is related to sediment supply. If a point is in the upper right-hand quarter of Figure 2-1, then "historical" factors will be more important than the effect of any given wave regime. It would be interesting to know how many coastal sediments are not in equilibrium with present local "energy" conditions. In summary, qualitative estimates of "energy" (and bathymetry) are commonly used, but quantitative methods are available, and should be more widely applied.

The most important notions in this section are (1) "energy" is used loosely and imprecisely by geologists; in its place, information about current strength should be given because this is what usually is meant; (2) 95 percent of energy loss is at surf base (usually at less than 5 m); and (3) all discussions of "energy" are meant to apply to equilibrium conditions, whereas in fact historical aspects may overwhelm local processes.

Grain-Size Distribution: Expectations

There is a very large literature on the use of statistical properties of sediments in order to differentiate the environment of deposition in general, and bathymetry as one environmental aspect. In this section, I will go into the anticipated results from the study of grain-size distributions, and the following section will present published examples. In this section, I will first consider a normal distribution of size frequency, and then will discuss perturbations from a normal

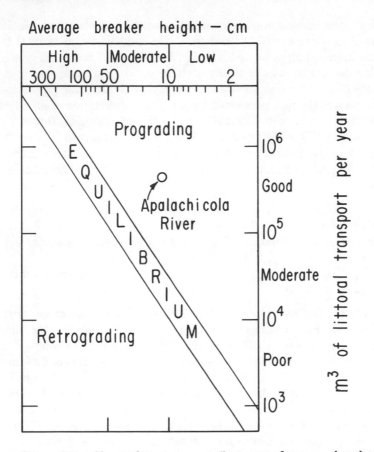

Figure 2-1 Chart of wave energy (in terms of average breaker height) versus sand transported. The circle plots the Apalachicola River delta, in which the amount of material transported to the coast is too much for the waves and currents to handle. Data for other coastal areas need to be obtained. After Tanner, 1961:93.

distribution and their significance. The basic notation of size-frequency distributions is given in the Appendix.

Many size-frequency distributions show no deficiencies in particular size fractions (Shea, 1974). In a significant number of cases, however, the size-frequency distribution of a sediment or rock does deviate from a normal or Gaussian or bell-shaped distribution (loosely speaking), and these instances provide added information on the origin of a deposit.

There are three general explanations for a significant deviation of a size-frequency distribution from a Gaussian distribution. (1) Direct breakage along joint or bedding plains should result in gravel, granular disintegration should result in sand, and chemical decay and platy disintegration should result in clay (Folk, 1966:81). These distinctive weathering processes may cause a deficiency of grains of intermediate sizes, approximately of the size of granules (0 to -2 $\emptyset$) and of finer silt (3 to 5 $\emptyset$). Spencer (1963) suggested that "all clastic sediments are essentially mixtures of three or less fundamental populations of log-normal grain sizes. The fundamental populations are:

(a) 'Gravel' with a median of -3.5 to -2 $\emptyset$ units and a standard deviation of 1.0 to 2.0 $\emptyset$ units
(b) 'Sand' with a median of 1.5 to 4 $\emptyset$ units and a standard deviation of 0.4 to 1.0 $\emptyset$ units
(c) 'Clay' with a median of 7 to 9 $\emptyset$ units and a standard deviation of 2 to 3 $\emptyset$ units."

The tendency for modal concentrations to be fixed at these three sizes owing to the type of source material has been termed the Sorby Principle, after Henry Clifton Sorby (1826–1908) who proposed it.

(2) Different breakdown processes result in particles of different shapes which can only be of certain sizes. Spalling and crushing of larger grains results in angular pieces. As material becomes smaller, abrasion increases in importance, resulting in rounding of grains, especially in the size range of 32 to 128 mm (Bluck, 1969:8).

(3) A change in mechanism of sediment transport will be reflected in different degrees of sorting and other statistical properties of sediments (Sagoe and Visher, 1977). This applies because nearly all sediments are composed of a mixture of two or more sizes of sediment, each of which is, by itself, normally distributed (Folk and Robles, 1964). As shown in Figure 2-2, the material that is moved by sliding or rolling, the *traction load*, chiefly affects grains in the size interval 0 to 2 $\emptyset$; material that is intermittently suspended, the *saltation load* of the literature, chiefly affects the size interval 2 to 3.5 $\emptyset$; and the continuous *suspension load* includes the still finer material. These categories are dependent upon the velocity of the water, and they refer in this instance to conditions encountered on "typical" beaches. The observed "breaks" in the slope of the size-frequency curve have been attributed to the shear velocity of the flow (Middleton, 1976). Indeed, Middleton believed that "the observed 'breaks' must be explained by hydraulic sorting."

Agents of sediment transport chiefly alter a sediment's size-fre-

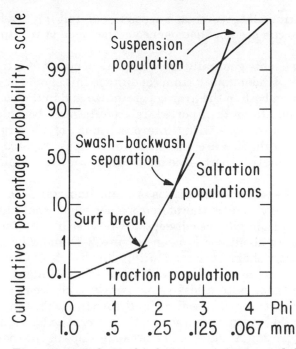

Figure 2-2   Relationship between grain size and the type of sediment transport. Note that the coarse and fine tails have a small amount of sediment over a wide range of sizes, and thus are more poorly sorted than are the intermediate saltation populations. After Visher, 1969:1079.

quency distribution at its finer end or its coarser end, and hence most of the interest in deciphering the origin of a sediment is directed at the nature of the tails of the distribution. Since diagenesis preferentially and significantly increases the amount of material of the finest sizes, the greatest interest reverts to characterizing the unaltered coarsest end of the distribution.

In addition to size-frequency plots, several statistical measures of the sediment as a whole often are used to infer the method of deposition. The measures which are computed most commonly are mean diameter, standard deviation, sorting, symmetry (skewness), and peakedness (kurtosis). The plot of C (the particle diameter of the coarsest 1 percentile) versus M (the mean grain size) has been used to isolate the mode of transportation (Passega, 1977). The principle is that as the mode of transport shifts from rolling to suspension, both the coarsest 1 percentile and the median diameter decrease in size. Thus if C and M are obtained for each of 20 to 30 samples representing all textures in a depositional unit (up to a few meters thick), one can see how homogeneous the mode of transport was for that unit.

Transporting agents generally are weaker in deeper water than in shallower water. Under equilibrium conditions, this should be reflected in a reduction both in the average grain size and in the maximum grain size able to be transported, relative to most beach deposits. However, just as with interpretations of "energy," interpretations of grain size are subject to limitations imposed by sediment *available* for transport—the historical factor in an equilibrium framework.

Recognition of shoreline deposits is of great importance in determining bathymetry. Let us therefore consider what would happen in typical conditions to a beach deposit whose sediment size is distributed normally about some mean (Figure 2-3) and which is composed of similarly shaped grains. Surf hits the beach, churns up the sediment, runs up the slope, partly sinks into the sediment, and returns down the beach with a less vigorous flow. The greater the wave height, the greater the sediment disturbance. The initial breaking surf should transport coarser particles from the beach into the surf zone (Figure 2-3, A and A'). The break in slope at the coarse end of the size-frequency diagram (Figure 2-2) has been referred to as the "surf break." Surf-zone sediments are thus negatively skewed and very poorly sorted.

Mid-beach sediments (Figure 2-3, B and B') lack most of the coarse mode and are better sorted than surf-zone sediments. When these mid-beach sediments are transported by wind onto the adjacent dune, their coarser particles are left behind. Dune sediments (Figure 2-3, C and C') should be extremely well sorted because medium- to high-energy surf previously winnowed out the fine fraction, and because the wind was unable to transport the coarse material, thus leaving the intermediate size classes intact. Further transportation of this dune material onto an eolian flat (Figure 2-3, D and D') should result in the gradual accumulation of fine material with a resultant positive skewness and concomitant gradual flattening of the size-frequency curve.

The degree of rounding of particles is expected to be related to the frequency and intensity of impact. Air is both less dense and more rapidly moving than water. Gentle breezes travel at a few hundred cm/sec (50 cm/sec $\simeq 1$ knot), whereas water rarely moves at greater than 100 cm/sec. Accordingly, grains from coastal dunes with onshore winds should be subject to a greater degree of impact, and thus be more rounded, than similarly sized materials in the ad-

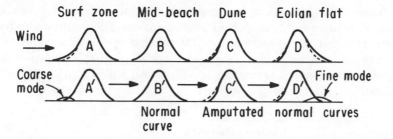

Figure 2-3   Explanation of changes in sizes of sediment in different environments according to an equilibrium model of moderate surf energy and an onshore wind. The dashed curve represents a normal distribution of constant shape and placed at constant position to serve as a reference. The solid-line curves in the upper row represent the type of frequency curve obtained upon analysis of actual samples; the lower set of figures is an attempt to explain why the actual sample curves depart from the reference curve. Mason and Folk proposed that a normally distributed nucleus population forms the great bulk of all samples, and that differences between samples (expressed as departures from normality) are caused by addition or subtraction of grains near the tails of this nucleus population. After Mason and Folk, 1958:225.

jacent beach. From offshore to swash zone to dune and beach ridge, rounding of grains increases by 10 to 20 percent (Shepard and Young, 1961; Waskom, 1958).

Transportation by rivers is very uneven from storm to storm, season to season, and year to year. Accordingly, river deposits may include extremely coarse materials (from the flooding season) and large amounts of clays and silts (deposited during the dry season). River sediments are expected to have significant tails in the coarse *and* fine fractions, and to be sorted much more poorly than beach deposits. Alluvial fans are closer to the original sources than are rivers and may be even less well sorted. Turbidites are deposited from material which was transported in a single pulse of energy, and therefore such deposits are anticipated to show no breaks in the slope of the line of the size-frequency distribution.

In summary, the main bathymetric expectation from analyses of grain-size distributions is to predict the occurrence of shoreline or beach deposits. The size-frequency distribution of nearly all sediments is a mixture of gravel, sand, and clay fractions, each of which originates and is transported in a different fashion. Coarse material originates from spalling and is transported in the traction load; moderate-sized material originates from mechanical abrasion and is transported by saltation; fine material originates from chemical decomposition and is transported in suspension. These three size fractions result in three sections on plots of size versus abundance. The degree of sorting and related statistical properties may reflect whether the deposit is marine or subaereal, and if marine, whether offshore or intertidal.

Grain-Size Distribution: Results

Some of the predicted trends regarding the size-frequency distribution of sediments have been confirmed. However, there are large differences in the degree of sorting and the size of grains, depending on the mechanism of transport for any given environment (Visher, 1969:Table 1).

Rivers transport a much wider range of sizes of materials than is found on beaches. This is reflected in plots of mean size versus standard deviation in which, for a given mean size, the standard deviation is much higher for river sand than for beach or coastal dune sands (see, e.g., Friedman, 1967). For gravel-sized sediment, stream and alluvial fan deposits have sorting coefficients three to five times higher than beach deposits (Emery, 1955).

Deposits subjected to surf action tend to be coarse and poorly sorted (Figure 2-4D), and to become better sorted but still have a significant coarse fraction in nearshore subtidal sands (Figure 2-4C). Dune sands are quite well sorted with a single saltation population (Figure 2-4B), whereas beach deposits typically have two populations in the size range which has been transported by intermittent suspension (Figure 2-4A). These two populations reflect water moving up a beach, sinking in, and flowing back down the beach.

The major discrepancy between the predicted and observed size-frequency distribution is in transects offshore. The continental shelves do not have bands of sediment which decrease in grain size from coarse nearshore to fine offshore, as is expected if sediment size followed strength of currents and settling velocity. Instead, the dominant pattern is that of fine sediments on the inner part of the continental shelf and coarse sediments on the outer part (Emery, 1968; Emery and Uchupi, 1972). The coarse sediments of the outer shelf record high currents—those of the shallow regions when sea level was many tens of meters below its present position. Those coarse sediments were left behind when sea level rose, and the currents became too weak to transport them. In a few places, those outer shelf coarse sediments are being reworked today (Milliman, 1972; Creager and Sternberg, 1972).

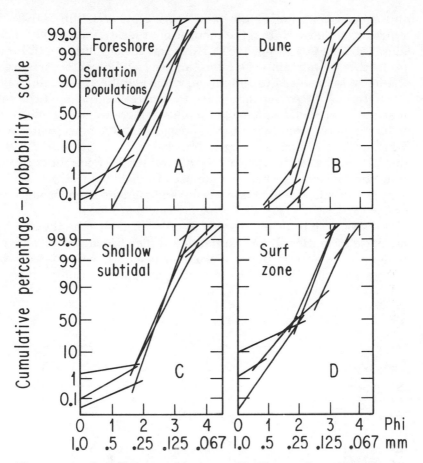

Figure 2-4 Size-frequency diagrams of sediments from different environments. *A*. Beach foreshore sands; note two populations of sediment carried by saltation. *B*. Beach dune sands; note single population of sediment carried by saltation. *C*. Wave-zone sands (shallow marine); note high proportion of coarse, poorly sorted population. *D*. Surf-zone sands; note very high proportion of coarse population. After Visher, 1969: Figures 7B, 8C, 9C, and 10D, respectively.

On a local scale, some areas behave "as expected," with finer sediments offshore (Pilkey, Trumbull, and Bush, 1978). Grain size generally decreases with water depth in the Gulf of Naples (Sindowski, 1957), but scatter in the data may be considerable, as for Bering Sea sediments (Sharma, Naidu, and Hood, 1972). Sediments from the outer continental shelf of the Bering Sea are more positively skewed (i.e., have more clay mixed in) than sediments from

the inner shelf. Sand, silt, and then clay occur in a regular
progression seaward from the mouths of at least seven major deltas
(data tabulated by Allen, 1967). However, the absolute depth limits
of the coarse sediment (2 to 15 m) and the shallowest occurrence of
the finest sediment (4 to 100 m) vary considerably, depending upon
the nature of the source materials, the energy regime, and the submarine slope. Shell fragments and platy minerals such as mica tend
to be winnowed from sandy littoral deposits and to be deposited seaward in finer sediments. At the Niger delta, a depth of 6 to 10 m
marks the shallowest depth of platy minerals. Sediments deeper
than the continental shelf are almost invariably silt-clay sized, except for turbidites and slump blocks and manganese-phosphate deposits.

Rocks must be disaggregated in order to determine the size-frequency distribution. Like sediments, lithified beach deposits have a
break in the intermittent suspension population compared with
marine deposits (Figure 2-5). And disaggregated turbidite deposits

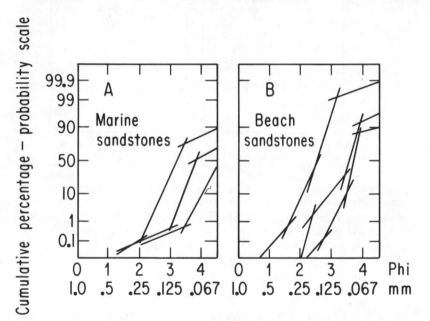

Figure 2-5 Size-frequency diagrams for disaggregated sandstones
from shallow water deposits. *A.* Sandstones from probable shallow
marine environment. *B.* Sandstones from probable beach environment. Note the two saltation populations in the beach sandstones.
Compare with Figures 2-2 and 2-4. After Visher, 1969:1098 and
1100.

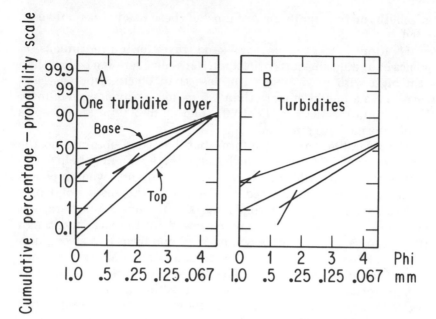

Figure 2-6 Size-frequency analyses of disaggregated sandstones from turbidite current deposits. Note how poorly sorted the material is. *A.* Variations within one 1.5 m turbidite bed. *B.* Turbidites from different ages. Contrast *A* and *B* with Figure 2-4. After Visher, 1969:1101.

are nearly linear (Figure 2-6) over several Ø units, just like modern sediments. Also the greater the inferred depth of water, the smaller the maximum grain size which was carried in the graded suspension (Passega, 1977:730).

Studies of grain-size distribution help one to determine the presence of the beach and shoreline. This is important because knowledge of the shoreline fixes the relative position of all adjacent stratigraphic units.

Other Sedimentological Criteria

Other sedimentological criteria of some use in gauging bathymetry include (1) bed forms, from ripple marks to sand waves, (2) lithologic characteristics, and (3) sedimentary sequences. Some of these same features are used to differentiate tidal deposits (Chapter 3). Although many sedimentary structures and directional features have been described in both modern and ancient sediments and rocks, the paleobathymetric significance of most of these has not been investigated.

(1) Ripples, megaripples, and sand waves form a continuous sequence of undulating surfaces. Ripples are considered to be less than 4 cm high with a wavelength of less than 60 cm, whereas sand waves have a height of more than 1.5 m and a wavelength of more than 30 m (after McCave, 1971:201) and up to 500 m. Megaripples are of intermediate size.

Considerable empirical information has been obtained on the environment of deposition of ripples, and these data may help one to identify the position of the shoreline. As many as seven empirical relationships have been used to determine the origin of ripple marks. Only two types of ripple marks are considered here, but these and many other data are reviewed by Reineck and Singh (1973). Current ripples are asymmetrical and form in a strong unidirectional flow. Wave ripples are symmetrical and indicate the lack of a strong unidirectional flow. The presence or absence of currents is of exceptional importance in distinguishing ancient tidal flats (low currents) versus lagoons (typically with steady currents) (van Straaten, 1954) and deeper water (see p. 85).

Some types of ripple marks characterize the intertidal region. Rhombohedral-shaped ripple marks typically form by the backwash of waves on beaches; the length/width ratio (made diagonally) changes from about 2.0 to 4.0 as the beach slope increases from 0°30′ to 1°45′. Flat-topped and terraced ripple marks form from typical ripples when the water level erodes crests of ripple marks. At least 11 categories and 18 varieties of ripple marks from shallow water (less than 1 m deep) are described. Finally "meta-ripples" (which are deposited on an erosion surface, and are themselves then eroded with no apparent relation to their internal form) indicate radically different current directions, as are typical of intertidal conditions (Imbrie and Buchanan, 1965).

Water depth (1 to 20 m) and wavelength (0.5 to 1 m) of ripple marks are directly related (Tanner, 1959:569). Similarly, water depth ($\simeq$1 to 100 m) and ripple height ($\simeq$0.1 to 10 m) are directly related (Allen, 1963). In addition, water depth, wave height, fetch and grain size are all related to ripple mark spacing. Following a series of studies over many years, Tanner (1971) empirically derived equations which relate all of these factors. The resulting predictions of the equations are summarized graphically in Figure 2-7 and explained in the caption. Under favorable conditions, water depth can be inferred from ripple marks.

The largest fields of sand waves exist in the North Sea between

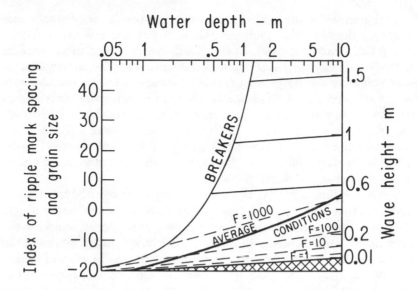

Figure 2-7 Plot of an index of ripple mark spacing and grain size, and of wave height, versus water depth. The left-hand vertical ordinate is the expression $0–97\ s – 3.72\ \ln g$, where s is ripple mark spacing in cm and g is grain size in mm. The curved line, rising to the upper right, represents the breaker zone; waves are not propagated under conditions represented to the left of the curved line. Moreover, waves are typically too small to form ripple marks in the lower right region where water wave heights are less than 1 cm. Thus the meaningful portion of the chart is the horn-shaped segment which opens toward the upper right; shallow water ripple marks plot in the narrow part of the horn, where uncertainty is modest. Dashed lines, sloping gently upward to the right, are marked in terms of fetch values (*F*, in km). Once values have been estimated for wave height and water depth, a general estimate of fetch can be read. As an example, assume one obtained a value of -10 on the left-hand vertical scale. For average conditions, water depth about 0.8 m, wave height about 0.25 m, the inferred fetch is less than a few hundred kilometers. After Tanner, 1971:84.

Holland and England, on Georges Bank off New England, and in the Gulf of Korea. In the North Sea, the field covers 15,000 km² (McCave, 1971) with wavelengths of 200 to 500 m, and the heights up to 7 m, in depths of 20 to 100 m. In sand waves from the Irish Sea, the trough height is usually 6 to 15 m, with 26 m a maximum. Linear shoals which are similar to these sand waves occur along the Atlantic coast of the United States from Florida to New England.

Structures of the size of sand waves must form in water considerably deeper than their height, and yet shallow enough that strong tidal currents exist. In the North Sea, surface tidal currents in the sand wave area are nearly 100 cm/sec. The attribution of sand waves to strong tidal currents is consistent with the observation that more than half of the 40 areas in which sand waves occur are at the tidal entrances to narrow embayments (Ludwick, 1970). The large-scale sand waves in the North Sea occur only in water shallower than 30 m where sufficient tidal current exists (Dingle, 1965). In areas of sand wave formation, tidal flow and sand supply are as important as depth per se.

In the geologic record, large sand waves apparently are not recognized, probably because large enough exposures are rarely if ever available. Sand waves may, however, be the source of sand for the large sheet sand deposits that are well known, although such deposits usually are attributed to the force of storms instead of to current or tidal influence (Goldring and Bridges, 1973). Smaller bar ridges of sand as high as a few meters in relief are known and are attributed to tidal deposits (Brenner and Davies, 1973, 1974).

The thickness of large-scale planar cross bedding also may be related to depth of water. Allen (1963) used this criterion to estimate minimum water depth over a range of about 0.5 m to 2 m shallowest depth in 11 examples (Table 2-3). Reineck and Singh (1973:12) graph the relationship between increasing current velocity and bed forms. For small ripples to megaripples, current (to 100 cm/sec) is a much more important variable than is depth (to 35 cm).

(2) Clues to the *relative depth* of deposition exist in several lithological features. For limestones, shallowest lagoonal conditions typically result in lithified lime muds, with the lime presumably of algal origin. Of particular interest have been the so-called birdseye structures, which indicate very shallow conditions. These structures represent open spaces larger than the average particle size, and may form according to a variety of physical, chemical, or biological causes (Deelman, 1972); many of the spaces may originate from gas bubbles. When birdseye structures occur, they commonly are the *only* diagnostic character of an otherwise nearly nondescript limestone.

Intertidal beach rock is known to be forming in many places at the present time (e.g., examples cited by Siesser, 1974), although the precise mechanism of cementation is not always certain. Only aragonite and high-Mg calcite are known as precipitates from normal marine salinities. Thus fully marine beach rock should have an

Table 2-3 Large-scale planar cross-stratal set thickness and predicted ranges of depth
of deposition (95 percent level). Data from Allen, 1963:213.

Formation	Median thickness (m)	Minimum depth range at 95 percent level (m)
L. Calcareous grit (Jur.), Berkshire	0.125	0.53–3.6
Pocono formation (Dev.), Appalachia	0.15	0.62–4.2
McNairy formation (Cret.), Mississippi-Arkansas	0.15	0.62–4.2
Chesterian sandstones (Miss.), Illinois	0.16	0.65–4.4
Ditton Series (Dev.), Shropshire	0.19	0.75–5.1
Lower Cretaceous sands, East Anglia	0.23	0.86–6.0
Woodbank Series (Dev.), Shropshire	0.23	0.86–6.0
Pennsylvanian sandstones, Illinois	0.24	0.90–6.2
Rough rock (Carb.) Pennine area	0.56	1.8–12.5
Folkestone Beds (Cret.), Farnham-Guildford area	0.59	1.9–13.0
Oil shale group (Carb.), Midland Valley	0.60	1.9–13.1

aragonite or high-Mg calcite cement (Bathurst, 1971:370) instead
of a low-Mg calcite content, which is typical of fresh-water cement.
Beach rock which forms in the zone of mixing of fresh and salt water
will have cement of mixed character (see p. 164). Since beach rock
forms rapidly (i.e., coke bottles a few years old are embedded), no
evidence of compaction should exist in the rock, in contrast to more
slowly cementing subtidal limestones. Mud cracks and salt casts
also, of course, identify the intertidal zone, as has been emphasized
for years. Owing to the great importance in bathymetry of determining shoreline, lithologic and other characteristics for carbonate tidal
flats are summarized in Table 2-4.

Shallow-water cements also differ from deep-water cements
(Milliman and Müller, 1977). Cements of beach rock, ooids, reef fill-
ings, etc., have $\delta^{18}O$ values between $+1$ and $-1‰$, and $\delta^{13}C$ values

Table 2-4 Relative frequency of occurrence of features in subenvironments within some modern and ancient carbonate tidal flats. A + indicates greater frequency of occurrence than an ×. Data from Kahle and Floyd, 1971.

Feature	Subenvironment		
	Supratidal	Intertidal	Subtidal
Celestite	+		
Halite cubes	+		
Carbonate levees	+		
Irregular laminations	+		
Lithoclasts	+		
General lack of fossils	+		
Mudcracks	+		
Bituminous films	+		
Eolianites	+	×	
Stromatolites	+	×	
Laminated sediment (algal)	+	×	
Nodular gypsum or anhydrite	+	+	
Dolomite	+	+	
Compressional ridges		+	
Blister structures		+	
Pelleted carbonate mud		+	
Restricted fossil assemblage		+	
Scour and fill		+	
Flat-pebble conglomerate		+	
Cross bedding		+	
Tidal channels		+	
Burrows		×	+
Oolites		×	+
Fossiliferous calcareous muds			+

chiefly between +3 and +5‰ (see Chapter 4 for explanation of isotope terminology); chemically, they contain more Sr in the aragonite (> 1 percent) and more Mg in the calcite (15 to 29 mole percent) than is expected from equilibrium inorganic precipitation. These chemical differences and the physical association with blue-green algae suggest that precipitation of the cements is biologically mediated. In contrast to these shallow-water cements, deep-water cements have a much wider range of $\delta^{18}O$ and $\delta^{13}C$ values, and the Mg content of calcite is much lower (9 to 13 mole percent), with each instance seemingly in equilibrium with specific deep-water habitats (muds in basins, or on volcanic seamounts, etc.). For deep-water cements there is no indication of blue-green algae, and the chemical compositions are within expectations for inorganic precipitation.

Some sedimentary sections which chiefly are carbonate are characterized by a cycle of a sharp increase in the frequency and degree of coarsening of quartz grains followed by a gradual decrease in grain diameter. This pattern represents successive shifts in supply related to successive rises in sea level (Wanless et al., 1957). This "index of clasticity" with quartz grains is useful because it remains unaffected by any dolomitization of the original limestone. Abundant "nodular limestones" (limestones with carbonate nodules surrounded by clayey streaks) originate offshore, but apparently are without any precise bathymetric significance (Dvořák, 1972), as are also carbonate sand bodies (Ball, 1967:583).

Many limestones of the fossil record were formed at depths of several hundred meters (Cook and Enos, 1977). J. L. Wilson (1969) characterized deep-water limestone as a pure lime mudstone, commonly dark (because of unoxidized organic matter), and laminated on a millimeter scale (because of a lack of burrowing). Grading of very fine sand to lime mudstone is observed over centimeter distances. Chert is often present. Beds are very uniform and planar, often 2 to 5 cm thick, and divided by thin shales. These limestones are sometimes interrupted by slump blocks. The limestone fauna is very restricted and consists chiefly of pelagic organisms. Many characteristics of deep-water limestones are found also in limestones from poorly oxygenated regions, independent of depth.

When sediments are eroded in bulk and then rapidly deposited, the largest particles settle first of course, followed by a graded sequence of finer sediments. Graded bedding is typical of turbidity currents (excellently reviewed by Bouma, 1972). Deposits of this type (the "Bouma sequence") form in marine areas of continental slopes and rises. Mudstones with alternating layers of mud and silt laminae also may form from turbidite flows.

Concentrically zoned calcium carbonate spheres called ooids and pisolites form in various sizes in marine, lake, and terrestrial (cave) conditions. An ooid is less than 2 mm in diameter; a rock or sediment composed chiefly of such particles is an oolite. A pisolite is a rock or sediment composed of such particles which are greater than 2 mm in diameter, but an individual grain is also called a pisolite (not pisoid, as symmetry would suggest). Marine ooids and pisolites originate as inorganic precipitates, and/or are closely controlled by the activity of bacteria and blue-green algae.

Kendall (1969) described a sequence of degrees of water agitation and calcium carbonate precipitation from lime mud through pellets to "grapestone" and pisolites, and then to botryoidal grains and oolites (Figure 2-8). Relatively agitated water is a prerequisite for these spherical grains, and this suggests (for marine regions)

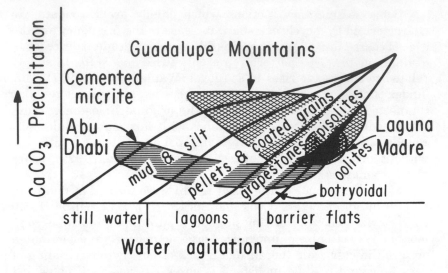

Figure 2-8   Diagram showing relationship of carbonate particles to water agitation and rate of calcium carbonate precipitation. Types of particles found in Abu Dhabi (Trucial Coast), Guadalupe Mountains (New Mexico), and Laguna Madre (Texas) are shown. Note that oolites occur in water of highest agitation. After Kendall, 1969:2521.

nearshore areas associated with the breaker zone (≈ 2 m). Grapestone forms under algal mats by the cementation of subtidal grains (Bathurst, 1971:319). The algal mat may rot and disappear, and the grapestone itself becomes a marker for the photic zone.

Ooids which form in hypersaline or fresh-water environments characteristically have radial instead of concentric growth (as is typical of marine ooids). The radial fabric is readily broken (in contrast to the concentric fabric), and therefore "broken ooids are considered significant indicators of unusual salinity if they comprise more than 1 percent of the grains in an oolite" (Halley, 1977).

(3) Sedimentary sequences can also be used to determine relative depth. In barrier island and deltaic prograding sequences, sea level can be placed at the base of cross-stratified beach sandstones. The sediment thickness from this level downward to the base of underlying offshore sediments may approximate the water depth into which the island or delta is prograding. This would apply only where compaction (if any) occurs very early, and where the offshore profile stays the same for a long time (i.e., subsidence or sea-level rise is not noticeable). Application of this method to the geologic record suggested water depths of 3 to 12 m offshore of ancient barrier

islands, 10 to 25 m offshore of cratonic deltas, and 30 to 90 m offshore of basinal deltas (Klein, 1974).

Characteristics of deltas per se can be combined to show probable *relative* depth, as in Table 2-5, for the Devonian Catskill delta. The important point is that *associations* of characters are needed rather than single characters.

In summary, there are a large number of anecdotal and particularistic criteria for bathymetry which are best learned by field experience. Intertidal conditions are often indicated by any combination of characters which suggests deposition by alternating ebb and flow

Table 2-5 Rock type and structures significantly abundant (+) or rare (−) in the nearshore paleoenvironments. Blank where not significant. Data from Sutton and Ramsayer, 1975:806.

Rock type or structure	Environments								
	Delta front	Distributary mouth bar	Estuary	Outer platform	Middle platform	Nearshore platform	Lagoon	Open shelf	Prodelta
Oscillation ripples	+	−	−			+	+	−	
Cross bedding	+	+						−	−
SANDSTONE	+	+		−		−	−	−	−
RUDITE	−	+		−	−			−	
Parting lineation	+	+							−
Current ripples	+							−	
Cross lamination	−								+
Laminae	−		−	+	+	−			+
SHALE	−	−	+		−		+	+	+
SILTSTONE	−	−			+			+	+
Cuspate ripples	−	−				+		+	+
Tracks and trails	−	−				+	+	+	+
Burrows	−	−						+	+
MUDSTONE	−		−		+	+		−	
Pillows					+		+	−	−
Load casts		−	−						+
Groove and scour		−	−			−	+		+
Prod and brush Laminae		−		−	−	+			+
Flute casts		−		−	−	−	−	+	+

(also see discussion of characteristics of tidal deposits, p. 79). Eriksson, Hobday, and Klein (1976) even suggested that "the recognition of bipolar-bimodal orientations of directional data appears to be the only method of distinguishing marine from non-marine depositional systems in the Precambrian." Also of diagnostic use are some types of ripple marks, but every field instance has to be judged on its own merits.

Backtrack Method and Carbonate Compensation Depth

The standard method of determining deep-sea bathymetry from the Jurassic to the present is to use the curve of age of normal oceanic crust versus oceanic depth, as shown in Figure 2-9. By using this curve, one can convert ocean bottom of any given age to the depth which it would have had at some earlier age, or indeed will have at some future time (method introduced by Berger, 1972). The relationship of depth to age is valid because it represents a simple cooling curve for a basaltic plate which is initially hot (and therefore relatively buoyant) and then moves laterally, cools, contracts (and therefore sinks in isostatic equilibrium). In general, new crust is introduced at 2.6 km, is at 3 km by 2 m.y., at 4 km by 20 m.y., at 5 km by 50 m.y., and at 5.5 km by 70 m.y. For this period of time, the cooling curve follows an exponential decay of $t^{1/2}$. For ocean basin

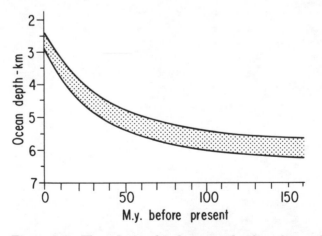

Figure 2-9 The relationship between depth and age of normal oceanic crust. Extensive data from the Atlantic, Pacific, and Indian oceans have been found to fall within the stippled envelope (600 m high). After Sclater, Abbott, and Thiede, 1977.

older than 70 m.y., the depth approaches an asymptote consistent with a basaltic plate which is being heated at a steady rate from beneath (Parsons and Sclater, 1977).

Use of the backtrack method to solve geological problems assumes that sea level has remained relatively constant and that the cooling curve for basalt has not changed over the past 200 m.y. In addition, a correction in the final bathymetry must be made for the effect of isostatic loading by sediments.

A first determination of deep-sea bathymetry using the backtrack method has been completed for the Atlantic and the Indian oceans for the Jurassic to the present. Sclater, Hellinger, and Tapscott (1977) reconstructed the bathymetry of the North and South Atlantic at 12 intervals between the Jurassic and the present, with each bathymetric chart approximately 30 m.y. after the preceding one. Figure 2-10 illustrates Atlantic bathymetry for 140, 110, 65, and 10 m.y. ago. Sclater, Hellinger, and Tapscott estimated that the backtrack method predicts the gross bathymetry for ocean crust of two-thirds of the Atlantic (i.e., younger than 80 m.y.) to be accurate to ± 300 m. Among the quantitative conclusions of this analysis are (1) the North Atlantic was a totally enclosed basin from the mid-Jurassic (165 m.y.) until the early Cretaceous (125 m.y.); (2) the South Atlantic was also a totally enclosed basin from the early Cretaceous until the late Cretaceous (95 to 110 m.y.); and (3) the deep-water connection between the North and South Atlantic probably could not have occurred before the late Cretaceous (80 m.y.) or early Tertiary (65 m.y.). In addition, by the lower Oligocene (35 m.y.), an essentially modern deep-water circulation was firmly established in the South Atlantic (Le Pichon, Melguen, and Sibuet, 1978).

Sclater, Abbott, and Thiede (1977) reconstructed the Indian Ocean bathymetry for 36, 53, and 70 m.y. ago (i.e., early Oligocene, early Eocene, and late Cretaceous). Distributions of calcareous and noncalcareous sediments were also plotted. Among the conclusions documented were that (1) between the middle Cretaceous and the Oligocene, the Indian Ocean was separated into three basins which were isolated from each other below 2,000 m; and (2) Oligocene calcareous sediments are not nearly as widespread as would be expected for the smaller, shallower ocean of Oligocene time.

Calcium carbonate shows systematic changes with depth in the ocean. It is abundant in many sediments shallower than 0.5 km in depth, is taken slowly into solution over the range of 0.5 to 3.0 km (Berger, 1976), and is especially susceptible to solution at 3.5 to 5.5 km, depending upon the ocean and the latitude.

The carbonate compensation depth (CCD) is the depth at which

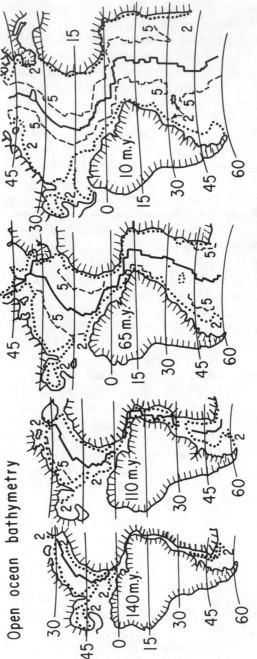

Figure 2-10 Development of bathymetry of the Atlantic Ocean at 140, 110, 65, and 10 m.y. ago. Shown are continent borders, 2 and 5 km contours, and the mid-Atlantic ridge crest (~2.7 km). At 165 m.y. ago, the whole Atlantic was closed. Note that as time passes, there is a separate development of deep basins in the North and South Atlantic, and an increase in amount of ocean that is 5 km or more in depth. After Sclater, Hellinger, and Tapscott, 1977.

the rate of carbonate addition is equal to the rate of carbonate dissolution (Bramlette, 1961), and is operationally identified by the great
reduction of calcium carbonate in sediments. In deep-sea sediment
samples, the presence of abundant carbonate is a conspicuous feature, and hence its near disappearance (say less than 10 percent) at
the CCD "forms the most obvious and significant facies boundary in
the deep-sea deposits" (Ramsay, 1977b:1374).

Figure 2-11 illustrates the approximate depth of the CCD for
calcite in sediments in the modern ocean. In regions where carbonate productivity and resulting sedimentation rates are high, such as
those near the equator, the CCD is depressed. The oceanic CCD for
calcite is now at 3.5 to 5.5 km, being deepest in the North Atlantic
and shallowest in the North Pacific. Within an ocean, the shallowest
depths of the CCD are adjacent to continents and may be at only
500 m off Antarctica (Kennett, 1966). Presumably, the oxidation of
organic matter in shallow sediments produces excess CO_2, which
acts to dissolve the carbonate.

The CCD for aragonite is the depth in the ocean at which shells
of Pteropoda disappear (pteropods are a group of planktonic mollusks). Most Atlantic values of the aragonite CCD are about 2 km,
and most Pacific values are closer to 1 km, with values of 0.5 km
common in regions adjacent to continental margins (Berger, 1978a).
Below about 5.5 km, all of the calcite also has gone into solution.
When all oceanic carbonate is gone, an association of organic, siliceous, and/or chitinophosphatic fossils is consistent with a depth of
formation equivalent to modern continental rises and abyssal
plains.

Samples from particular depths in cores can be referred to their
original depth of deposition by use of the backtrack method of bathymetry. By noting the presence or absence of carbonate, and the
original depth of deposition, the sample can be placed above or
below the CCD for the time of formation. In this way, carbonate
cycles through the late Mesozoic, and Cenozoic have been related to
the CCD and its bathymetric history.

Over the past 100 m.y., the first-order pattern of change in the
CCD has been from the shallowest in the middle Cretaceous
(2.5 km, Indian Ocean; 1.5 km, South Atlantic), to a deepening in
the late Cretaceous and early and middle Tertiary (most values near
3 to 4 km in the Indian Ocean, North and South Atlantic, and non-
equatorial Pacific), and then to a further deepening from the late
Miocene to the present (about 5 km) (Sclater, Abbott, and Thiede,
1977; Ramsay, 1977b; Le Pichon, Melguen, and Sibuet, 1978).
Superimposed on this general pattern is an apparent shallowing in

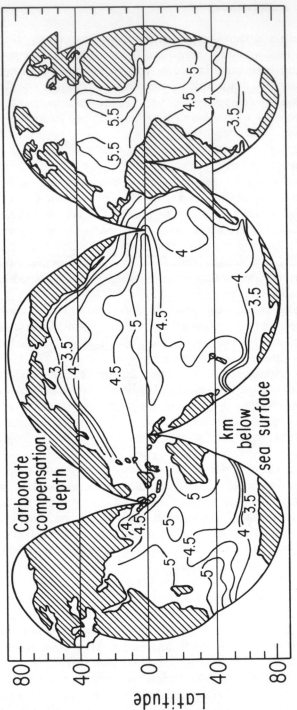

Figure 2-11 Topography of the carbonate compensation depth (CCD) for surface sediments. The CCD is defined by interpolating between sediments with abundant carbonate and sediments with no or only a few percent of carbonate. Numbers on contours denote kilometers below sea surface. Note that the CCD is deepest in the Atlantic, shallowest in the Pacific, and is lowered beneath tropical zones of high carbonate production. After Berger and Winterer, 1974:12.

the Eocene and again in the middle Miocene (in each instance to about 3.5 km). Very limited data indicate an upper Jurassic CCD of approximately 3.5 km. Within this general framework, different oceans have different patterns of CCD variation. Reasons for this are very speculative at present, and are left for the reader to pursue.

The 100 m.y. pattern of deepening in the CCD parallels cooling in the deep water from about 14° (middle Cretaceous) to 2 to 4°C (present) (see Figure 4-13). A gradually deepening CCD is what is predicted from a cooling ocean. Variations in both the CCD and in temperature are major oceanographic changes which have in common that they are communicated to all oceans by the general oceanographic circulation.

The depth of rapid increase in solubility of $CaCO_3$ is about 0.5 to 1 km shallower than the CCD, and is known as the lysocline. The concept of the lysocline was introduced "to denote a surprisingly well defined facies boundary zone between well-preserved and poorly preserved foraminiferal assemblages" (Berger, 1974:222). The sedimentary lysocline is what is charted in the shallowing lysocline around oceanic margins, and is what is seen in ancient sediments. (In contrast, the hydrographic lysocline is the depth in the water column at which occurs a dramatic increase in dissolution; the hydrographic lysocline is more uniform in depth than the sedimentary lysocline, being depressed only in regions where carbonate production is high, such as equatorial zones.)

Several alternatives have been proposed to account for the formation of the lysocline, and the question of its origin is not entirely settled. However, different water types of different carbonate ion concentration show that the lysocline has its greatest depth where water of high carbonate ion content extends to the bottom (North Atlantic), has an intermediate depth where water of Antarctic origin is adjacent to the bottom, and a shallow depth where nutrient-rich water of low carbonate ion content overlies water of Antarctic origin, as in the North Pacific (see also Chapter 3). Broecker and Takahashi (1978:93) contended that "to within 200 m the depth of the oceanic lysocline can be explained in terms of the carbonate ion concentration distribution alone."

In addition to the natural variation in depth of the lysocline, which is attributable to the mechanism of its formation, one must also consider the longer-term changes in depth due to position on a tectonic plate. Sediments which are high in carbonate may form on a ridge which is subsequently transported over time into deeper water. Accordingly, any model postulating a close coupling of depth and carbonate dissolution also must take into account historical factors of source region and subsequent tectonic history.

If there is independent evidence for the depth of the CCD and the lysocline (for example, by use of species with different susceptibilities to solution), bathymetric changes can be estimated. Some 12,000 years ago, the calcite lysocline is thought to have been about 1 km shallower than it now is (Berger, Johnson, and Killingley, 1977).

Deep-sea bathymetry for periods of time older than the Jurassic, or for deep-sea sediments now on land, can also be estimated. To begin with, the height of ocean ridge where basalt is released is uniform over the whole ocean, at a depth of 2.6 km ($\pm$ 200 m) (Veevers, 1977). Accordingly, where ophiolite of the ridge system is found, this depth can be assigned to it, assuming no systematic change through time. Aragonite and calcite would accumulate at this depth, but go into solution as the depth increases beyond the CCD. Silica persists yet deeper. The "red clay" at 4,500 m and deeper "is uniquely restricted to the deep-sea environment" (Berger, 1974:217). (It is possible that prior to the development of carbonate-secreting phytoplankton in the mid-Mesozoic, cherts would have formed at shallower depths if deep-sea carbonates, being less abundant, did not dilute them.)

A sequence of ophiolite, aragonite, calcite, quartz, and finally clay can be interpreted as a sequential deepening owing to lateral movement of a plate below the various compensation depths. Since approximately 20 m.y. are (now) required to move below the carbonate compensation depth, deep-sea calcareous layers would not normally exceed 200 m at standard deep-sea sedimentation rates (< 10 m/m.y.).

In the Jurassic Alps, the sedimentological sequence is reversed (ophiolite covered by siliceous rocks covered by carbonate). Bosellini and Winterer (1975) and Hsü (1976) believed this sequence reflected an initial very shallow CCD (shallower than 2,500 m to permit dissolution of all carbonate and accumulation of siliceous radiolarite directly on the ophiolite), followed by a deepening CCD (which first permitted calcitic cephalopod jaws [aptychus] to accumulate, followed later by a yet deeper CCD (which allowed the aragonitic cephalopod shells to be preserved).

Geochemical and Mineralogical Criteria

A few geochemical or mineralogical criteria exist for gauging absolute depth, and several criteria are used for estimating relative depth. These various attempted methods include observations on (1) lava, especially vesicle size, (2) iron minerals, (3) pressure-sensitive mineralogical phases, and (4) a hodgepodge of other geochemical trends.

(1) Perhaps the most utilized indication of general bathymetric change among geochemical criteria is shown by lava erupted into the ocean. Sea level is marked by the transition from sheet lava to pillow lava and breccia (Jones and Nelson, 1970:15). With increasing depth (pressure), the volume and diameter of individual vesicules decrease in quenched margins of pillow basalts, resulting in an increase in basalt density (Figure 2-12). Aumento (1971), however, suggested that vesicularity must also be interpreted in terms of the saturation of the magma chamber.

Following the vesicularity criterion, a sequence of Ordovician pillow lavas in Wales was suggested to have been extruded in increasing depth over a range of tens to perhaps 2,000 m (Jones, 1969). And similar deposits in Nevada were believed to have been extruded at 600 m and at 4,000 m (Wrucke, Churkin, and Heropoulos, 1978). The depth of eruption from the crest of the Reykjanes Ridge of Iceland was deduced (by this criterion) to have shallowed from 2,500 m 35 m.y. ago to 600 m 2.7 m.y. ago (Duffield, 1978). Garrison (1974:378) summarized other applications of this bathymetric indicator.

Varioles are pea-size spherules which occur in basic igneous rocks, and are thought to form by liquid immiscibility in the basaltic magma at the time of eruption. These are frequently reported in pillow lavas which formed under deep-water conditions (1,600 to 5,000 m), and are unknown in subaerially erupted basalts (Furnes,

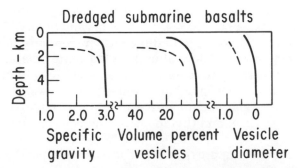

Figure 2-12. Increasing specific gravity, volume percent vesicles, and average vesicle diameter (in mm) plotted against increasing ocean depth, for dredge hauls of submarine basalt. Solid curves are for rocks dredged off Hawaii; dashed curves are for rocks dredged from Revillagigedo (west coast of Mexico). After Moore, 1970:340

1973). Gas cavities or vesicules are sparse in or absent from basalts with varioles, and their absence also suggests a deep-water origin.

(2) A gradient exists in some places in the sequence of iron minerals extending seaward on the continental shelf (Porrenga, 1967a). Nearest shore (0 to 10 m), detrital geothite ($HFeO_2$) may appear. In burial and compaction, it may be dehydrated and become hematite (Fe_2O_3) if a high Eh can be maintained (i.e., very little organic matter). Alternatively, with considerable organic matter and under reducing conditions, geothite will disappear and pyrite (FeS_2) is the most abundant species. Indeed, the occurrence of pyrite in sedimentary rocks means almost without exception that organic matter was present initially. In the sense of indicating organic matter, pyrite is a very good "fossil."

Offshore, the greenish-black minerals chamosite and glauconite appear: chamosite at 10 to 50 m and glauconite at 125 to 250 m, off the Niger delta (however, Nikolaeva [1972] considered that glauconite displays climatic zonality and occurs north and south of an equatorial "nonglauconite" region). Chamosite is the iron-rich member of the chlorite group of silicates with a formula of $(Mg, Fe, Al)_6(Al, Si)_4O_{10}(OH)_8$. Glauconite has a structure like muscovite, with a formula of $K(Fe, Mg, Al)_2(Si_4O_{10})(OH)_2$. Chamosite and glauconite appear to form by alteration of preexisting clay minerals subject to interstitial fluids much higher in iron than in the environment from which the clay minerals weathered. The iron appears to be released by the reduction of iron oxides in sediments owing to the reducing conditions and lower pH (< 7) that follow decay of organic matter.

The sequential arrangement of the iron minerals geothite to chamosite to glauconite has been reported for a Tertiary sedimentary sequence of the eastern United States (Owens and Sohl, 1973), and is logically interpreted as a gradual increase in depth. In addition, in the Tertiary example, low alumina glauconite was associated with sediments attributed to deeper water, and high alumina glauconite with deposits believed to have formed in shallower water (although the converse relationship with alumina was reported by Distanov and Sorokin, 1975; and this indicates the need for further study).

(3) Pressure increases approximately 1 atmosphere per 10 m. Accordingly, chemical reactions for authigenic minerals which are pressure sensitive over the range 1 to 1,000 atmospheres might yield evidence of paleobathymetry, if a nonreversible reaction took place. Prime suspects for such reactions would be in phases of man-

ganese, phosphate, or sulfide minerals, but pressure-sensitive phases of these minerals have not yet been reported in natural materials. Lowenstam (1972) found no depth-dependent changes in aragonite (to 6,200 m), calcite (to 8,900 m), opal (to 5,200 m), and dahllite, amorphous phosphate minerals, magnetite, geothite, fluorite, and weddellite (to 3,500 m).

Phases of manganese oxides which occur with iron and phosphate in nodules have been attributed to pressure effects, but thermodynamic calculations now show that pressure would have a negligible influence (Glasby, 1972a, 1972b). The principal manganese phase in nodules from open ocean localities is δMnO_2 (most abundant at 0.5 to 3 km; Barnes, 1967). For unknown reasons (but apparently related to redox potential and not to pressure), 10 Å manganite (Mn·OH, todorokite) is the principal mineralogical phase of continental margin manganese nodules (and most abundant at 3.5 to 5.5 km). Manganese nodules are known from more than a dozen deposits from the Cambrian to the present (Jenkyns, 1977), and are best described from the Jurassic of Sicily and the Cretaceous of Timor. Fine-grained deep-sea sediments chiefly composed of iron and manganese oxides ("umbers") are known from many regions (Robertson and Hudson, 1974), but their mineralogy has not yet been described from the point of view of evaluating pressure-related (i.e., bathymetric) changes.

(4) There are several general geochemical trends that roughly are related to increasing depth. Some of these trends parallel changes in sediments from coarse to fine (i.e., sands to clays), and therefore the geochemical trends may be merely second-order correlations with sediment type. Other trends parallel a depth-dependent factor such as temperature. Distinguishing the environment of origination from the present environment in which the mineral is found is often very difficult (Calvert, 1976).

Owing to the complex interrelationships among some minerals, transportation, local depositional environment, and diagenetic rates and processes, binary plots of element abundance or concentration versus depth may obscure basic relationships. More likely to be helpful in many circumstances are multivariate statistical analyses. Spencer, Degens and Kulbicki (1968) used multivariate methods to identify several factors which were responsible for elemental distributions. They identified five principal factors which were correlated with elemental distribution: carbonate, quartz dilution, oxidation-reduction potential, illite-glauconite, and montmorillonite. None of these factors showed a simple relationship with depth, though in any given section, any of them could be correlated with depth. Thus data on elemental distributions are substantiating rather than primary evidence for bathymetric analysis.

Füchtbauer (1963) reported that under conditions of lower pH and strong oxidation, reddish brown biotite (a clay mineral) and olive-green tourmaline (a boron and aluminum silicate) have been found only in brackish marine sediments, whereas brown and blue-green tourmaline is "enriched relatively in limno-fluviatile rocks of the same age."

Some additional mineralogical and geochemical trends related to bathymetry include the following.

(1) In carbonates at 0 to 100 m, compared with those at 3,200 m, Pilkey and Blackwelder (1968) reported the following trends: the amount of low-Mg calcite increased from about 35 percent to 95 percent; the amount of aragonite decreased from 50 percent to 2 percent; and high-Mg calcite decreased from 15 percent to 3 percent. Chilingar (1963) had concluded previously that the Ca/Mg ratio increased with depth and distance from shore.

(2) Concentrations of several trace metals change with depth (Wedepohl, 1959; Tourtelot, 1964; Nicholls, 1967) and parallel an increase in finer-grained sediments.

(3) Any temperature-related trend (see Chapter 4) will parallel bathymetry. For example, oxygen isotope values of ahermatypic corals parallel temperature, and therefore depth (Weber, 1973a).

In summary, geochemical and mineralogical criteria for determining ancient bathymetry are far easiest to use in investigating Pleistocene and Tertiary deposits. Perhaps of greatest use in Paleozoic deep-sea deposits is the physical evidence of change in basalt vesicularity as a function of the depth of eruption.

Biological Criteria

Biological criteria are more explicit and of more general use for estimating bathymetry of sedimentary rocks than any of the other approaches thus far presented. After one determines whether a sedimentary deposit formed seaward of the shoreline, the next general decision is whether or not it formed in the photic zone.

Four basic approaches allow biological data to be used to determine bathymetry: (1) Some biochemical adaptations (such as the ability to photosynthesize) are depth (i.e., light) dependent; (2) some mechanical adaptations (such as shell strength) are depth (i.e., pressure) related; (3) data on depth of living taxa can often be applied to their fossil relatives; and (4) some changes in general properties of organisms, such as their diversity or size, are related to depth.

Table 2-6 gives selection of depth-related aspects of organisms useful in bathymetry that are not discussed in the text. The accuracy of a given depth is probably ± 10 to 50 m (depending on the method) for shelf depths, and ± 200 to 500 m for estimated depths of a kilometer or more. Relative depth is much easier to establish, and to contour, than is absolute depth.

(1) The outstanding physical or chemical property of organisms which is depth dependent is the limitation of photosynthesis to depths generally shallower than the upper 100 m (and an absolute depth limit of approximately 200 m). Light penetration is a function of the amount of suspended matter in water (Manheim, Hathaway, and Uchupi, 1972:Figure 4) so that near shore, photosynthesis may be limited to the upper 5 m. *The distinction of whether a fossil deposit was within or below the photic zone is the most important decision that can be made regarding the depth of deposition of shelf deposits.* Encrusting algae, including blue-greens (which form layered deposits called stromatolites) and reds (which form rocklike encrustations called rhodolites), usually indicate in situ conditions. Many of the green algae disintegrate on death and thereby contribute an immense volume of calcareous sediments, some of which may be transported to deeper water and would be recognized as such owing to their inclusion in stratigraphic sections with graded beds and turbidites.

The effect of light on limiting photosynthesis to shallow depths is also evident in the distribution of reef-building (hermatypic) corals with their symbiotic zooxanthellae (dinoflagellates), and this may apply also to the Cretaceous rudistid bivalves (Vogel, 1975). Foraminifera with a radial hyaline wall structure also harbor algae, and therefore often have a greenish appearance (Haynes, 1965). These foraminiferal-algal symbiotic relationships occur in both planktonic and benthic species of the photic zone.

Another adaptation is dependent upon the absence of light, namely, blindness. The incidence of occurrence of degenerate, small, pigmentless eyes in modern crustacea increases markedly in forms from depths greater than 500 m (Clarkson 1967:372). Presumably, this adaptation obtained also for deeper-water trilobites during the Paleozoic, particularly the small phacopid and proetid trilobites of the upper Devonian and early Carboniferous.

(2) Some mechanical adaptations related to structural support may be associated with depth (i.e., pressure), and one example is

Table 2-6 A selection of depth-related aspects of organisms of use in paleobathymetry, other than those discussed in the text. The selection was made to emphasize the diversity of aspects covered.

Reference	Group and bathymetry	Possible limiting cause
Playford, et al., 1976	Stromatolites: up to 100 m below sea level	Available light
Swinchatt, 1969	Algal borings: probably less than 20 m; none below 50 m	Available light
Adey and Macintyre, 1973	Crustose coralline algae: depth distribution	Available light
Byrnes, 1968	Receptaculitids: if these fossils are dasyclad algae, then they may not occur deeper than 12 m	Available light
Phleger, 1964	Foraminifera faunal assemblage changes at depth of 13 to 18 m	Boundary of intense water turbulence
Orr, 1967	Foraminifera: within a species, wall thickness increases as the form assumes a deeper water habitat (>500 m)	Rate of growth slower, but live to be very old(?)
Murray, 1971	Foraminifera: shelf seas of normal salinity have a much higher diversity than hyposaline or hypersaline nearshore shelf seas	Less speciation in less typical habitats
Berger, 1969	Abundance of both benthic and planktonic living foraminifera decreases with depth	Decreasing food?
Bandy and Arnal, 1969	Increase in maximum size of individuals at depth greater than 200 m	Live to be older?
Bandy, 1960b	In uvigerinid foraminifera, sculpture changes with depth: striate (100 to 200 m) to costate (200 to 1,500 m) to spinose (1,500 to 2,000 m) to papillate (2,000 to 5,000 m)	Unknown
Delaca and Lipps, 1972	Firm and close attachment of foraminifera to substrate in shallow water	Permit occupation of sites where current or wave action is great
Bandy, 1960a	Spindle-shaped types of foraminifera are characteristic of 20 to 80 m depth: many other depth-correlated structural changes are indicated	Unknown

Reference	Observation	Explanation
Stevens, 1969	Fusilinids: adult fusilinids reproduced in water deeper than 12 m	Unknown
Ross, 1968 and earlier	Fusilinids: different species characteristic of region from surf base to the lower part of the photic zone; most Pennsylvanian and Permian species characteristic of the region shallower than surf base, usually in less than 4 to 5 m depth	Unknown
Reid, 1968	Sponges: calcarea decrease and hexactinellida increase with depth	Carbonate less available and silica more available at depth?
Abbott, 1973	Stromatoporoid shape and environmental "energy" are closely related	Unknown
Schopf, 1969	Bryozoans: ratio of erect to encrusting species increases with depth	Fewer substrata for encrusting species
Schopf, 1969	Bryozoans: among erect species, rigid forms occur in greater abundance in deeper water, whereas flexible forms occur at any depth	Flexible forms can "shed" sediment whereas rigid forms cannot
McKee, Chronic, and Leopold, 1959	The number of species of spores, pollen, dino-flagellates, diatoms, microforaminifera and sponge spicules per gram of sediment, increases with depth from 0 to 80 m	Quieter, deeper water allows smaller particles to settle out
McAlester and Rhoads, 1967	Bivalve adaptations for deep vertical burrowing are common only in intertidal and shallow subtidal environments; bivalve adaptations for near-surface horizontal burrowing predominate in deeper subtidal environments	Highest stress occurs in shallowest water
Collins, 1978	Coiled cephalopods in vertical position limited to less than 10 m depth	Buoyancy lost at greater pressure
Howard, 1972	Crustacean burrow: shoreline types	Unknown
Foerste, 1930	Color markings much more abundant in shallow-water species	Various

given here. Implosion of cephalopod shells is related to the strength of the shell, and ability to withstand pressure is a function of the ratio of wall thickness to radius of curvature. The greater the wall thickness and/or the lower the radius of curvature, the more shell strength increases. Most cephalopods apparently were limited structurally to depths of less than 300 m, although shells of some species may not have imploded until depths of more than 1,000 m (Westermann, 1973). Laboratory experiments showed that *Nautilus* seemingly could not live below 700 m (Saunders and Wehman, 1977), and living *Nautilus* placed in cages imploded at depths of 750 to 900 m (P. Ward, personal communication, 1979).

(3) For fossil species or genera with living relatives, the depth range of the fossil assemblage often can be inferred. Figure 2-13 il-

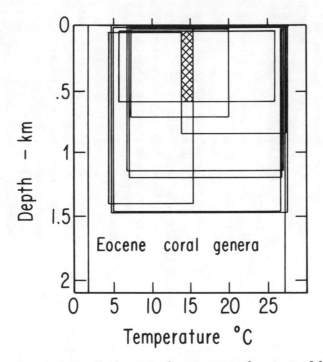

Figure 2-13 Each rectangle represents the range of depth and temperature for one particular genus of ahermatypic coral (middle Eocene, Gulf Coast of U.S.). The probable depth and temperature range for the entire deposit is indicated by the dotted region close to 15°C and 50 to 600 m. With this narrow temperature range, the depth could be more precisely estimated by learning the typical depth of 15° water for tropical temperature-depth curves. After Wells, 1967:362.

lustrates this technique applied to an Eocene deposit with nine genera of solitary corals. Although individual genera may range to
2,000 m, and exist in a temperature range of 0° to 27°C, data for all
genera limit the probable depth from 50 to 500 m, and the temperature to approximately 15°C. For species of foraminifera, 13 paleobathymetric zones were established between the depths of 0 to
2,000 m "by noting the modern depth distributions for species of the
Miocene and comparing modern and fossil homeomorphs" (Bandy
and Arnal, 1969:787). Examples of this general method of estimating bathymetry are numerous.

In some cases, the similarity between the recent organism and
its fossil counterpart is morphologic in only a general sense. For
foraminifera in water depths of 0 to 150 m, the relative proportions
of arenaceous benthics, calcareous benthics, and planktonics change
so that the proportions are different for every depth (Figure 2-14).
The proportion of planktonic to benthic species was found to increase systematically from 10 percent near the shore to 50 or 60 percent at the shelf break (Stehli, 1966). Studies a decade later (Murray, 1976) confirmed the general pattern for shelf depths. In deeper
water, however, the planktonics may undergo considerable solution,
the calcareous benthics probably less, and the arenaceous benthics
do not dissolve at all. Accordingly, ratios of different types of foraminifera in deeper-water deposits must be carefully evaluated for

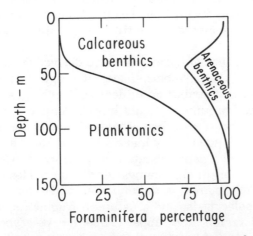

Figure 2-14 Variation in percentages of present-day arenaceous
benthic, calcareous benthic, and planktonic foraminiferans as a function of water depth in a traverse normal to the shore and extending
offshore to a depth of 150 m. Note that there is a unique assemblage
associated with every depth. After Stehli, 1966:224.

diagenetic effects (Berger, 1973). Deep-water deposits containing only agglutinated tests indicate a depth of deposition in excess of the regional calcium carbonate compensation depth (Prince et al., 1974).

Even with so-called trace fossils (all fossils are only traces of life), characteristic types occur in different depth zones (many examples cited in Crimes and Harper, 1970, 1977; and Frey, 1975). In shallow water, animals dig vertical burrows into which they retire as the tide goes out. In deeper water, horizontal burrows are more common and tend to be associated with food-gathering activities, with different species and burrow types occurring at different water depths.

(4) Both stratigraphic and lateral sequences of ancient faunal and floral "associations" (i.e., the communities of much of the paleontological literature) can be established from field evidence. From such reconstructions, the diversity of specific groups can be plotted. For both the upper Ordovician and lower Silurian, brachiopod diversity has been documented to increase offshore (Anderson, 1971), paralleling a gradually increasing depth of water toward the continental margin. This trend has been noted in benthic faunas of many geologic ages (McKerrow, 1978:58, 88, 121, 148). Modern foraminifera display a similar trend, with diversity increasing seaward over the shelf. Maximum foraminiferal diversity occurs when environmental conditions stabilize, which is at water depths of about 15 m off Georgia (Sen Gupta and Kilbourne, 1974) and at the shelf edge at the Bay of Biscay (Schnitker, 1969). In general, foraminiferal diversity increases seaward away from the shore (Gibson, 1966).

Even though many notions of bathymetry are derived from observations of living species, the paleobathymetry of a region must of course be inferred from the distribution of death assemblages. An analysis was made of dead organisms found in sediments collected at 60 stations distributed across the continental shelf and upper continental slope off New England (Wigley and Stinton, 1973). The results (Figure 2-15) indicated a depth zonation in prominent groups with centers of concentration in shallow waters for barnacles and the common sand dollar of the area, in the mid-shelf and outer-shelf regions for gastropods, pelecypods, and crustacean fragments, and in the outermost-shelf region and upper slope for pelecypods, scaphopods, a common echinoid, and an abundant cup coral. In deeper water, fish otoliths and cephalopod beaks were the most characteristic hard parts.

Hertweck's (1971) excellent studies of shallow-shelf sediments

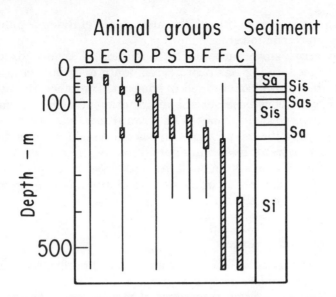

Figure 2-15 Plot of the distribution of hard parts in bottom sediments off New England, arranged according to water depth and sediment type. From left to right, animal groups are B (barnacles), E (*Echinarachnius,* the sand dollar), G (gastropods), D (decapods), P (pelecypods), S (scaphopods), B (*Brisaster,* an echinoderm), F (*Flabellum,* a cup coral), F (fish), and C (cephalopods); sediment abbreviations are Sa (sand), Sis (silty sand), Sas (sandy silt), and Si (silt). Depths of greatest abundance may not correspond to depth of greatest abundance of living population. After Wigley and Stinton, 1973.

and fauna of the Mediterranean documented in detail the fact that depth ranges for the living assemblages are much more restricted than depth ranges for the shells. The reason for this is that shells are widely distributed owing to the influence of extreme weather conditions. Brackish and fresh-water clams are extremely abundant as much as 400 to 500 m seaward from the shore. A paleontological interpretation would assign these deposits to a lagoon or to a brackish water region when that emphatically is not the case.

Along the Atlantic margin of the United States, the depth zonation of shell debris on the full shelf is also not precisely identical to the position of living populations (Milliman, 1972). In general, molluscan debris is so much more abundant than any other type of shell debris that it swamps all other sources. Therefore, these other shell types (barnacles, bryozoa, echinoids, foraminifera, algae, etc.) achieve prominence only where molluscan debris is sparse, gen-

erally in deeper water, even though the main depth of living populations may be much shallower.

To summarize, since each of the hundreds of thousands of
marine species now living has a preferred depth distribution, one
may presume that this also obtained in the past. Therefore, the use
of arguments by analogy based on the distribution of modern forms
may be excellent, especially if based on several taxa of a given assemblage. Certainly the use of algae allows one to be certain a deposit was formed in water less than approximately 100 m deep, and
very likely less than about 20 m deep. As with other methods, however, it more often than not is impossible to distinguish 5 m from,
say, 25 m, or 50 m from 250 m.

General Bathymetric Models

Conclusions on bathymetry for individual stratigraphic sections
lead directly into considerations of regional depositional models. I
will consider here two contrasting types of models for continental
platforms: (1) the *epeiric sea* model and (2) the *marginal sea* model.

(1) Shaw (1964) and Irwin (1965) emphasized a depositional
model of epeiric seas in which a very shallow nearshore "low-energy" zone up to "hundreds of miles" wide was followed seaward by a
"high-energy" belt "tens of miles" wide, leading further seaward
into a "low-energy" zone "hundreds of miles" wide (Figure 2-16).
Such an epeiric sea was envisaged to have been 600 to 1,000 km
wide, and to parallel the continental margin for at least this distance. Modern counterparts closest to the presumed magnitude of
ancient epeiric seas are Hudson Bay (1,000 km in radius), the Persian Gulf (850 × 300 km), and the region of the South China Sea,
Java Sea, Timor Sea, and Arafura Sea (if sea level were raised 60 m
by melting of glaciers, there would be two regions, one east of Malaysia, the other north of Australia, each of which was 1,000 km ×
1,000 km). In the ancient epeiric seas, salinity was presumed to
have increased gradually from the margin toward the interior, and,
owing to the shallowness, the water was to have been isohaline from
top to bottom. In modern seas, this type of situation occurs in Shark
Bay, Australia (where salinities inshore eventually reach 65‰), and
west of Andros Island in the Bahamas (where salinities range to
about 45‰). However, in these modern examples, the distance over
which this salinity gradient is maintained is only about 80 km in
each case—not the hundreds of kilometers hypothesized by Shaw
and Irwin for epeiric seas.

In North America the most extensive shallow seas—the most

Epeiric sea sedimentation – Shaw & Irwin Model

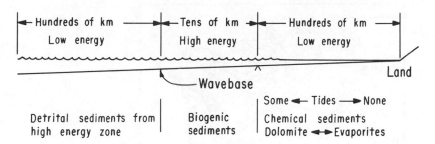

Figure 2-16 Schematic representation of presumed energy zones in epeiric seas, as used by many authors. Note the scale. After Irwin, 1965.

truly epeiric—existed in the early Paleozoic. In the maximum transgression of the late Cambrian *Saukia* zone, "algal shoals" are suggested to have occupied an area 800 km wide and 1,000 km along depositional strike (Lochman-Balk, 1971), and the shoreline is drawn an additional 100 to 500 km landward of the shoals. All of this area ($\simeq$ 1,300 km $\times$ 1,000 km) would be in the photic zone. This late Cambrian sea would fit the scale postulated by Shaw and Irwin.

For the late Cambrian and early Ordovician carbonates, the epeiric sea model has been used in another context because evaporation was expected (in the right climatic zone) over a considerable area. These early Paleozoic limestones occurred near the continental margin with contemporaneous, subtidal dolomites extending landward (Figure 2-17). The idea was that evaporation over the central part of the epeiric sea increased salinity, with Mg, in particular, increasing in interstitial fluids, thus enabling dolomite to form. In this interpretation, many stromatolite horizons were also viewed as subtidal. Some thousands of square kilometers of carbonate as much as 1,000 m thick have been dolomitized. This model seems to require that a fairly restricted set of climatic conditions be in existence for millions of years, over vast distances.

The model of epeiric seas has been utilized by workers for many regions for rocks of the upper Proterozoic, and virtually every period of the Phanerozoic. No matter how heuristic the epeiric model may be, however, some aspects of it are open to question.

(a) Because of the presence of winds and storms, the extreme view of an inner band "hundreds of miles wide" which was "low energy" seems quite improbable.

(b) The presumed slope of the land surface is such that it would

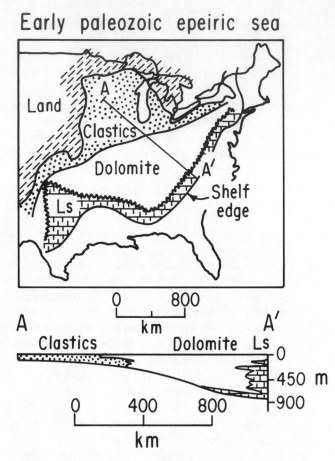

Figure 2-17 Generalized paleogeographic map of late Cambrian and early Ordovician time showing regional distribution of facies of the Knox Group and equivalent rocks for an ancient epeiric sea. The inferred slope is about 1 m/km. After Harris, 1973:64.

require a nearly horizontal surface over hundreds of kilometers. Even though modern continental shelves have a very gradual slope, the shelf is still 150 m deep within approximately 100 km of the present shore, and in some places is a good deal deeper. Nearly flat surfaces of the magnitude predicted for epeiric seas originate only by sedimentation, as in basins, not by erosion. Therefore, a sea nearly everywhere less than 30 to 50 m deep over distances of 500 to 1,000 km seems physically improbable.

(c) An alternative explanation for the deposits of the lower Paleozoic is, as already suggested by Shinn, Lloyd, and Ginsburg

(1969:1227), that "such wide spread tidal-flat rocks represent many transgressive and regressive tidal-flat packages spread out through time and gradual subsidence." A second alternative explanation for these widespread lower Paleozoic carbonates is that they represent deposits of the highest intertidal region similar in tidal position, but not in regional climate, to the Ranns of Kutch (see p. 82). There, an essentially flat surface extends inland for approximately 300 km along 100 km of the coast.

(d) Improper paleogeographic reconstructions for the early Paleozoic (and many other times) may have accentuated the presumed size and distance from the ocean of these shallow seas.

(2) The model for *marginal seas* is much more easily visualized (Figure 2-18) than is the model for epeiric seas because of the existence of numerous modern analogues. The average width of the present inner continental shelf (depth of 0 to 65 m) is 16 km, with 68 percent of the world continental shelf between 3 and 80 km wide (Hayes, 1964). The outer shelf (depth 65 to about 130 m) is about 50 km wide, on the average. Thus the average overall distance from shoreline to shelf break is now approximately 70 km. Landward of the shelf is of course the intertidal region, but this cannot extend too far. Shinn, Lloyd, and Ginsburg (1969:1227) summarized their views of intertidal environments by writing that "with tides and currents as we know them, it seems impossible for a single tidal flat to be much more than 10 miles wide at any one time." Thus modern

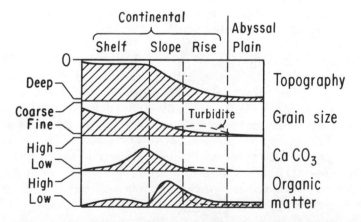

Figure 2-18   Generalized diagram showing idealized continental margin and character of sediments in each subdivision of the margin. Dashed lines are for interbedded turbidites. After Emery, 1965:1381.

marginal seas, including the adjacent intertidal region, are characteristically 100 km or less in width.

Many geologic situations appear to represent deposits from seas less extensive than those envisaged in epeiric sea models, but without a discernible shelf break, as always occurs at the seaward edge of marginal seas. In a paper of 1937, M. K. Elias illustrated the type of paleobathymetric analysis which was to become important a quarter of a century later (Figure 2–19). By following the sequence of lithologies and of faunas and floras, and by carefully noting modern counterparts, Elias established a pattern of depth zonation in the Permian seas of the mid-continent of North America. The ecologic significance of some of the evidence has subsequently been reexamined (see Elias, 1964), but the innovative method of analysis which Elias pioneered has now become widely used.

Many regional bathymetric studies have been undertaken. Especially notable for its combination of sedimentological, biological, and geochemical criteria is Sellwood's (1972) reconstruction of the bathymetry of Jurassic shelf deposits of England. That Cretaceous chalk deposits of northwestern Europe accumulated at outer-shelf depths (∼100 to 250 m) is described in the excellent paper of Håkansson, Bromley, and Perch-Nielsen (1974). Several paleobathymetric analyses have been conducted on reef facies because of their potential in petroleum exploration. For example, shallow-water paleobathymetry for a Carboniferous reef of England was presented by Broadhurst and Simpson (1973:378). The last example I cite of general bathymetric studies is the very instructive debate of Gibson (1967, 1968) and Leutze (1968) on the depth of the Miocene shelf off the southeastern United States, in which faunal evidence is critically examined.

In addition to interpretions of paleobathymetry for a given moment in geologic time, changes in bathymetry through time are also of interest. As is well developed in stratigraphy textbooks, the investigations of Johannes Walther (1860–1937) demonstrated that a sequence of environments that follows itself vertically must also have existed laterally (Middleton, 1973). Therefore, if bathymetry can be inferred for successive time planes, the net subsidence or uplift can be mapped (Bandy and Arnal, 1960). The result is a tectonic map which provides evidence of the controls over patterns of sedimentation. Similarly, a rapid change in bathymetric indications of faunas suggests that the intervening intermediate depths were removed by faulting, erosion, or another type of drastic environmental change.

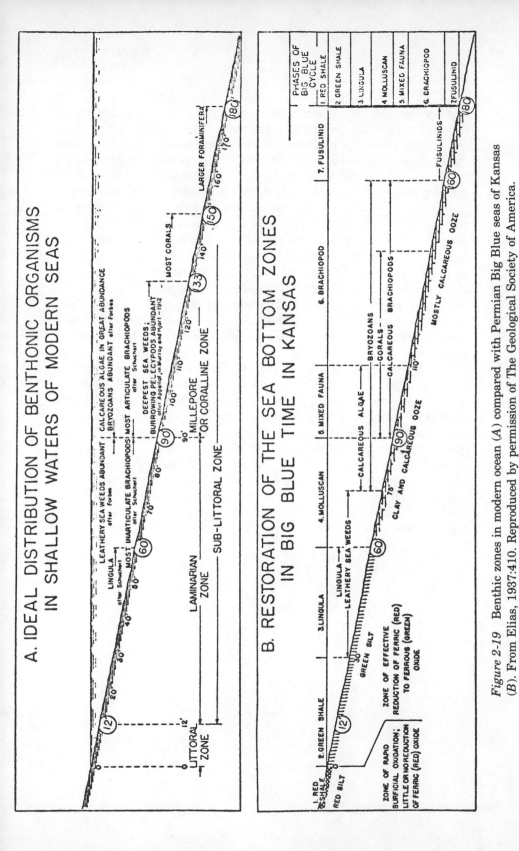

Figure 2-19 Benthic zones in modern ocean (*A*) compared with Permian Big Blue seas of Kansas (*B*). From Elias, 1937:410. Reproduced by permission of The Geological Society of America.

Summary

In this chapter I have summarized sedimentological, geochemical, mineralogical, and biological criteria for the establishment of bathymetry. The only bathymetric decisions which can usually be made are in relation to (1) the shore, (2) the photic zone (usually 0 to 30 m), and (3) the depth of the ophiolite ridge ($\simeq$2,600 m). Relative depth (rather than absolute depth) is much easier to estimate in sedimentary sequences, and sedimentary units can often be tied to the shoreline, the photic zone, and the ridge crest. Because of the precise relationship between age of ocean crust and its depth, we have the remarkable conclusion, completely unforeseen a few years ago, that deep-sea bathymetry for the Jurassic to the present is more easily and more accurately determinable (say, within 10 percent) than is shallow-water bathymetry. This approach and the now widespread appreciation of the use of the hypsometric curve as a way to gauge sea-level changes combine to give paleobathymetry a sound basis.

The two types of continental platforms which have been of greatest interest in sedimentary geology are marginal seas and epicontinental "epeiric" seas. These two settings provide very different models of bathymetric relationships. Only the marginal seas are well represented at the present time, and it appears to me that epeiric seas on a scale envisaged by Shaw and Irwin may not have existed. Many inferred seas of the geological past seem to be on a much smaller scale than the epeiric seas. Deep-sea lithologic sequences go from ophiolite at ridge crests, to aragonite and calcite, to only calcite, and then to quartz and clays as the oceanic plate moves laterally into deeper water and sinks below the lysoclines.

3. WATER STUDIES

Oceanography is such a young science that it is not possible to tell whether these water masses of the world ocean are in complete equilibrium with present climatic conditions or whether adjustment is still taking place.

L. V. Worthington, 1968

Grouped in this chapter are three rather disparate topics—tides, currents, and circulation—united only by their common concern with water movement. For each of these I first present the modern distribution or causes, then proceed to discuss how it is that each of them can or could be recognized in the geologic record, and finally consider their geologic record.

Tides

With respect to tides I will consider their (1) origins, (2) occurrence on continental shelves, (3) geologic indicators, and (4) occurrence over geologic time.

(1) Ocean tides are largely dependent upon earth-moon interactions, as shown in Figure 3-1A. At E, the *center* of the earth, each kilogram is attracted to the moon with a gravitational force (which happens to be 3.38 mg weight). This is countered for the whole of the earth by a centrifugal force which is equal in magnitude (3.38 mg) but opposite in direction. Such a system is in equilibrium when the sum of the forces is equal to zero. The surface of the side of the earth which is closer to the moon has a different set of forces acting on it than does a point in the center of the earth (which is truly at equilibrium) or a point on the opposite side of the earth. At point Z in Figure 3-1A, which is on the side of the earth close to the moon, there is a higher than normal gravitational attraction (equal to 3.49 mg). If the earth's centrifugal force (of 3.38 mg) is subtracted from the gravitational attraction, there remains a net absolute difference (of 0.11 mg), which accounts for the tidal bulge toward the moon.

On the opposite side of the earth at N (Figure 3-1A), there is a lower than normal gravitational attraction (equal to 3.37 mg). As before, there remains at N a net absolute difference (of 0.11 mg) between the gravitational attraction and the centrifugal force

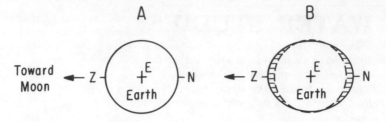

Figure 3-1 Diagram showing generation of tidal forces; letters *Z*, *E*, and *N* are referred to in the text. *A*. Equilibrium between the total attractive force of the moon and the total centrifugal force of the earth. *B*. Deformation of a continuous oceanic envelope caused by tide-generating forces. After Defant, 1958:27, 41.

(3.38 mg). Thus any water at point *N* will be held less closely to the earth's surface than is water at points nearer to the moon, and therefore water will "escape" from the earth's surface, thus creating a tidal bulge on the side of the earth opposite the moon.

According to Newton's law of gravitation, bodies are attracted in proportion to the product of their masses and inverse proportion to the square of their distance apart. In the case of the earth, moon, and sun, the much closer distance between the earth and the moon compensates for the much greater mass of the sun, which is much farther away. The tide-generating force of the sun is approximately half that of the moon, and so when the sun and moon are lined up, their combined force results in tides one and a half times normal tides. Similarly, if the sun and moon are 90° apart, their cancelled forces result in tides of about half of normal magnitude. Unusually high tides are known as spring tides (from the German *springen,* meaning to leap) and unusually low tides as neap tides. Spring and neap tides thus march around the earth (in theory at least) as the position of the moon (and sun) change with respect to the earth, as in Figure 3-1*B*.

The further prediction of tides depends upon a knowledge of the resultant forces generated by the sun and moon as each travels in an elliptical orbit with varying declination and distance from the earth. The most important tidal constituents are listed in Table 3-1. The main lunar (M_2) and solar (S_2) components account for more than 80 percent of the tidal force.

This generalized view of tidal behavior requires additional information before coastal tides can be predicted. As ocean water over the deep sea meets and is piled up on the shallower continental shelf, the speed of the water over the shelf is sharply reduced. Because of differences from place to place in water depth on the shelf

Table 3-1 The most important constituents of the tide-generating forces.
Abbreviations are M for moon; S for sun; O refers to oscillations associated with the
moon's declination; P refers to oscillations associated with the sun's declination; and
K refers to the sun-moon interaction with regard to declination; N refers to lunar
variation with distance. Subscript (1) indicates a diurnal tide; (2), semidiurnal; and
(f), fortnightly. Data from Defant (1958:48).

Tide	Symbol	Period in solar hours	Amplitude $M_2 = 100$	Description
Semidiurnal tides	M_2	12.42	100.00	Main lunar (semidiurnal) constituent
	S_2	12.00	46.6	Main solar (semidiurnal) constituent
	N_2	12.66	19.1	Lunar constituent due to monthly variation in moon's distance
	K_2	11.97	12.7	Soli-lunar constituent due to changes in declination of sun and moon throughout their orbital cycle
Diurnal tides	K_1	23.93	58.4	Soli-lunar constituent
	O_1	25.82	41.5	Main lunar (diurnal) constituent
	P_1	24.07	19.3	Main solar (diurnal) constituent
Long-period tides	M_f	327.86	17.2	Moon's fortnightly constituent

and in shelf width, shoreline localities a few hundred kilometers
apart may have the time of high tide differ by several hours, as
along the east coast of the United States (Emery and Uchupi,
1972:231). From a paleotidal point of view, this difference in time of
high water would make co-time maps showing crest lines of high
tides difficult to prepare. For many geological situations, however,
the relative time of high tide is not nearly so important as is recognizing relative height of the high tide.

(2) Tides of continental shelves are higher than tides of the open
ocean. Over much of the ocean, the tidal oscillation has an elevation
of approximately 0.8 m over distances of thousands of kilometers.
However, the tide at the shore is modified by the co-oscillation of
water over the continental shelf (Redfield, 1958b). The initial standing wave engendered by tide in deep water builds up as water shal-

lows from 3.5 km to 0.2 km over the short distance of the continental rise and slope. The tide is augmented further by any geographic
features which funnel water shoreward.

For the Atlantic margin of the eastern United States, as shown
in Figure 3-2, there is a close positive relationship between the
width of the continental shelf and the height of the tide at the shoreline. On a worldwide basis, a definite linear relationship obtains between tidal height and shelf width for eight regions up to 300 km
wide, and one region 500 km wide (Cram, 1979). The absolute
values of tidal height, however, are more a function of local geography. At a shelf width of 250 km, the tidal range was 1 to 5 m,
with a mean near 2 m.

The general conclusion that semidiurnal tides of extreme magnitude occur in areas of the widest continental shelves is of basic importance for paleoceanography. Figure 3-3 shows by inspection that
the coastal areas with tidal ranges of more than 3 m are associated
with wide shelves. Such regions are also characterized by sand
ridges and high current velocities. Redfield (1958b:447) computed
for the east coast of the United States (Figure 3-2) that "the greatest
amplitude of the M_2 tide occurring at any width of shelf does in-

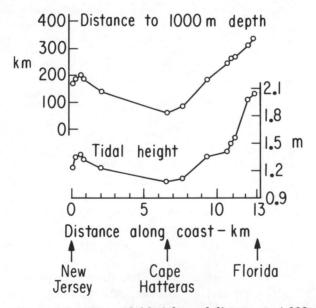

Figure 3-2   Mean tidal height and distance to 1,000 m contour along
the eastern coast of the United States. Note that tidal height increases as shelf width increases. After Redfield, 1958b:435.

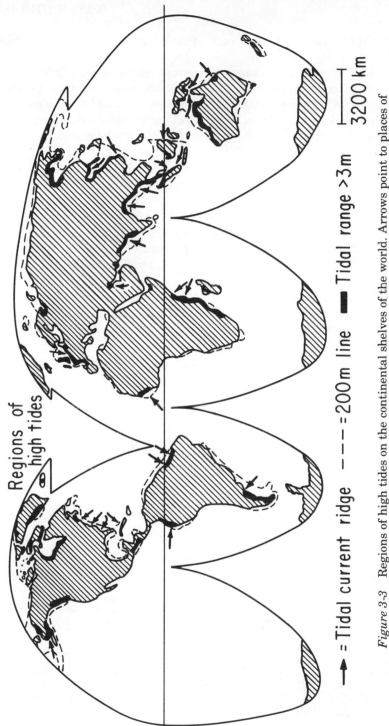

Figure 3-3 Regions of high tides on the continental shelves of the world. Arrows point to places of tidal current ridges. After Off, 1963:332–333, who lists the specific localities.

Regions of high tides

➝ = Tidal current ridge - - - = 200m line ▬ = Tidal range >3m

3200 km

crease with the width at a rate of about one foot per 20 nautical miles."

The question can be asked to what extent would the effect of the main lunar component of the tide be changed if the continental shelf were narrow versus wide. The natural free period (T) of the tidal oscillation will be a function of mean depth (h) and of distance from the shelf break to shore (l), according to the relationship

$$T = 4l/\sqrt{gh} \tag{3.1}$$

where g is the acceleration due to gravity. Note that the period T is directly proportional to length l and inversely proportional to the square root of depth, h. In equation (3.1), if $T = 12$ hours ($= 43,200$ sec), and we assume a mean shelf depth, h, of 40 m, with $g = 10$ m/sec², then l is equal to 216 km. Figure 3-4 shows the distance from the shelf break for the greatest amplitude of the lunar component of tide for various average depths of water over the shelf. For depths of 50 to 100 m, the greatest effect of the lunar tide is 200 to 300 km from the shelf break. This means that the effect of the main component of the tide (the lunar component) will be felt a great distance from the shelf break (assuming a shelf depth of 50 to 100 m,

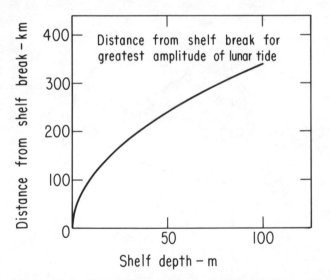

Figure 3-4 Plot of distance from shelf break for greatest amplitude of lunar (M₂) tide, for different average shelf depths. For average shelf depths of 50 to 100 m, the greatest amplitude of the lunar tide is at a shelf width of 200 to 300 km.

and that the shelf is continuous with the open ocean). This relationship says little about the *absolute* tidal height at the shore. Absolute tidal height is governed mostly by the coastal geography, the highest tides at the places into which water is funneled.

In predicting tides in ancient seas, the *first consideration* must be given to paleogeographic setting. Shallow seas that are separated from the open ocean have very low tides because such seas are separated from the resonance of the ocean. Lake Chiem in Bavaria is approximately 13 km wide, and is the smallest lake in which a tide has been demonstrated (1 mm). Lakes on the order of 400 to 1,000 km in the longest dimension may have tides of approximately 6 to 8 cm, and the Mediterranean, with nearly separate eastern and western basins, each approximately 2,000 km long, may have tides 10 to 15 cm (Defant, 1958:Table 4). Brosche and Sündermann (1977) attempted a first approximation of tidal height for the open ocean of the Permian using a numerical simulation method. More work of this type is badly needed with special attention to the continental margins.

(3) Geologic indicators of tidal height occur in both clastic and carbonate sediments.

(a) In clastic sediments, *tidalites* (Klein, 1971) is the name which has been given to deposits which form in the intertidal zone and include the vertical sequence of a surf zone coarse sand (traction load), tidal flat medium sand (swash zone), and tidal flat fine muds (suspension load). Mean low water is indicated by sedimentary structures that result from the runoff at lowest tide. Tidal deposits represent a sequence of "fining-upward" sediments, as shown in Figure 3-5. Examples of tidal deposits from the Precambrian to the Pleistocene are given by Ginsburg (1975) and Klein (1977).

The most important criterion for ancient tidal deposits is evidence of periodic changes in current direction or velocity—the flood and ebb of the tide. This evidence for tidal deposits can take many forms, including (1) alternation of deposition and erosion (so-called reactivation surfaces), (2) alternation of coarse bedload and fine suspension load (which is trapped by ripple troughs at slack water and leaves a texture to the alternating clay and ripples known as *flaser* [*vein,* in German] bedding [if little clay], wavy bedding [≈ 25 percent clay], or lenticular bedding [≈ 50 percent clay] [Reineck and Wunderlich, 1968]); (3) alternation of cross stratification with modes 180° apart ("herringbone" pattern) owing to the ebb and flood of currents, and (4) alternation of exposure and coverage by water (with mudcracks, etc.). Additional features indicative of shorelines were presented in Chapter 2.

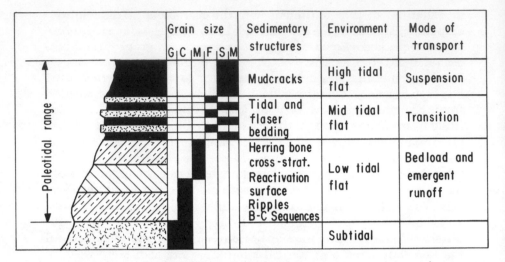

Grain size G|C|M|F|S|M	Sedimentary structures	Environment	Mode of transport
	Mudcracks	High tidal flat	Suspension
	Tidal and flaser bedding	Mid tidal flat	Transition
	Herring bone cross-strat. Reactivation surface Ripples B-C Sequences	Low tidal flat	Bedload and emergent runoff
		Subtidal	

Figure 3-5 Fining-upward sequence of tidal zone deposit showing vertical changes in modal grain size, sedimentary structures, and environments. From left to right, grain size abbreviations are G = granule and pebble gravel; C = coarse-grained sand; M = medium-grained sand; F = fine-grained sand; S = silt; M = mud. Tidal bedding consists of even layers of sand and mud; B-C sequences are alternations of (B) low tidal flat or sandbar environments, with (C) late stage emergence changes in flow structure; other terms are identified in the text. After Klein, 1972:540–541.

Two criteria must be met in order for deposits to be preserved and recognized in the stratigraphic record: (1) "moderate to rapid rates" of progradation are essential, and (2) a variety of sediment textures and structures must occur (Klein, 1972). If these two criteria obtain, the fining-upward tidal deposits can be distinguished readily from other types of fining-upward deposits, namely, those of meandering alluvial channels and those of turbidites. In general, alluvial channels differ from intertidalites in lacking sedimentary structures typical of alternating suspension and bed-load processes (such as flaser bedding) and in being underlain by a bottom-channel lag deposit. Turbidites have pronounced graded bedding, linear sedimentary structures, and exhibit considerable vertical variability in both grain size and heavy mineralogy. Deep-sea sediments deposited by bottom currents lack graded bedding, and are rather uniform in both grain size and heavy mineralogy. Thus intertidal fining-upward sequences are distinctive.

Several sand deposits of the geologic record have been ascribed to intertidal conditions (Klein, 1970, listed 46 such studies), although Visher (1969) earlier suggested that beach sands per se were rarely preserved as ancient sands. More recently, given precise criteria for intertidal deposits, specific comparisons between modern and ancient deposits of the intertidal have been drawn (Goodwin and Anderson, 1974; Klein, 1977; Ginsburg, 1975). Quartz sands now are believed to be nearly always intertidal in origin (Klein, 1977; Klein and Ryer, 1978).

The maximum value of the tidal range may be indicated by the thickness of a continuously deposited sequence between emergence (e.g., root zones or mudcracks at the top) and submergence (e.g., marine faunas at the bottom). For parts of the Devonian "Catskill Delta," this thickness is 2 m or less, which was therefore suggested as the maximum tidal range (Walker and Harms, 1975). Some authors have claimed that the thickness of a coarse sedimentary unit (e.g., gravels) indicates the tidal range, but in a transgressive sea (or subsiding basin), thick nearshore deposits might accumulate below tidal depths. Laterally extensive sheet sands of the geological past may be no more than ancient nearshore deposits (or areas of sand waves) such as are forming at present (Chapter 2). A summary of the tidal range for different geologic periods is presented below.

(b) In carbonate sediments, intertidal deposits are characterized by specific small-scale features, as shown in Figure 3-6. On Andros Island in the Bahamas, the mean neap tidal range is 0.17 m, and the mean spring tidal range is 0.41 m, with extreme values to nearly 1 m under unusual wind conditions. Stromatolites of blue-green algae are regulated by the extent of tidal cover. Their morphology, together with the development of different types of mudcracks in different degrees of exposure and the type of infauna, results in characteristic patterns of bioturbation and sedimentary structures, depending upon the exposure (Figure 3-6). The occurrence of these structures defines an "exposure index" (Ginsburg and Hardie, 1975).

In addition to the Bahamian type of intertidal carbonate deposit, Ginsburg (1975:268) identified as a second type the very widespread laminated, thin-bedded, and stromatolitic limestone of the upper Cambrian and lower Ordovician. These deposits lack the repeated, small-scale vertical sequences of lithofacies and structures which are typical of Bahamian intertidal carbonates.

This lower Paleozoic type of carbonate deposit represents an environmental regime no longer widely available—if at all—and it may have originated as follows. Before the Silurian, in the absence of land plants, nearshore areas would not have had their banks nearly as well defined as today (see p. 217). Hence the debouching of

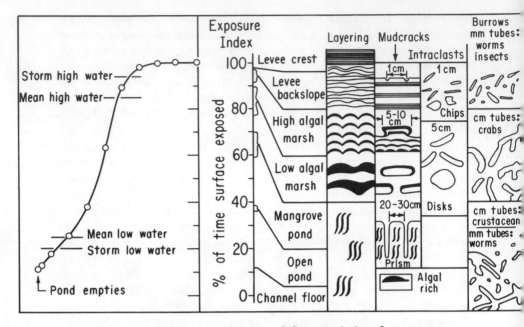

Figure 3-6 Diagrammatic view of characteristics of an exposure
index, based on observations in the tidal flats of Andros Island, Bahamas. Symbols are ∫∫ = moderately bioturbated, ∫∫∫ = strongly
bioturbated, ⌐⌐⌐ = knobby algal stromatolite, ▬ = algal stromatolite. After Ginsburg and Hardie, 1975.

water into the sea may have covered a wide area which was flat for
some tens to hundreds of kilometers, such as exists today in the
Ranns of Kutch, south of the Indus River along the northwestern
coast of India. In the Great Rann (which extends inland for 300 km
and along the shore for 100 km) and the Little Rann, approximately
30,000 km² are flooded seasonally by the southwest monsoon (Glennie and Evans, 1976). This immediately supratidal environment
today is accumulating a layer of gypsum. With a nonmonsoonal hydrography, a region of this type might accumulate very widespread,
thin-bedded stromatolitic limestones.

    As for clastics, tidal range for carbonates may be estimated by
finding the vertical distance between emergent land and fully
marine conditions. Individual sinusoidal laminations of stromatolites may have their highest point when wetted at high tide, and
their lowest point when just covered at low tide, thus yielding the
tidal range (Cloud, 1968). This occurs in some stromatolites in

Shark Bay, Australia. However, dome-shaped stromatolites (which should be useful in inferring tidal range) have been shown to occur subtidally in Devonian reef facies to depths of 45 to 100 m below Devonian sea level (Playford et al., 1976). Thus stromatolites make uncertain tide gauges.

(4) Tidal height through geologic time has long interested both astronomers and geologists. Since the largest component of the present tide is the lunar (M_2) input, and since the moon is receding slowly from the earth, tides of the geologic past may have been greater than they are now. Tidal friction and core accretion in the earth also have slowed the rate of the earth's rotation and decreased the frictional effect of zonal winds, thereby presumably decreasing the height of tides. Estimates of the rate of recession of the earth from the moon are derived from astronomical data and from measurements of an increase in day length based on banding in fossil shells (reviewed by Pannella, 1972; Rosenberg and Runcorn, 1975).

The argument has been put forth that the moon approached from its present distance of 60 earth radii to within a few earth radii during the past 2 b.y., possibly as late as 700 m.y. ago, and may have been captured at that time (Olson, 1972, and earlier; Merifield and Lamar, 1970). Lunar capture cannot have been that late, however, because stromatolites "with developed tidal bands" show that the moon has been associated with the earth for at least 3 b.y. (Pannella, 1976:680).

On the one hand, three lines of evidence are cited for considerably higher Precambrian tides.

(a) Precambrian sheet sands may have been molded into large sand waves similar to those which exist today in regions of high tidal currents (reviewed in Chapter 2; see also Merifield and Lamar, 1968). However, for two specific late Precambrian sheet sand deposits, "it is not possible to conclude that the earth-moon distance was considerably smaller at the time these rocks were deposited, because both occurrences could be explained by local conditions, such as those prevailing today in the Irish Sea and English Channel" (Merifield and Lamar, 1970:38–39). The 2.5 b.y. old gold-bearing conglomerates of the Witwatersrand have also been interpreted as indicating high tidal velocities that would suggest a closer earth-moon distance than exists today (Hargraves, 1970). The careful analysis of Vos (1975), however, referred the deposits to a "braided alluvial plain and lacustrine environment," not a tidal environment.

(b) A 6 to 15 m amplitude in height of stromatolite laminations may indicate high Precambrian tides (records cited by Cloud, 1968, and Walter, 1970, but these authors specifically did not agree with the interpretation of enormous late Precambrian tides).

(c) Deposits which are consistent with a tidal range of 12 to 25 m occur in the 3 b.y. old Pongola Supergroup in South Africa (Mason and von Brunn, 1977). Data from a single geographic region, however, do not indicate a general pattern.

On the other hand, most evidence suggests that tidal height for at least the past 1 b.y. has not been any greater than it currently is.

(a) Klein (1977) applied the fining-upward method of determining tidal deposits to 428 occurrences and found the following paleotidal ranges: late Precambrian, 0.3 to 13.0 m; Cambrian, 1.1 to 7.9 m; Ordovician, 0.3 to 1.4 m; Silurian, 3.3 to 6.1 m; Devonian, 2.0 to 8.9 m; Jurassic, 0.8 to 4.1 m; Cretaceous, 0.5 to 8.0 m; and observed Recent range, 0 to 17 m.

(b) The number of days per synodic (= lunar) month has decreased from slightly in excess of 31 in the Ordovician to approximately 29 at present (Pannella, 1972). Accordingly, the earth-moon system has been essentially stable for about 500 m.y. Stromatolites with banding typical of those which form intertidally today occur in Gunflint deposits approximately 2 b.y. old (Pannella, 1972). Gunflint stromatolites, however, have slightly fewer than 40 bands per month or 445 bands per year (Figure 3-7), whereas theory predicts 520 bands. Perhaps this discrepancy is because intertidal stomatolites always are eroded and thus leave far fewer bands than there are days per year. On the other hand, some experienced workers accepted the Gunflint data. It is essential to have this argument resolved.

(c) Stromatolites with small amplitude "tidal" bands occur in rocks 2 b.y. old. The stromatolites with the greatest amplitude (6 to 15 m) are latest Precambrian to Cambrian in age (Walter, 1970) and are not in older rocks, which should be the case if tides were higher in yet more ancient times.

In summary of tides, the lunar component of earth tides has twice the effect of the solar component, and the two factors together account for more than 80 percent of the magnitude of normal ocean tides. On continental shelves, tidal height is amplified as the shelf width increases for at least approximately 300 km from the shelf break, depending upon the average shelf depth. The most important factor in evaluating the existence of tides in the geological past is paleogeography. Regions with high tides must have extensive connection with the open ocean.

Deposits attributable to the intertidal zone occur in clastic and

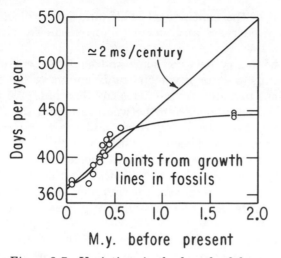

Figure 3-7 Variations in the length of the year since Gunflint time (≈2.0 b.y.). The straight line is the predicted pattern for an earth slowing at a rate of ≈2 ms/century. The curved line fits the data. Note the major discrepancy between theory and data. After Pannella, 1972:236.

in carbonate deposits. Many criteria have been developed for the recognition of "tidalites" in the geologic record. Over the past 2 b.y., tidal height appears to have been as variable from place to place as it now is; there is no evidence for a systematic increase (or decrease) in tidal height as one goes back into geologic time.

Currents

Currents are caused by tides and winds, with tides much the more important on continental shelves. However, bottom currents at 60 m have been attributed to surface winds (Lavelle, Keller, and Clarke, 1975). In the past few years, considerable attention has been paid to paleocurrents, two aspects of which are of special paleocean-ographic interests: (1) magnitude and (2) direction. Patterns derived from their measurements yield an indication of ocean circulation and are considered in the next section.

(1) With respect to magnitude, the speed of modern-day ocean currents can be classified as slow (< 20 cm/sec), moderate (20 to 100 cm/sec), and fast (> 100 cm/sec). Twenty and 100 cm/sec correspond approximately to 0.4 and 2 knots, respectively. Low current velocities are typical of the continental slope and rise, with moder-

ate velocities typical of the continental shelf, and high velocities characteristic of tidal currents and storm conditions in restricted areas.

The suggested boundary between slow and moderate currents (20 cm/sec) corresponds to a mean velocity at 15 cm above the bottom which is sufficient to transport particles smaller than 62.5 μm (Postma, 1967:158)—the size boundary between sand and silt. The suggested boundary between moderate and fast currents (100 cm/sec) is sufficient to erode sediment that is 2 mm in diameter. Although the absolute numbers for current speed are somewhat arbitrary, the numbers chosen do have physical significance, and the desirability of having a standard for the commonly used qualitative terms slow, moderate, and fast is sufficiently great to justify this approach.

Geostrophic currents that follow contours of the continental slope and rise have been characterized as "steady, low-velocity (2-20 cm/sec)" (Heezen, Hollister, and Ruddiman, 1966). In deeper water, seven of eight deep-sea (2,5000 to 4,600 meters) direct-current measurements made from 0 to 600 m above the bottom were between 6 and 12 cm/sec (Heezen and Hollister, 1964). These values of less than 20 cm/sec are typical of the many later measurements of current strength in deep water. In contrast, currents on the continental shelf are commonly higher in strength, on the order of 25 to 100 cm/sec, depending on the tidal phase. Thus, according to the convention adopted here, currents deeper than 200 m are commonly low, whereas those of the continental shelf are commonly moderate. These generalizations do not apply in special circumstances, such as the influence of western boundary currents (e.g., Gulf Stream), even at depths of 1 to 2 km (e.g., Blake Plateau; Pratt, 1963).

Fast currents (> 100 cm/sec) are associated with boundary currents (where a large body of water is geostrophically piled up and must pass through a narrow region), or with tides in shallow water (where a large mass of water is forced over a shallow region).

The relationships among current strength, particle size, and sediment deposition and erosion have been widely applied (e.g., by Ledbetter and Johnson, 1976). For any given particle size, there is a characteristic settling velocity. Obviously, if the settling velocity of a particle through the medium is faster than the velocity of the medium, deposition should occur.

The current velocity necessary to cause erosion decreases as sediment size decreases, but only to a point. It is at a minimum for particles 0.3 to 0.6 mm in diameter (fine sand), as shown in Figure 3-8, the Hjulström diagram (Hjulström, 1939). Particles smaller

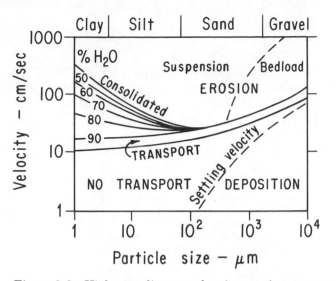

Figure 3-8   Hjulström diagram showing erosion, transportation, and deposition velocities measured 15 cm above bottom for different grain sizes and water content of clay and silt. Note that the curve for erosion is "U-shaped," with clays of low water content requiring very high velocities for erosion. After Postma, 1967:158; Allen, 1965:109, 1967:433; and Gibbs, Matthews, and Link, 1971:11.

than 0.3 mm tend to adhere to each other by surface forces, and a much higher velocity is required to erode them than is required to transport them. This accounts for the "U-shaped" pattern of the diagram.

The water content of fine sediments is critical in determining the velocity necessary for erosion (Figure 3-8). For example, a current of 20 cm/sec is sufficient to erode a clay with 90 percent water, but a current of 200 cm/sec is required to erode a clay with only 50 percent water.

The situation at any given place on the shelf is always changing, and depending on the state of the tide and time of year, the same sediment can be deposited, transported, or eroded (McClennen, 1973). The observed rock record, which can be inferred from sediment size, represents only the last event, i.e., deposition. Once deposited, material is much more difficult to erode than it is to transport.

Current velocities inferred from sediment sizes represent *minimum* values. For example, if sediments of the Asiatic continental shelves were to become rocks, most of the region would be composed

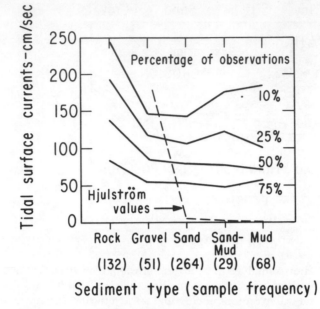

Figure 3-9   Median surface current velocities on East Asiatic continental shelves over various types of bottom. For example, over mud
bottoms, 10 percent of the recorded velocities are more than about
175 cm/sec, 25 percent more than 100 cm/sec, 50 percent more than
60 cm/sec, and 75 percent more than 55 cm/sec. Note that the observed current is far above the value required for transport of sand,
sand-mud, and mud (according to the Hjulström diagram; see Figure
3-8). After Shepard, Emery, and Gould, 1949.

of shales and siltstone and fine-grained sandstones (Figure 3-9).
However, median velocities of tidal currents on these shelves reach
70 cm/sec 50 percent of the time. That velocity would easily suffice
to *transport* this fine material elsewhere. This is presumably not
happening since the material is still there, and thus the material is
probably not available for transportation. Evidently, once the fine
sediment was deposited (under historically different hydrographic
conditions?), it became largely decoupled from the overlying transporting medium.

   Despite these difficulties in inferring current strength from particle size, use of the Hjulström diagram to estimate paleocurrents is
not uncommon. One recent example in Antarctic deep-sea work is
based on the absence of sediments of age 0 to 3 m.y. over an area 3 ×
10^6 km^2 between Australia–New Zealand and Antarctica. This
region is considered to have had an increase in the velocity of bot-

tom currents from less than 10 cm/sec to more than 10 cm/sec more recently than the past 3.5 m.y. (Watkins and Kennett, 1972), and either to have eroded the underlying sediments or to have prevented their deposition during that time.

Manganese nodules and phosphate pavement and nodules form very slowly: an average of 1 to 10 mm per *million* years, in contrast to deep-sea rates of sedimentation of millimeters to centimeters per *thousand* years (Ku, 1977). Thus they can originate only if sediment is not being deposited in amounts sufficient to swamp out the nodule formation. In the deep water of the Pacific, currents of less than 10 cm/sec permit the deposition of calcareous material. Currents of more than 10 to 15 cm/sec prevent this sedimentation and thus permit the formation of maganese nodules (Kennett and Watkins, 1975).

Currents of reasonable magnitude can occur if the water is aerated or anoxic. But in the absence of currents, water stagnates, and thus lack of circulation is *one* of the reasons basins become anoxic. Under anoxic conditions, bacteria convert sulfate to iron sulfide in the presence of organic matter. Pyrite accumulated in the Sea of Japan during glacial periods (Kobayashi and Nomura, 1972) because the sill depth prevented circulation with the open ocean.

In a sea where the slope of the bottom is *very low,* energy is theoretically able to be dissipated so that waves never break. Thus a solitary wave 2 m high would dissipate its total energy over a gradually shoaling distance of 12 km (Table 3-2). Under these conditions, waves might leave no measurable effect on shore sediments. The ex-

Table 3-2   Distance from shore needed to dissipate energy from waves of a given wave height so that the waves reach the shore with zero height and thus do not break. Note the enormous distances required for even small waves. Data from Keulegan and Krumbein, 1949.

Wave height (m)	Distance from shore (km)
4.0	40.0
3.0	24.0
2.0	12.0
1.0	3.5
0.5	1.0
0.2	0.21
0.1	0.06

ercise is important because it shows how extreme the conditions must be before one would expect wave energy to be a trivial factor in sorting sediments. Such conditions probably have never prevailed in the geologic record.

Currents in shallow seas may be especially sensitive to barometric conditions, such as exist along the western shore of Lake Michigan and the Texas Gulf Coast. Fox and Davis (1976, and earlier) reported that when a low-pressure system approaches the coast, barometric pressure drops and there is an increase in wind velocity, breaker height, and long-shore current velocity. When the low pressure leaves the coast, barometric pressure rises and there is a reversal of wind direction together with a reversal of long-shore current direction. In response to barometric changes, sandbars oscillate back and forth along the shore. Given this example, and the continually changing weather, it seems quite likely that wind currents responding to barometric pressure may have been an important *day-to-day* process for sediment movement in ancient seas.

(2) Indications of current direction are sufficiently numerous that it is possible to derive a pattern of bottom currents over an entire basin or continental margin (Potter and Pettijohn, 1977). Criteria of use include those of both biological origin (any elongate fossil is suitable under some current circumstance) and sedimentological origin (especially cross bedding, ripple marks, turbidite structures, and sediment fabric).

In shallow-water areas, paleocurrents may be influenced by runoff from land, by wind, by tides, and by ocean currents. Accordingly, there is no necessary relationship between direction of bottom currents and paleoslope. Only by analyzing data from a large region can the significance of current indicators be deciphered.

Specimens which have been oriented by waves can be distinguished from those which have been oriented by currents (Figure 3-10). Unidirectional flow is typical of *currents* and results in shells and other current indicators having the same orientation, i.e., being directed into the current. In contrast to current-oriented objects, *wave*-oriented shells, etc., display two maxima, at 180° and perpendicular to each of the directions of wave movement. This comparison of current and wave orientation has been demonstrated in both field and laboratory experiments, and has also been documented in stratigraphic studies (Nagle, 1967). As expected, wave-oriented material is not as typical of deep-water deposits as it is of shallow-water deposits (Jones and Dennison, 1970).

In summary of currents, slow (<20 cm/sec), moderate (20 to 100 cm/sec), and fast (>100 cm/sec) currents are respectively char-

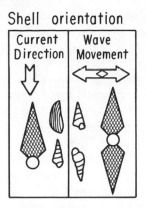

Figure 3-10 Generalized patterns of orientation of shells for current versus wave motion. Actual orientation patterns may show some asymmetry. Note that current-oriented shells have one maximum, but wave-oriented shells have two maxima, at approximately 180° to each other. After Nagle, 1967:1125.

acteristic of (1) the continental slope and deeper regions, of (2) the continental shelf, and of (3) major ocean surface currents, storms, and tidal currents. Estimates of current speed can be derived from the Hjulström diagram—a plot of sediment size versus velocity required for erosion, deposition, and transportation. Winds and change in barometric pressure cause enough motion in shallow water for one to be sure that all parts of shallow seas of the past were at times strongly influenced by currents.

Circulation

Circulation will be discussed according to the topics (1) principles and modern patterns and (2) patterns over geologic time.

(1) A knowledge of principles and modern patterns of ocean circulation is the most important ingredient in predicting the distribution of biogenic sediments, faunal provinces, and the global heat balance. Yet, as we will see, a tremendous amount remains to be learned. Indeed, experienced physical oceanographers are still debating when and where (and on what time scale) water is being added to the deep ocean.

With realization that any configuration of world geography is "correct" for only a short period of geologic time, the subject of paleoceanographic circulation must come to rely more on the general principles causing that circulation than on information from specific present-day patterns. Five major aspects of oceanic circulation will be treated below: (a) the zonal distribution of windstress, the chief factor allowing prediction of surface currents (upper 1 km) in the world ocean; (b) whether evaporation or inflow is dominant in a given basin; (c) the density distribution within the ocean, the main factor determining the flow at depths greater than 1 km; (d) upwelling; and (e) the oxygen minimum layer.

(a) In the zonal (i.e., latitudinal) distribution of winds, air is heated in equatorial regions and therefore rises. In order to replace rising air, surface air is brought toward the equator from low latitudes, as shown in Figure 3-11 and Table 3-3. The air and the oceans are on a rotating sphere, i.e., this is geostrophic flow, and a relative motion is imparted to the air and the water. This relative motion is known as the Coriolis effect and is recognized by the fact that both air and water currents are deflected to the right of an observer looking downstream in the Northern Hemisphere (and to the left of an observer in the Southern Hemisphere). Owing to the Coriolis effect, the surface wind in equatorial latitudes is deflected so that to an observer standing on the equator, the chief winds come from the northeast and from the southeast.

Patterns of zonal winds are modified when land surfaces near the equator are heated exceptionally. For example, in India in late

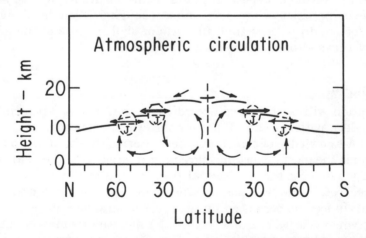

Figure 3-11 Diagrammatic cross section of atmospheric circulation. Note the upward movement of air in the tropics, sinking at 30°, rising at 60°, and sinking at the poles. Cross circulation and horizontal exchange are shown by double arrows in the region of the tropopause and jet streams (J). After Flohn, 1969:85.

Table 3-3 Summary of atmospheric characteristics of latitudinal zones. Data from Flohn, 1969:106.

Hemisphere		Air-pressure belt	Wind systems	Upper boundary
South	North			
90°	90°	Polar high		
			Polar east winds	2 to 3 km
65°	65° to 75°	Subpolar low-pressure trough		
			Extratropical westerly winds	Summer: 18 to 22 km Winter: 80 km
25° to 30°	30° to 40°	Subtropical high-pressure trough		
			Tropical easterly winds	Ocean: 6 to 10 km Continent: at least above 20 km
	0 to 10°N	Equatorial low pressure trough		

summer, land warms considerably more than the adjacent water, so that the driving force of air (and water) circulation becomes the air rising from the land, which then draws in air off the adjacent ocean. As the seasons change and the sun moves into the Southern Hemisphere, the air circulation reverses, with accompanying reversal in ocean winds and ocean currents. This is what is responsible for the monsoon region of the Indian Ocean between India and Africa, as shown in Figure 3-12. Correct paleogeography is clearly a prerequisite for understanding the monsoon type of oceanic circulation.

Air which rises in equatorial regions spreads to higher latitudes, where it cools and sinks, thus causing a high-pressure zone at approximately 30°N and 30°S, as shown in Figures 3-11 and 3-12 and in Table 3-3. On reaching the earth's surface, the air travels either equatorward (with the consequences reviewed above) or poleward. If the air goes poleward, it is again deflected to the right, and thus forms the westerlies of the mid-latitudes. This air is subsequently warmed at the surface, and rises at approximately 60°N (Figures 3-11, 3-12), forming a low-pressure zone. At elevation, the air spreads equatorward to sink at 30°N, or spreads poleward, where it sinks forming the polar high and the clockwise circulation of the North Polar Sea (=,Arctic Ocean).

The surface circulation of the ocean parallels rather well the

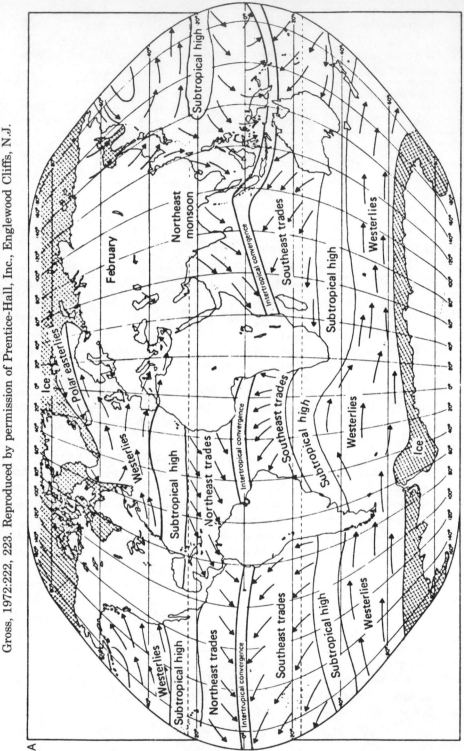

Figure 3-12 Generalized surface winds over the ocean in February (A) and in August (B). From Gross, 1972:222, 223. Reproduced by permission of Prentice-Hall, Inc., Englewood Cliffs, N.J.

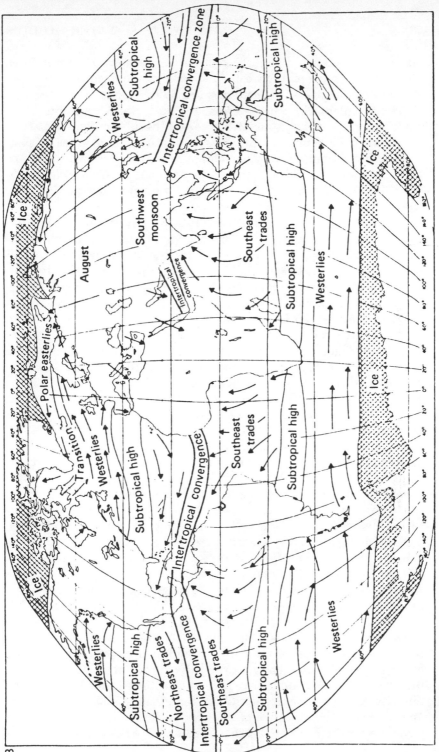

B

surface wind distribution just reviewed, as is shown in Figure 3-13. However, currents on the western sides of ocean basins are intensified and narrowed—Gulf Stream, Kuro Shio (= Japan Current), and their equivalents in other oceans. As these warm currents leave their boundary margins and move across the adjacent ocean basins, they spread out (Figure 3-13). Thus mid-latitude land areas on eastern sides of oceans have a warmer climate than comparable latitudes on western sides of oceans; for example, Land's End in southwest England has palm trees, but Newfoundland, also at 50°N, has icebergs.

In addition to the western boundary currents, the equatorial countercurrents (Figure 3-13) also may not be obvious from observing the wind patterns. These countercurrents run west to east in narrow, high-velocity "jets." They result (in part) from the disequilibrium caused by the equatorial wind piling up water on the eastern side of an ocean. The combination of equatorial currents and countercurrents affords a mechanism for larvae of tropical species to be distributed back and forth across the ocean.

Some of the patterns of oceanic circulation outlined above have been studied by physical or simulation modeling. Physical experimental models have made use of rotating "dishes" (von Arx, 1957) or sphere against sphere experiments (Baker, 1970). In the rotating dishes, a hemisphere is seen as though from an immense distance, with the pole in the middle and the equator at the circumference. Continental and oceanic configurations for a hemisphere can be assembled on the dish. This method is particularly suitable for examining the flow of surface currents. However, flow *between* hemispheres is prevented, and hence some observed patterns may be misleading. Middle Cretaceous seas have been modeled by using rotating dish experiments (Luyendyk, Forsyth, and Phillips, 1972). Sphere against sphere experiments permit flow between hemispheres, but they have not progressed to the point of including effects of the geography of land masses.

An additional method of charting ocean circulation owes its origin to the fact that ocean currents are the chief means of dispersal of organisms (see Chapter 7). Currents can themselves be barriers to organism dispersal; Cifelli and Smith (1969), for example, wrote that "the importance of the Gulf Stream as a major biogeographic boundary for plankton has yet to be fully appreciated." On continental shelves, biogeographic provinces appear to be formed by the combination of a specific current and the restrictions of local geography, and to lead to the evolution of indigenous floras and faunas. This is also true for the deep sea, in that "present-day benthonic foramini-

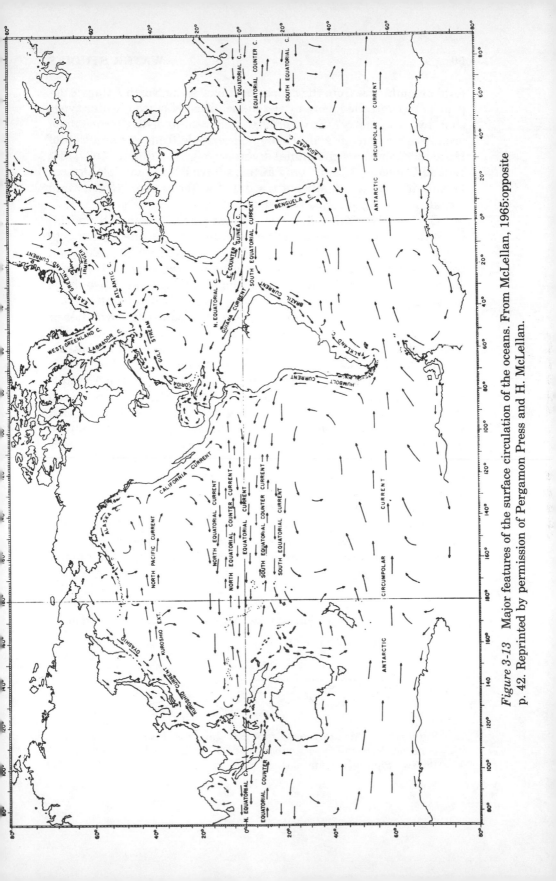

Figure 3-13 Major features of the surface circulation of the oceans. From McLellan, 1965:opposite p. 42. Reprinted by permission of Pergamon Press and H. McLellan.

feral assemblages from the deeper portions of the North Atlantic appear to be controlled more by the distribution of bottom water types than by bathymetry" (Streeter, 1973). Recent and Pleistocene open ocean surface currents are clearly discernible from examination of faunal distributions and faunal diversity (e.g., Ruddiman, 1977:151; Bé and Duplessy, 1976). Many authors have found that "planktonic foraminiferans are significant tracers of water masses" (Bandy and Arnal, 1969:803; see also Bradshaw, 1959).

Surface circulation in 39 of 40 lakes, seas, estuaries, and lagoons is counterclockwise in the Northern Hemisphere owing to wind stress (Emery and Csanady, 1973). If the pattern of circulation could be determined in ancient seas and large coastal bodies of water (as, for example, by the direction of sandspits), this would be an indication of the hemisphere of the ancient body of water.

(b) A second important factor in ocean circulation is whether or not evaporation or net inflow is dominant for surface waters of a basin (Berger, 1970). As shown in Figure 3-14, this is the contrast between arid Mediterranean type circulation and humid estuarine type circulation. In the Mediterranean type, just as in the Mediterranean Sea, water flows in at the surface from the open ocean and then evaporates and sinks as it moves toward the head of the bay; water exits as a deep flow beneath the inflowing surface current. In estuarine type circulation (called lagoonal by Berger), fresh water flows in from the head of the bay and flows out on the surface; deep water flows in from the mouth of the bay and mixes upward. In Mediterranean type circulation, runoff from the land is meager and low in nutrients (P, N, Si), and therefore basin water is high in oxygen; basin sediments are rich in $CaCO_3$ but poor in organic carbon. Estuarine type circulation has the opposite characteristics, with runoff from the land being high in nutrients and basin water being

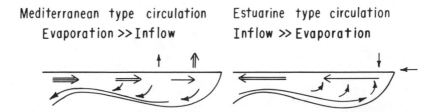

Mediterranean type circulation   Estuarine type circulation
Evaporation >> Inflow   Inflow >> Evaporation

Figure 3-14   Mediterranean type circulation and estuarine type circulation. Note the opposite movement of surface and deep water in the two types of circulation.

low in oxygen (depleted by oxidizing organic matter); sediments are poor in $CaCO_3$ (dissolved away) and rich in organic carbon.

The North Atlantic is a Mediterranean type of ocean and has been since the late Eocene (Ramsay, 1977b:1412). In it, dense water is sinking at its northern boundary, as is conclusively demonstrated by the transport of bomb-produced tritium (over only 10 years) to depths of 5 km by cold, dense high-latitude water (Östlund, Dorsey, and Rooth, 1974). This deep water then flows southward (and is called North Atlantic Deep Water: NADW).

In contrast to this Mediterranean type of circulation of the North Atlantic, the Pacific has been an estuarine type of ocean, also since the late Eocene. No dense water is forming in the North Pacific, and indeed the North Pacific bottom water owes its origin to water which has flowed all the way from Antarctica. Table 3-4 contrasts the salinity, oxygen, phosphate, and nitrate in the North Atlantic and North Pacific, and Table 3-5 shows the resulting differences, in percentage covered by calcareous oozes, siliceous oozes, and red clay. The low-oxygen, high-nutrients, and carbonate-poor conditions of the Pacific contrast with the high-oxygen, low-nutrient, and carbonate-rich Atlantic conditions. These same contrasts are seen qualitatively in comparing much smaller bodies of water, such as the Norwegian fjords (estuarine type) and the Persian Gulf (Mediterranean type). The same type of comparison probably can be made for geological deposits between basins with black shales and those with calcareous muds.

The chief implication of large-scale evaporite deposits from the point of view of this chapter (also see p. 198) is that circulation with the open ocean must have been severely restricted (Kinsman, 1975), and must have been of the Mediterranean type. The existence of an

Table 3-4 Comparison of water characteristics in the North Pacific and the North Atlantic ($\approx$50°N and $\approx$3,000 m depth). Data from Berger, 1970.

Characteristic	North Pacific	North Atlantic
Temperature (°C)	1.5	2.25
Salinity (‰)	34.65	34.95
Oxygen (ml/l)	2.8	6.3
Phosphate-P (μgat/l)	2.9	0.9
Nitrate-N (μgat/l)	34	13

Table 3-5 Comparison of Pacific and Atlantic sediments: percent areas covered by main types of pelagic sediments. Data from Berger, 1970.

Sediment	Pacific	Atlantic
Calcareous oozes	36.2	67.5
Siliceous oozes	14.7	6.7
Red clay (percent manganiferous)	49.1(87)	25.3(43)

enormous number of major salt deposits (Kozary, Dunlap, and Humphrey, 1968) indicates the importance of Mediterranean circulation. This is true whether one accepts a model of total basin evaporation (Lucia, 1972), as probably occurred over most of the Mediterranean during the Miocene (Hsü et al., 1977), or a model of evaporite precipitation from a layered sea (Schmalz, 1969; Sloss, 1969; Stuart, 1973), or a model of coastal *sabkha* (an Arabic word for salt flats) (Kinsman, 1969; Leeder and Zeidan, 1977).

(c) A third factor controlling ocean circulation is the density of sea water. Increasing density of sea water is a function both of increasing salinity and of decreasing temperature. At normal ocean salinities, a change of 7°C has approximately the same effect on density as a change of 1‰; thus at the present time the global latitudinal temperature change is from approximately 28 to 0°, but the salinity change is only 2 to 3‰. Hence formation of dense water is chiefly a function of low temperatures at high latitudes, given modern climate. (Temperature and salinity also differentially effect the fine-tuning of density. At 0° and 34‰, to raise temperature 1°C is equivalent in its effect on density to a salinity decrease of only 0.035‰; at 20°C, a similar 1°C increase would be equivalent to a much larger decrease of 0.39‰ [Weyl, 1968].) During the Cretaceous (and perhaps other times as well, see Chapter 4) when the latitudinal temperature gradient was from approximately 28 to 14°C, mid-latitudes might have been around 21°C. These mid-latitudes correspond to intense evaporation (and salinity approximately 2‰ higher than equatorial water). If that climatic regime obtained, then the densest water may well have corresponded to this salinity maximum instead of to a higher latitude temperature minimum. Thus for periods of time in the geologic past when water at high latitudes was much warmer than at present, temperature would have decreased in importance and salinity increased in importance in determining the density distribution of the ocean.

The difference in density of water from different sources accounts for the water masses of the deep ocean, as is shown for the Atlantic in Figure 3-15. Dense water forms today both from relatively highly saline water—as in Mediterranean Water (MW) of S $\geq 36‰$ flowing into the North Atlantic and in Norwegian Sea water at S $\geq 34.9‰$ going into the North Atlantic and forming North Atlantic Deep Water (NADW); and from very cold water—as in Antarctic Bottom Water (AABW) at potential temperature $\leq 2°C$. On top of NADW in the South Atlantic is Antarctic Intermediate Water (AAIW). At S $\leq 34.8‰$, AAIW is as fresh as AABW but is much warmer (potential temperature is $\simeq 5°C$) and therefore much lighter than AABW.

Whether or not significant amounts of Antarctic Bottom Water are also *currently* forming adjacent to the Weddel Sea is a matter of dispute (Worthington, 1977), although tritium at a depth of 2 km indicates that "some deep water must have come from the surface in the last 10 to 20 years" (Michel and Williams, 1973). The time it takes a parcel of water to leave the ocean surface, sink, and resurface has been estimated to be from approximately 500 years to

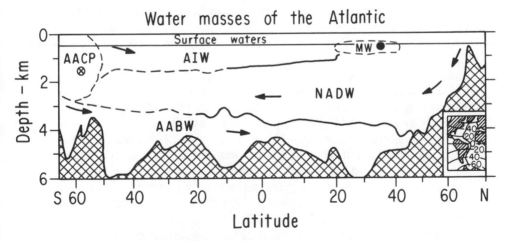

Figure 3-15 Schematic relationship of water masses of the Atlantic Ocean. Saline North Atlantic Deep Water (NADW) flows southward over cooler Antarctic Bottom Water (AABW) and below fresher Antarctic Intermediate Water (AIW). Two meridional flows are shown: Antarctic Circumpolar Polar (AACP) flowing out of the diagram (west to east), and Mediterranean Water (MW) flowing into the diagram (east to west). Based on McLellan, 1965:48; Berger, 1970:1386; and Broecker, Takahashi, and Li, 1976:1085.

1,500 years, and may be taken as on the order of 10^3, with shorter times for the Atlantic and longer times for the Pacific.

(d) The fourth aspect of circulation emphasized here is upwelling. Upwelling is the name given to the transport of subsurface water (usually less than 200 m deep) to the surface. It chiefly occurs in subtropical latitudes within 100 km of the shore where local winds blow from the north along the western side of Northern Hemisphere continents, or from the south along the western side of Southern Hemisphere continents. The wind shear results in each case in the transport of water westward (offshore), following the Ekman spiral (see McLellan, 1965:Chapter 12). Accordingly, upwelling is characteristic of western sides of continents, although special occurrences of upwelling along the eastern sides of continents are known (Summerhayes, de Melo, and Barretto, 1976). Locally, upwelling can also occur through a mixing process caused by evaporation, as off Ghana (Pople and Mensah, 1971). Coastal upwelling is important because large amounts of nutrients are thereby recycled upward into the surface water. These nutrients are the basis for the world's largest fisheries (off Peru), and are the source of phosphorus for the formation of phosphorite deposits.

Phosphorite is the main geological indicator of coastal upwelling; the phosphate occurs as one or another variety of the mineral apatite, $Ca_3(PO_4)_2$. Considerable literature (Kazakov, 1937, and his followers, many cited by Gulbrandsen, 1969, and Tooms, Summerhayes, and Cronan, 1969) has emphasized the importance of upwelling as a renewable source of phosphate, and has presumed a direct chemical precipitation from sea water. The source of phosphate is certainly the organic matter (phytoplankton and zooplankton), of which approximately 0.1 percent is phosphate. Rather than have precipitation of apatite directly from sea water (which is a rare phenomenon if it occurs at all), let us consider what goes on in interstitial fluids. Oxidation of organic matter during the sulfate reduction releases phosphate to the interstitial fluids (Burnett, 1977). Especially in organic-rich sediments, calcium is also released by the solution of foraminifera. At a pH of 7.0 to 7.8, phosphate combines in interstitial fluids with the much more abundant calcium to yield calcium phosphate (Brooks, Presley and Kaplan, 1968).

Manheim, Rowe, and Jipa (1975) wrote that the formation of sedimentary phosphates requires (1) a continual resupply of organic matter, as in areas of upwelling along the outer continental shelf; (2) a low rate of detrital sedimentation (as in arid regions), because a high rate would prevent interstitial fluids from reaching high phosphate concentrations; (3) an oxygen minimum zone impinging on the sea floor to prevent immediate oxidation of organic matter; and (4) a sufficient but not an excessive supply of carbonate. Excess denitrification leading to a decrease in the marine N:P ratio below its present value of 16:1 also would leave excess P, and thus contribute to phosphate precipitation (Piper and Codispoti, 1975). Phosphorite is forming today in organic-rich sediments below upwelling areas off southwestern Africa (Summerhayes et al., 1973) and off Peru (Burnett, 1977).

An area of nonupwelling in which phosphorite pavements are apparently being formed is on the Blake Plateau off the southeastern United States (summarized by Emery and Uchupi, 1972:111). The source of phosphate is a previous lag deposit of phosphate nodules, which is now exposed. Strong currents prevent dilution by sediments (see p. 89), and the phosphate nodules are being cemented into a phosphate pavement. Similar phosphorites are forming in strong currents on the Agulhas Bank off South Africa, far from shore where the only animals are suspension feeders: "in some areas virtually the only animals present were bryozoans, which must have grown forest-like on the sea floor" (Parker and Siesser, 1972:439). In general, the occurrence of phosphate nodules indicates (1) a source of phosphate (often upwelling), (2) a concentration mechanism, and (3) a lack of dilution (often owing to considerable bottom current).

Open ocean latitudinal zones of divergence of currents are also marked by upwelling when water at depth is brought to the surface (as in the Arctic, Antarctic, and tropical divergences). The high nutrients and silica of these zones lead to abundant growth of diatoms and radiolaria, whose siliceous tests in turn fall to the bottom and record these divergences as latitudinal bands of silica-rich sediments (Lisitzin, 1972 and earlier). Oceanic fertility is also reflected in high diatom abundances of the Eastern Pacific compared to those of the Western Pacific (Mikkelsen, 1978).

Diatom and radiolarian tests are composed of opal-A (amorphous hydrated silica), which is readily dissolved in interstitial water. The SiO_2 is chiefly reprecipitated as opal-CT (low-temperature cristobalite-tridymite, terminology after Jones and Segnit, 1971), and rocks consistently composed dominantly of opal-CT are called porcelanite. Opal-CT often occurs as spherical microcrystalline aggregates of bladed crystals (lepispheres of Wise and Kelts, 1972), which are explained as regularly interpenetrating intergrowths according to a tridymite twinning law (Flörke et al., 1976). Opal-CT is rarely found in rocks less than 10 m.y. old, and by the Jurassic it has all recrystalized into quartz (Calvert, 1977; Riech and von Rad, 1979). The rate of conversion of opal-CT into quartz seems chiefly a function of age, but is modified by the facies of accompanying sediments. Laboratory experiments indicate a faster conversion in carbonates than in clays (Kastner, Keene, and Gieskes, 1977), but for reasons not yet understood, this does not seem to be closely mirrored in nature (Riech and von Rad, 1979). Rocks consisting mostly of quartz are called chert, and all sedimentary siliceous rocks older than the Cretaceous are now chert. Probably nearly all of the Phanerozoic cherts once ascribed to volcanic source materials are one variant or another of the nodular and bedded cherts which today form from biogenous silica.

Open ocean zones of divergence, often related to upwelling, are marked by silica-rich deposits. Drewry, Ramsay, and Smith (1974:545) plotted the paleolatitude of oceanic cherts and found that 66 of 69 occurrences were between 30°N and 30°S. By the same reasoning, the *absence* of deep-sea cherts in some deep-sea drilling sites can be explained by the fact that their sediments were not deposited in "the high-productivity equatorial zone" (Keene, 1975).

Oceanic zones of convergence are also regions of high primary productivity of calcite-secreting organisms. They occur along the warmer side of the Antarctic and Arctic convergences and in the equatorial current system. As the zones of convergence of high-latitude currents migrate northward or southward with climate, the underlying bands of carbonate debris should also migrate latitudinally. As shown in Figure 3-16, deposits of carbonate debris from coccolithophorid tests will be diachronous, just like ice-rafted debris (see p. 204). These changes in the concentration of carbonate debris may also parallel shifts in the biogeography of temperature-sensitive species. The equatorial current system has been hypothesized to yield higher production of calcium carbonate (from foraminifera and coccoliths) during cold-water conditions because of presumed intensified atmospheric and oceanic circulation (Arrhenius, 1966, and earlier).

(e) Another factor important in ocean circulation is the oxygen minimum layer, an oceanic "layer" of water 200 to 500 m thick which is reduced in oxygen content relative to water above and below it. The oxygen minimum occurs at a depth of a few hundred meters (~ 0.5 to 2 km) in most parts of the ocean. The really low values of oxygen in the water column (< 20 μg-at/l—"or even below the limits of detection," Deuser, 1975) occur in only a few places in the ocean, each of which is associated with upwelling and high organic productivity. These regions are off the western coast of the Americas from Peru northward to Baja, California, in the Arabian

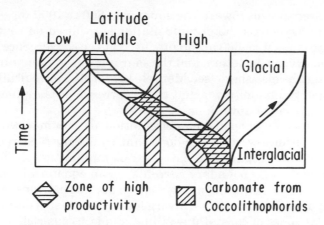

Figure 3-16 Anticipated migration of the high-productivity zone of coccolithophorid algae through an interglacial-glacial climatic cycle. They are depicted as seen in the carbonate sediment less than 74 μm in size in cores from different latitudes underlying the zone of migration. Note that the high-productivity zone crosses latitudes as climate changes. After McIntyre, Ruddiman, and Jantzen, 1972:64.

Sea between Saudi Arabia and India, and in very small areas on the eastern margin of India and adjacent to part of Antarctica. The eastern Pacific oxygen minimum layer has enormous lateral extent, and continues westward in 2 prongs (centered at approximately 15°N and 6°S). The northern prong extends westward for 1,500 km, and is some 360 km wide when it leaves the continental margin. The southern tongue of oxygen minimum water extends seaward only half as far as the northern tongue probably because southern Pacific surface water has more oxygen and is more vigorously transported (Reid, 1965). (In marine basins with restricted circulation, oxygen of course can be absent, as in the Black Sea; characteristics of the known, mostly small, 33 anoxic marine basins are given by Deuser, 1975).

The oxygen minimum layer owes its origin to notably high nutrient production in upwelling regions near the shore, with subsequent sinking of organic matter to the sediments, oxidation of the organic matter, followed by movement offshore of this oxygen-stripped water (Menzel, 1974; Suess, 1976). Where this occurs, the sediments are a trap for nutrients and are organic rich (see also pp. 102–104). It now appears unlikely that nutrient-rich waters from the Arctic and Antarctic convergences are the main sources of the oxygen minimum layer, as earlier seemed most plausible (Redfield,

1942). This water spreads toward the equator at 800 to 1000 m, and it is reduced in oxygen as the organic matter is oxidized en route. However, there is insufficient carbon produced at the convergence to account for the oxygen depletion, and the source of the oxygen minimum water must be elsewhere (see Menzel, 1974:671). Geologically, both the oxygen minimum layer and the occurrence of phosphates are indicative of coastal upwelling.

General principles of open ocean circulation are summarized in Figure 3-17 for surface waters. The dominant wind-driven currents are westward in the tropics, and eastward at mid-latitudes. They are tempered by western boundary currents, by an equatorial countercurrent, by coastal upwelling along eastern oceanic margins, and by open ocean upwelling at oceanic divergences. Phosphates may accumulate under zones of coastal upwelling, and siliceous-rich sediments under zones of oceanic divergences. At the pole, exceptionally cold water forms and sinks to fill the deep ocean.

The first assignment in determining paleoceanographic currents is to decide upon the paleogeographic setting. Given that, the method used to discern patterns of ancient oceanic circulation is to compare these with modern patterns, sometimes amplified by use of simulation models. Evidence for circulation patterns is recorded

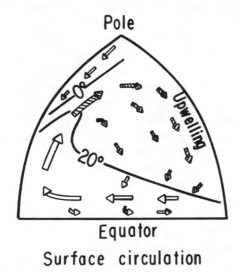

Figure 3-17 Idealized and simplified current and temperature (°C) patterns for an ocean. Open arrows indicate warm water; dark arrows indicate cool water. Note the asymmetry on different sides of the ocean. After Emery, 1968:447.

in several types of geologic deposits, including phosphorites, silica-rich sediments, evaporites, organic-rich sediments, and carbonate-rich sediments.

(2) Knowledge of ocean circulation over geologic time is greatly handicapped by lack of information on paleogeogrpahy. Perhaps the best that can be done at present is to summarize the suggested changes in circulation with the knowledge that much more is to be learned.

If no water in the upper kilometer becomes sufficiently dense to displace deeper water, then the deep ocean would not be mixed above the permanent thermocline (500 to 1,000 m depth). In particular, if the ocean were cool (as it is at present) and the earth were to warm up, then high-latitude water would become warmer and less dense, and deep-ocean water would not be as vigorously renewed. Instead, heating of deep water from below by crustal heat flow would serve to cause density instabilities, as may have happened in the Pleistocene (Worthington, 1968).

Sharp discontinuities in temperature and salinity occur and are somehow maintained between modern water massess. Density differences as fine as 0.0001 are maintained (Worthington, 1968), and water masses are not simply stirred into a homogeneous mixture (Broecker, Takahashi, and Li, 1976). It is difficult to go from this view of seemingly strong maintenance of a steady-state condition to the view that historical climatologic conditions may, in reality, be the chief reason why water masses of given characteristics are where they are. Yet that is a distinct possibility. Both steady-state and historical explanations need be kept in mind (although not always mentioned) in the ensuing discussion.

Cloud (1973, 1974a) suggested that the deep ocean was stagnant during the period prior to the change in climate which resulted in the late Precambrian glaciation, and that this climatic cooling caused dense water to form and rapid ocean turnover. Major times of ocean stagnation have also been suggested for parts of the Ordovician, Silurian, and Devonian because of the widespread occurrences of black shales. And Hallam (1975:60) wrote that during the Jurassic almost the whole ocean tended toward stagnation, a view which is also invoked for various intervals of the Cretaceous to account for extensive organic-rich layers in the Atlantic (Hallam, 1977a).

The only instance for which there is evidence of extensive cooling of the deep ocean is from the Cretaceous to the Recent, a 10°C change over a time scale of 100 m.y. (see Chapter 5). As high-latitude water is cooled, dense water is formed, and this contributes to deep-ocean circulation, and the renewal of oxygen. In contrast, once

an ocean is cooled, it is very hard to replace the deep water, especially if surface climate is itself becoming warmer. If so, one should
expect to find in the geological record deep-sea deposits of oxygenated water for several tens of millions of years *before* major glacial
periods. Deep oxygenated water would characterize the later Precambrian, the Ordovician, the Carboniferous, and the late Cretaceous to Recent. Intervening times would be characterized by a deep
ocean whose oxygen was sporadically renewed, a condition that could
lead to extensive organic accumulations. These intervals of time include the Cambrian, the Silurian and Devonian, and the middle
Permian to middle Cretaceous. As accurate paleogeographic maps
become available, it will be possible to test this idea. For the Jurassic and Cretaceous anoxic events, the question to ask may *not* be
why one finds anoxic sediments, but rather why, during those times,
one finds oxygenated deep-sea sediment.

Preliminary maps depicting oceanic surface currents have been
prepared for major portions of the world ocean for every geologic
period (some 25 papers are known to me). These maps are derived
from first principles of where ocean currents ought to appear (as discussed previously in this chapter and in a paleogeographic context
in Chapter 7). The accuracy of the charts is chiefly a function of the
degree to which adequate paleogeography is known. Their main
value has been as a heuristic device to consider the distribution of
fossils and of climate (especially usefully done by Ross, 1976, for the
Ordovician; Ziegler, Hansen et al., 1977, for the Silurian; Berggren
and Hollister, 1974, for the Mesozoic and Tertiary; and Ramsay,
1977b, for the Tertiary). The key to reliable charts is paleogeography.
But unfortunately, even for the past 100 m.y., estimates of bathymetry for the Atlantic Ocean "are not sufficiently precise that they
can be used directly for studies of the past circulation of water
masses" (Sclater, Hellinger, and Tapscott, 1977:548), and the same is
said to hold true for the Indian Ocean (Sclater, Abbott, and Thiede,
1977:51).

Upwelling has been inferred often from geologic evidence. The
Permian Phosphoria Formation of the western United States is the
most famous deposit whose origin is attributed to upwelling (Sheldon,
Maughan, and Cressman, 1967:51, Plate 11). The amount of phosphate accumulated in the sediment is more than six times greater
than the amount of the whole ocean today (Piper and Codispoti,
1975). Upwelling even has been used to account for disjunct distributions of organisms. The penetration of northern species into
subtropical waters was attributed to upwelling for, e.g., modern
foraminifera (Cifelli and Bernier, 1976) and Cretaceous ammonites
(Einsele and Wiedmann, 1975).

Now that paleogeography has been mapped for the South Atlantic, the depth of origination of organic-rich deposits was identified as 500 to 2,500 m. These sediments correspond closely with modern deposits formed in oxygen-poor waters, specifically the oxygen minimum layer (Thiede and Van Andel, 1977). Also in deep-sea deposits, a change from light to heavy $\delta^{13}C$ and then back again has been interpreted as indicating an expanding and then contracting oxygen minimum layer in the late Cretaceous paralleling changes in phytoplankton production (Scholle and Arthur, 1976; see discussion of $\delta^{13}C$, p. 236). In much older deposits, an offshore organic-rich belt of shallow oxidized sediments has been charted for the Cambrian. Fortey (1975) interpreted this as the expression of the oxygen minimum layer impinging on the continental shelf, although this view must be examined in light of latitude and side of the ocean.

In general, the density stratification of the ocean has fundamental consequences for geologic events (Gartner and Keany, 1978; Thierstein and Berger, 1978). On a time scale of a few thousands to a few millions of years, less saline water released to the open ocean in large amounts from an isolated basin (as is hypothesized for the Arctic Ocean during the latest Cretaceous) would act as a freshwater lid; conceivably this could have profound effects on the plankton (and dependent species) of the photic zone. In addition, very dense water released to the open ocean has the effect of stabilizing the deep ocean, thus leading to oxygen depletion and deposits typical of anoxic conditions (as occurred during the Jurassic and Cretaceous, whose paleogeography is known, and possibly during the Devonian—and other times—whose paleogeography is not yet so well known). The interplay of small and large ocean basins of variable salinities, set in a correct paleogeography, should lead to a much clearer understanding of geologic history over the next few years. The estimate of water volume and of rates of exchange *must* be quantitative, however, in order for hypotheses to be tested reasonably.

The greatest interest in determining oceanic circulation for any geologic interval has been for the Tertiary (Ramsay, 1977b:1436) and the Pleistocene. This effort has been greatly aided by the recognition that there are regional hiatuses in deep-sea sediments and that the CCD has nearly worldwide changes, but the general history of ocean circulation is still not understood (p. 108). In the Southern Hemisphere, as Australia moved northward from Antarctica, southern ocean circulation through the Tertiary responded to the changing geographic patterns (Kennett et al., 1972). The circum-Antarctic current could not develop until the middle to late Oligocene, when Australia and Antarctica fully separated. Possibly the exten-

sive erosion or nondeposition in the South Atlantic from the late Eocene to middle Miocene (generally 50 to 70 percent hiatuses in sampled sections) is related to the initiation of flow from Antarctic
Bottom Water (Ramsay, 1977b:1424). During this same interval of
time, the North Atlantic shows fewer and fewer histuses, as though
water currents were diminishing in strength. Hiatus patterns for
the Indian Ocean are like those for the North Atlantic, whereas the
hiatus patterns for the eastern-central and southwestern Pacific
show no particular time-dependent trends (30 to 50 percent hiatuses
for any given interval sampled [Ramsay, 1977b: Figures 31-32]).

One idea for Pleistocene circulation for the North Atlantic is
summarized in Figure 3-18. By mapping modern and ancient faunal
associations, Schnitker (1974) found that bottom water formed in
the subarctic rather than in the Arctic or the Antarctic. He related
this finding to the proposal of Weyl (1968, 1972) that sea-ice formation is favored by having a larger amount of lower-salinity (less
dense) surface water flowing into the North Atlantic than presently
does. A fresher northernmost Atlantic would mean a steeper salinity gradient with depth, and therefore the more stable formation of
sea ice. If these conditions obtained in the Pleistocene, sea ice may
have extended farther toward the equator than it now does, and typical present sources of North Atlantic Deep Water, such as the Norwegian Sea, would then have been ice covered. This may have led to
the formation of deep water at subpolar gyres at 50° to 60° latitude
instead of at higher latitudes, as is currently the case. Weyl's theory
received additional confirmation from oxygen isotope data on planktonic foraminifera (see Chapter 4). The Norwegian Sea bottom
was found to have had a higher temperature in the Pleistocene than
at present (Duplessy, Chenouard, and Vila, 1975). That is, with
warmer, lighter water, the Norwegian Sea could not have been a
source of deep water.

Worthington (1968) calculated that at the time of greatest extent of Wisconsin glaciers, the mean salinity of the ocean would
have been about 35.7‰, or one part per thousand more saline than it
now is (see Chapter 5). This very dense water would have filled the
deep-ocean basins. Worthington also calculated that heat flow
through the ocean floor over 15,000 years is sufficient to raise the
temperature of a 1 km column of water 7°C, enough to reduce its
density to modern values. Worthington therefore attributed the
present deep-water masses of both the Pacific and Indian oceans to
sinking of denser, more saline water during the Pleistocene. The
deep water of the Atlantic Ocean is, however, much younger than
that of the other oceans, and Worthington suggested that "most of it

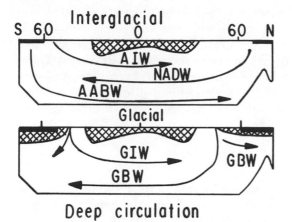

Figure 3-18   Schematic north-south diagram of thermohaline circulation in Western Atlantic Ocean for interglacial and glacial conditions. The stipled pattern is low-density surface water; the solid pattern is sea ice. Note that the region of formation of deep water is pushed much further toward the equator in glacial times, owing to the movement of sea ice. AIW, Antarctic Intermediate Water; NADW, North Atlantic Deep Water; AABW, Antarctic Bottom Water; GIW, Glacial Intermediate Water; GBW, Glacial Bottom Water. Based on Weyl, 1968:Figure 30; and Schnitker, 1974:386.

was formed during the period A.D. 1600–1800 when the northern climate was more severe than at any time since the Wisconsin period"—during the "Little Ice Age."

Worthington's and Schnitker's views on the importance of historical events bring to mind the basic question of how clearly ocean circulation at any given time can be understood by using steady-state rather than historical explanations. What effect, for example, does a decade of very cold winters have on creating water of extraordinary density, and what is the volume of that water compared to the volume of the reservoir into which it is flowing? In waters of the upper kilometer, the effects of even a single intensely cold winter can be quite noticeable. During the severe winter of 1976–77, a very much larger amount of water than was "normal" was cooled to 18°C at the southern edge of the Gulf Stream. This water sank, flowed southward in the Sargasso Sea past Bermuda, and is known as 18° water (Worthington, 1959). During that winter, the thickness of 18° water was increased by 100 to 150 m, indicating a very much larger volume addition at the source area (Leetmaa, 1977).

In general, Worthington's view emphasized the importance of punctational events in understanding what one measures at any given moment, in contrast to steady-state models, which have been "the traditional bulwark of ocean modeling" (Broecker, Takahashi, and Li, 1976) and which regard historical events as local perturbations. This is the same contrast evident in models of sediment origins, biological species diversity, and indeed in all subjects with a basis in natural history.

Summary

Tidal height increases inland approximately 400 km from the shelf break on modern shelves under favorable conditions. Although many geologic deposits are intertidal in position, few of these can be attributed definitely to tides per se. To the extent that it presently can be determined, tidal height has not changed markedly for at least 2 b.y. The best—but weak—key to judging past current velocity is the application of the Hjulström diagram. Ocean circulation of both surface and deep waters of ancient oceans is now a subject of great interest, and is based on an understanding of why modern oceans have the patterns they do.

Oceanic circulation is now coming into its own as a topic to which geologists can contribute actively. As paleobathymetry for the Jurassic to the present becomes known precisely, circulation will be plotted, sediment associations noted, and these in turn will provide the clues for going back beyond the Jurassic. There is no other part of paleoceanography which has as promising an immediate future as does the subject of determining ancient ocean circulation.

4. TEMPERATURE

Why we think the sun should be any of these
[i.e., perfect, or, if not perfect, constant, and if in-
constant, regular] when other stars are not is more
a question for social than for physical science.
J. A. Eddy, 1976

In this chapter I first develop the idea that the temperature of the open ocean may have 33°C as its approximate upper limit, owing to negative feedback mechanisms. Geologic, biologic, and chemical methods for recognizing ancient temperatures are then presented; these include isotopic methods, other chemical methods, taxonomic and morphologic methods, and faunal and floral gradients in diversity, age, and assemblages. Any *single* method is likely to yield results which are accurate to within only 2° to 3°C of the true temperature. The section on analytical methods is followed by one on possibilities for changing the atmospheric and solar ground rules that seem to set the earth's temperature, and this leads to a discussion of temperature over geologic time. The chief constraint on the earth's temperature from about 3.2 to 0.7 m.y. ago is that liquid water always existed ($T = 0°$ to $100°C$), and from 0.7 m.y. to the present is that metazoan life has flourished much as we know it today (chiefly 2° to 30°C).

Principles

As shown in Figure 4-1, the reflectivity (= albedo) at the earth's surface is dependent upon latitude. At high latitudes, the same amount of incoming solar energy is distributed over a larger surface area of the earth than at low latitudes because of the curvature of the earth's surface. Once at the surface, heat is maintained generally within belts parallel to latitude by the winds and ocean currents (Figure 4-2). Some of the excess heat of the tropics is transported poleward either as sensible heat (*directly* as warm air and water masses) or as latent heat (*indirectly* in the form of water vapor which has evaporated and then released heat on condensation). For the heat flux transported poleward of 30°N and 30°S, latent heat is responsible for 25 to 33 percent, sensible heat in the atmosphere for

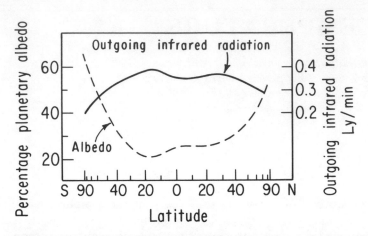

Figure 4-1   Mean meridional profiles of outgoing infrared radiation (ly/min, solid line) and planetary albedo (percentage, dashed line) (1 langley/min = 1 cal/cm²/min). After Vonder Haar and Suomi, 1969:667.

another 25 to 33 percent, and sensible heat in the oceans for the remaining 33 to 50 percent (Kellogg and Schneider, 1974).

The temperature range in the world ocean is from approximately 60°C in a few isolated coastal embayments of tropical latitudes to −2°C in high latitudes, but these extremes give a distorted view of the usual temperature distribution. The normal range is 2° to 30°C, and ocean water is in fact quite chilly on the average. Most of the water in the ocean is deeper than 2 km in depth, and most of this water is cooler than 4°C because it originated at the surface in high latitudes. If water at high latitudes were not so thoroughly cooled, deep waters would be correspondingly warmer.

At the upper range of temperature, surface waters appear to be constrained so as not to exceed 33°C, for the following reason. The heat balance between a large body of water and the atmosphere can be expressed as:

$$R = H + LE + G \tag{4.1}$$

where R is the net radiation at the water surface, H the vertical heat flux into the air, E the rate and L the amount of the latent heat of evaporation, and G the heat flux into the water (Priestley, 1966). With constant solar radiation, G is rather uniform. Hence changes in surface temperature are a function of H and LE.

Priestley (1966) showed that over a continuously moist surface,

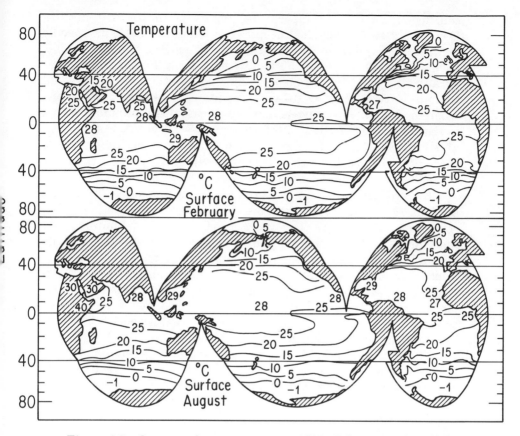

Figure 4-2 Ocean surface temperature (°C) in February and August. Simplified and redrawn from Sverdup, Johnson, and Fleming, 1942.

the cooling effect caused by evaporation of the water (LE) increases faster than the warming effect caused by the vertical heat flux (H). This leads to an equilibrium temperature which in theory is 32° to 34°C, and which in fact is approximately 33°, as is substantiated by data on open ocean tropical water temperatures.

The same principle applies to any moist surface: "vegetated surfaces are usually cooler than bare ground because much of the adsorbed solar energy is used to evaporate water" (Charney, Stone, and Quirk, 1976). Indeed, Linacre (1967) wrote that leaf temperature tends to come to equilibrium at 33°C by this same process, provided the plant is well watered and is exposed to bright sunshine. If a negative feedback system operates to limit ocean surface temperatures to approximately 33°C, then this leads to a very important conclusion about the temperature history of the earth's surface. Owing to a simple physical property of water, namely, its specific heat, and for as long as solar heating at the surface of the earth has been essentially the same, 33°C may have been the maximum temperature for water of the open ocean. As Priestley wrote (personal communication, May 8, 1974), "Exceptions could of course occur where horizontal advective effects complicate the local balance, e.g., in narrow or very shallow seas bordering deserts or where tides regularly draw water into contact with very warm rocks, or where there are other peculiarities of local circulation, but over the open oceans the same limit generally appears to hold."

If there is a natural upper limit to oceanic temperatures, it may be no accident that tropical marine species have an exceedingly narrow range of 34° to 37°C as their lethal limit (Kolehmainen and Morgan, 1972). This is not a lethal limit for comparable forms of life on land, however. In hot springs, metazoans (ostracods of the phylum Arthropoda) live and reproduce in temperatures of 50°C (Wickstrom and Castenholz, 1973). Thus the evolutionary potential exists for living in high temperatures, although it may never, for the reasons indicated above, have been realized in oceanic species. The evolutionary implications of this "limit" are discussed in Chapter 7.

There is yet another negative feedback mechanism to maintain a low oceanic water temperature. Adam (1975) combined the fact that hurricanes can form only where the oceanic surface temperatures are in excess of 26°C with the observation that these storms release immense amounts of energy. The net effect (through mixing, etc.) is to *decrease* the ocean surface temperature. Adam therefore concluded that this negative feedback mechanism may act to keep tropical temperatures not "far in excess of 26°C for long periods of time."

If an upper limit is set for tropical temperatures, we may ask if there are similar constraints on temperatures at higher latitudes. The mean global heat balance is based chiefly on the latitudinal variation in incident radiation. At present, latitudes lower than about 35° have a net surplus of incoming radiation, and higher latitudes a net deficit. Since equatorial temperatures, as well as the latitudinal distributions of heat, are set, it appears that markedly warmer higher latitudes would result only from perturbing the system from the outside—as by a change in the solar constant (which will be discussed later).

Temperature also varies systematically as a function of ocean depth (Figure 4-3). Seasonal warming and cooling is limited to approximately the upper 100 m (the seasonal thermocline), but an-

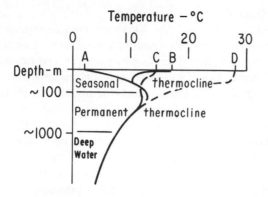

Figure 4-3   Temperature structure in the ocean at temperate latitudes to show seasonal thermocline, permanent thermocline, and
deep water. The transition from *A* to *D* represents the development
of the seasonal thermocline. *A*. Extreme winter cooling. *B*. After
spring warming with light winds. *C*. Condition *B* after vigorous wind
stirring. *D*. Extreme summer condition. After McLellan, 1965:29.

nual effects sometimes are barely discernible to as much as 600 m
for winter-cooled water (Pocklington, 1978). Below the seasonal
thermocline, extending to about 2000 m, the ocean temperature
gradually declines, and this depth interval is known as the permanent thermocline. Worthington (1976) divided the Western Atlantic
into five temperature depth zones: (1) greater than 17°C, which is
the lower limit of Gulf Stream water, followed by three temperature
divisions of the permanent thermocline: (2) 12° to 17°C in the upper
thermocline; (3) 7° to 12°C in the mid-thermocline; (4) 4° to 7°C in
the lower thermocline; and ending with (5) water less than 4°C in
the deep ocean. These temperature zones may be generally applicable in periods when oceanic circulation is well developed, as at
present.

Geologic Recognition

The ease of recognition of paleotemperature is strongly dependent
upon the environment and geologic age of the sampled strata. The
most precise temperature data come from planktonic taxa (especially foraminifera) of Pleistocene and Tertiary age. Temperatures
may be based on comparisons of temperature limitations of recent
taxa as "transferred" to their fossil relatives, or on oxygen isotope
data. Data on temperature for the Paleozoic through the Jurassic
generally are beyond the range of any assistance of isotopic data for

carbonates (owing to diagenetic effects), and largely rely upon latitudinal zonation of animals and plants. Data on temperatures for
Precambrian time are derived chiefly from oxygen isotopic data in
cherts, but whether these accurately reflect original environmental
temperatures is still a moot question (as will be discussed). In general, the only way in which a given estimate of paleotemperature
can be evaluated is by using more than one method. Thus it is extremely valuable to put together results from as many different
methods as possible. Four types of methods are discussed below: (1)
isotopic methods, (2) other chemical methods, (3) taxonomic and
morphologic methods, and (4) faunal and floral gradients. A comparison of isotopic and faunal methods concluded that each is beset
with errors which are on the order of 2° to 3°C for any given estimate
(Berger and Gardner, 1975).

(1) Isotopic methods include the use of (a) oxygen, (b) carbon, (c)
deuterium, and (d) combinations of these.

(a) In 1947, H. C. Urey reported that temperature effects on par-
titioning of stable isotopes might be used as a guide to paleotemperatures. For the reaction

$$H_2{}^{18}O + \tfrac{1}{3}C^{16}O_3{}^- \rightleftharpoons H_2{}^{16}O + \tfrac{1}{3}C^{18}O_3{}^= \qquad (4.2)$$

K, the equilibrium constant, is

$$K_{(t)} = \frac{[(C^{18}O_3{}^=)/(C^{16}O_3{}^=)]^{1/3}}{(H_2{}^{18}O)/(H_2{}^{16}O)} \qquad (4.3)$$

$K_{0°C} = 1.0176$, and $K_{25°C} = 1.0138$, or a temperature coefficient of
0.016 percent K per degree C. As temperature increases, the lighter
isotope (^{16}O) perferentially moves and therefore is incorporated into
the carbonate.

Since the numerical difference in mass is very small, the convention is to multiply all isotopic deviations by 1,000, in the form:

$$\delta^{18}O‰ = \left[\frac{(^{18}O/^{16}O) \text{ sample} - (^{18}O/^{16}O) \text{ standard}}{(^{18}O/^{16}O) \text{ standard}} \right] \times 10^3 \quad (4.4)$$

For example, the expression $\delta^{18}O = +10$ means that the isotopic
ratio of the sample exceeds that of the standard by 10 parts per
1,000, or 10 per mil, or 1 percent. The precision of the mass spectro-
metric technique is on the order of ± 0.1 percent (which is the same
as a temperature change of slightly less than ± 0.5°C). Changes in
$\delta^{18}O$ with respect to temperature over the range 0 to 30°C are shown
in Figure 4-4. A change of 1 per mil corresponds to an apparent effect of 4.5°C.

The final empirical equation for the carbonate-water isotopic

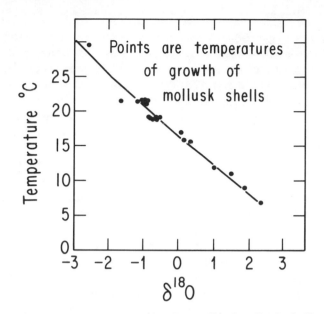

Figure 4-4 Temperature of growth of mollusk shells versus the $^{18}O/^{16}O$ ratio expressed as $\delta^{18}O$ relative to the PDB standard. After Epstein et al., 1953:Figure 9.

temperature scale in the range 0° to 30°C is given here in the form advocated by Shackleton (1974);

$$t(°C) = 16.9 - 4.38(\delta c - \delta w) + 0.10(\delta c - \delta w)^2 \qquad (4.5)$$

where δ terms for carbonate (c) and water (w) refer to the true δ values for CO_2 from these sources (when obtained from carbonate by acid extraction with H_3PO_4, etc., and from water equilibrated at 25°C, etc., both analyzed against the same mass spectrometer standard).

The original CO_2 standard against which comparisons were made was CO_2 derived from the calcite of a cephalopod belemnite (*Belemnitella americana,* upper Cretaceous) of the Peedee Formation (named for exposures along the Great Peedee River, in southeastern United States); the standard was referred to as PDB-1. This is still the standard for $^{18}O/^{16}O$ ratios in carbonates.

At the present time, values of $\delta^{18}O$ for *water* are given in terms of the SMOW (standard mean ocean water) standard, which is close to the true mean for ocean water. On the SMOW scale, $\delta^{18}O$ of seawater is −0.28 per mil. The SMOW standard is defined relative to the National Bureau of Standards (NBS-1) isotopic water standard

(Craig, 1961). The PDB-1 carbonate standard is $+30.6$ per mil on the SMOW scale (Craig, 1965). The need for better interlaboratory calibration led to the development of a second water standard called SLAP (Standard Light Antarctic Precipitation). The recommended δ value for SLAP relative to V-SMOW (Vienna Standard Mean Ocean Water) is: $\delta^{18}O$ (SLAP) $= -55.5\text{‰}$. V-SMOW was obtained "by mixing distilled ocean water with small amounts of other waters in order to bring its isotopic composition as close as possible to that of the defined SMOW" (Gonfiantini, 1978).

In practice, the oxygen isotope method is widely used for foraminifera (e.g., Emiliani, 1955, 1966; Shackleton and Opdyke, 1973), but it is also applied to calcareous nannofossils (Margolis et al., 1975), to many groups of metazoans, and to a wide variety of sedimentary, metamorphic, and igneous rocks.

The equation for isotopic equilibrium (without further corrections) is based on several assumptions: (i) an absence of a species-specific "vital" effect, (ii) a known isotopic composition of sea water, (iii) an absence of diagenetic changes in the sample, (iv) isotopic equilibrium between sea water and the organic calcium carbonate precipitated from sea water, and (v) an absence of any effect caused by a change in depth or sea-water density. Berger and Gardner (1975) and Savin and Stehli (1974) evaluated these biases for temperature from foraminifera. They concluded that the combined effect of routine bias is on the order of $2°$ to $3°C$ for any given value. The first three of these assumptions are now discussed.

(i) Isotopic fractionation values have been reported which are species-specific and which are distinct from values of local waters. In foraminifera they may account for "errors" from true temperatures of from $2.5°$ to $5°C$ (Duplessy, Lalou, and Vinot, 1970). Workers are now urged to limit isotopic analysis "to monospecific samples" (Emiliani, 1977), although multispecies assemblages probably yield reasonable average values (Savin, Douglas, and Stehli, 1975). Symbiotic algae occur with some taxa (e.g., many corals, and foraminifera [Bé, 1977]) and contribute ^{16}O-rich metabolic CO_2 to the internal pool of CO_2 in the carbonate-secreting host. Consequently, carbonate ^{18}O values from these organisms are not in equilibrium with ambient temperatures and will yield temperatures displaced upward by as much as $5°C$ (Erez, 1978).

(ii) Measured isotopic values are sensitive to initial sea-water values (the δw terms in equation 4.5). Evaporation of sea water preferentially removes the lighter ^{16}O isotope, thus enriching sea water in ^{18}O (Craig and Gordon, 1965). (For the same reason, evaporation will enrich sea water in ^{34}S, relative to ^{32}S, in deuterium, and in ^{13}C

relative to ^{12}C, although for $^{12}C/^{13}C$ fractionation, the influence of organisms is a much more significant influence than is evaporation.)

The present open ocean (with a salinity range of 33.0 to 38.0‰) has a variation in $\delta^{18}O$ of about 1.5 per mil (Berger and Gardner, 1975: Figure 6). The upper limit of enrichment in $\delta^{18}O$ in brines owing to evaporation of water may be 6 per mil (Lloyd, 1966). In a transect from open ocean to fringing reef to shore, an enrichment of $\delta^{18}O$ may occur, perhaps because of evaporation near shore on a small volume of water (Lloyd, 1964). Fresh water is greatly depleted in $\delta^{18}O$ since it forms from evaporated sea water.

Before the Plio-Pleistocene development of glaciers, the isotopic composition of ocean water would have been − 1.0 per mil, SMOW standard (Shackleton and Kennett, 1975). In contrast, for the open ocean during the time of maximum removal of fresh water to glacial ice, a value just over 1‰ isotopically more positive "is now widely accepted" (Shackleton and Opdyke, 1973), and may be as high as 2.7‰ in the Mediterranean (Grazzini, 1975). If so, then Pleistocene changes in $\delta^{18}O$ record changes chiefly in ice volume (and thus sea level) rather than in temperature (Shackleton, 1967). This has been proved to be the case for the past 120,000 years because the patterns of variation in $\delta^{18}O$ of planktonic and benthic foraminifera parallel each other, independent of surface and bottom temperatures. For the Pleistocene, van Donk (1976) believed that "at least 90 percent of the changes in the isotopic composition are attributable to variation in the isotopic composition of ocean water, which is due to the waxing and waning of large continental glaciers." Thus Pleistocene data from oxygen isotopes record major climatic changes—not in temperature but in ice storage.

In addition to regional variation in $\delta^{18}O$, there can be significant local variation. Daytime sea-water samples are enriched in ^{13}C relative to water collected at night because of CO_2 fractionation by phytoplankton during the day (Weber and Woodhead, 1971). Thus the ratio of local biomass to local water mass can be important in interpreting isotopic values for species from restricted bodies of water.

(iii) Correction of $\delta^{18}O$ values is required for skeletal samples which have had significant changes because of partial solution (Berger and Gardner, 1975) or postdepositional isotopic exchange (Veizer, 1977). Isotopic exchange is more likely for shells than for the sedimentary matrix (Tan and Hudson, 1971). In general, in any given geologic deposit, if aragonitic shells are "unaltered," the even more stable calcitic shells that occur with them may yield reliable isotopic data. Such deposits are most likely to be in bituminous or asphaltic sediments, although oxygen isotopic values of petroleum brine waters appear to have remained unchanged since the Paleozoic (Degens et al., 1964). Brine waters, like asphaltic sediments, generally are removed from the influences of ground water.

Asphaltic sediments are rare, however, and it appears that nearly all geologic deposits would yield carbonate material whose isotopic values must be corrected before isotopic temperature determinations can be made. The extent of postdepositional isotopic exchange can be estimated because the equilibrium and/or "typical" amounts of Sr, Mg, and other trace elements are well known for carbonate minerals from sea water and fresh water. For example, Mn, Fe, and Mg will be increased greatly in marine carbonates which subsequently have been "washed" in fresh water, whereas Sr will be systematically depleted. Therefore, systematic changes in these (or other) elements in skeletal material provide an index to diagenetic change in coexisting isotopic values (excellently employed by Veizer, 1974, who confirmed that belemnites were low-Mg calcite, and that the specimens with which he was working had had about a 10 percent postdepositional exchange with fluids of the enclosing sediment).

Although the general warning of *caveat emptor* applies particularly to isotopic data, an amazing degree of sensitivity can be realized from single specimens. Samples from successive septa in a Jurassic ammonite recorded seasonal changes, as shown in Figure 4-5, and thus permitted the inference that this animal was born in the spring, and died in the winter of its second year.

In addition to the widespread use of oxygen from carbonates for $\delta^{18}O$, other oxygen carriers, in the form of phosphates, sulfates, and silicates, have been evaluated as paleotemperature indicators. A second source of ^{18}O should provide an independent isotopic test of equilibrium conditions between sea water and the mineral phases. In the case of phosphates, the isotopic fraction for oxygen is practically identical with that of carbonates (Longinelli and Nuti, 1973). Thus temperature and environmental isotopic composition cannot be determined *independently* by using the phosphate system. However, in the absence of carbonate, phosphate itself has the potential for yielding reliable isotopic temperatures.

With regard to sulfate, isotopic equilibrium of ^{18}O with surface waters is not obtained, seemingly because of a very low exchange rate (Longinelli and Craig, 1967; Lloyd, 1967). In addition, tremendous scatter in isotopic data from shells of both living and fossil forms indicates that shell sulfate probably is not involved in metabolic processes, since such processes usually exhibit fractionation

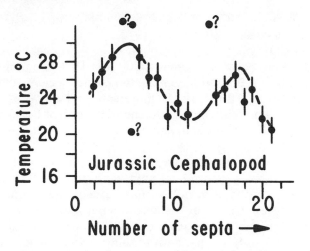

Figure 4-5   Temperatures of formation of the septa of an ammonite of Jurassic age. Variation with age is interpreted as seasonal, with the animal having been born in the spring, and having died in the winter of its second year. After Stahl and Jordan, 1969:177.

with temperature. Possibly postdepositional processes easily change initial sulphate isotopic values. Whatever the cause, sulfate is not usable for paleotemperature analysis.

With regard to silicates, oxygen isotopic fractionation in SiO_2 of marine diatoms and sponge spicules (Labeyrie, 1974; Mikkelsen, Labeyrie, and Berger, 1978) is very similar to carbonate fractionation, since t (°C) $= 5 - 4.1$ ($\delta_{SiO_2} - 40 - \delta_{H_2O}$). Silicate $\delta^{18}O$ should provide an independent check on temperature for at least the Pleistocene. Over a longer time scale, chert may be an appropriate source of silica for temperature determinations (Knauth and Epstein, 1976), and results based on chert are presented below. Kolodny and Epstein (1976) stressed, however, that chert paleothermometry "cannot, even at its best, compete with the fine resolution capabilities of the carbonate paleothermometer."

(b) In a manner entirely analogous to that of oxygen isotopes, data for carbon isotopic values are reported as

$$\delta^{13}C\text{‰} = \left[\frac{(^{13}C/^{12}C) \text{ sample} - (^{13}C/^{12}C) \text{ standard}}{(^{13}C/^{12}C) \text{ standard}} \right] \times 10^3 \quad (4.6)$$

The standard for $^{13}C/^{12}C$ ratios is PDB-1, referred to above. Values of $\delta^{13}C$ are subject to the same types of controls as are specified for $\delta^{18}O$, with the added complexity that different biosynthetic pathways fractionate ^{13}C to different degrees in terrestrial plants (C_3

versus C_4 plants; Lerman, 1974). A wide variety of organisms have been investigated (Degens, 1969; Smith, 1972).

In the system CO_2 (gas): CO_2 (aqueous): HCO^-_3 (aqueous), the carbon isotopic fraction between gaseous CO_2 and HCO^-_3 changes from -10.8 per mil to -7.4 per mil over the temperature range 0° to 30°C, at sea-water pH of 8.2 (Mook, Bommerson, and Staverman, 1974). This fractionation occurs chiefly in the hydration of CO_2 and not in the transfer of CO_2 from atmosphere to water (Deuser and Degens, 1967).

The temperature fractionation may be quantitatively lower if the rate at which carbon is utilized by phytoplankton exceeds the rate at which isotopic equilibrium can be maintained. The extent to which this disequilibrium occurs in the ocean is not known. During photosynthesis, phytoplankton partition carbon isotopes at a rate of approximately -19 per mil, independent of temperature, as long as excess bicarbonate exists (Deuser, Degens, and Guillard, 1968).

In general, very negative $\delta^{13}C$ values ($\simeq -25$ to -28) are consistent with low temperatures and large excesses of dissolved CO_2. In contrast, relatively positive values ($\simeq -9$ to -15 by one account, $\simeq -21$ by another account) are indicative of warmer waters with less dissolved CO_2. Values of -28 per mil are known from plankton in the cold South Atlantic and of -13 per mil from plankton of the warm Gulf Stream (see Deuser, Degens, and Guillard, 1968; and Sackett et al., 1965). During the Pleistocene, sediment organic carbon $\delta^{13}C$ values changed from $-22‰$ to $-26‰$, and this has been correlated with warm and then cool (interglacial and then glacial) periods (Rogers and Koons, 1969). Alternatively, taxon-specific fractionation also exists, and this "problem" prevents any simple causal relationship with temperature, although these separate taxa (and their $\delta^{13}C$ values) do track water masses (Fontugne and Duplessy, 1978).

If excess bicarbonate is not available, phytoplankton fractionation is less than -19 per mil. CO_2 with a greater $^{13}C/^{12}C$ ratio (i.e., heavier carbon) is incorporated into the organic matter. This process may continue until not enough CO_2 exists to sustain growth, approximately at a fractionation of -6 per mil relative to the CO_2 source (Deuser, Degens, and Guillard, 1968). Values of $\delta^{13}C$ from limestones reflect changes in the oceanic $\delta^{13}C$ pool, which itself is influenced by the phytoplankton fractionation (see p. 236).

(c) The deuterium (D) content of precipitation decreases with decreasing temperature (Dansgaard, 1964) to approximately the same extent as $\delta^{18}O$ changes with temperature (Figure 4-6). The SMOW standard is used to express D/H ratios in water and other materials. In terms of the new standards (Gonfiantini, 1978; see

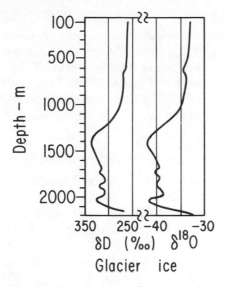

Figure 4-6   Oxygen $\delta^{18}O$ and hydrogen δD isotope ratios in glacier ice from a deep core hole at Byrd Station, Antarctica. Note the virtual identity between the patterns. After Epstein, Sharp, and Gow, 1970:1571.

above, $\delta^{18}O$ (V-SMOW) = $-0.2‰$ SMOW, and δD (SLAP) = $-428‰$ V-SMOW. Therefore, δD per mil is suitable as a paleotemperature indicator (thoroughly discussed by Craig and Gordon, 1965). It has been used in that way on peat (Schiegl, 1972), and potentially is useful on cellulose (Epstein, Thompson, and Yapp, 1977).

   (d) Both $\delta^{13}C$ and $\delta^{18}O$ are affected by salinity (see Chapter 5), but to different degrees. Therefore, if a series of carbon and oxygen isotopic values are known from shells presumed to have lived under varying salinity but the same temperature regime, a unique solution exists for the temperature of formation (Mook, 1971). Shell carbonate is in isotopic equilibrium with aqueous bicarbonate, and therefore extrapolation of a $\delta^{13}C/\delta^{18}O$ plot of data from shells of estuarine species to intercept the oceanic bicarbonate value would yield the approximate growth temperature (Mook and Vogel, 1968). This method effectively removes any fractionation caused by salinity.

   (2) Other chemical criteria have also been used for paleotemperatures. Grouped in this section are summaries of paleotemperature methods based on (a) racemization of amino acids, (b) noble gas solubility, and (c) trace-element analysis.

(a) Racemization of amino acids is a newly developed paleotemperature technique. Only L-amino acids usually occur in proteins, but after death the α-carbon of the amino acid changes stereochemistry so that with time, an equivalent amount of L- and D-amino acid is produced (Schroeder and Bada, 1976). This reaction is a function of time, temperature, and other factors. For the two stereoisomers of any amino acid, the conversion can be expressed by:

$$\ln \left\{ \frac{1 + D/L}{1 - D/L} \right\} - \text{constant} = 2\,kt \qquad (4.7)$$

where t = time, D/L is the amino acid enantiomorphic ratio, and k is the first-order rate constant for interconversion of the L- and D-enantiomeres (Bada and Schroeder, 1975). The constant k reflects any initial D-enantiomere plus any produced during the isolation procedure.

The half-life for racemization of different amino acids varies. For isoleucine racemization at 25°C, it is approximately 50,000 years, but for aspartic acid it is only 3,500 years (Bada and Schroeder, 1975).

Either the time or the time-averaged temperature must be known in order to apply the racemization technique to fossil material. Thus this method is especially useful in environments where temperature has been essentially uniform over very long periods of time, such as the tropics, the deep sea, or caves.

As with isotopic methods, taxon-specific bias occurs (Wehmiller, Hare, and Kujala, 1976), with as much as a factor of 2 difference in apparent racemization rates (Hare, 1977). This results because each species has unique shell proteins each of which has its own characteristic kinetic curve (Kriausakul and Mitterer, 1978). In situ leaching also can be a significant source of bias (King, 1978).

(b) The amounts of the noble gases Ar, Kr, and Xe which are dissolved in water until air saturation is reached are dependent upon temperature. For a temperature increase from 10° to 60°C, the solubility decreases by a factor of 2.0 for Ar, 2.3 for Kr, and 2.8 for Xe. If the system containing the gases remains closed until the time for analysis, then the paleotemperatures at the time of formation can be calculated. Mazor (1972) applied this method to trapped "fossil" ground water in North Africa.

(c) The influence of temperature on the mineralogic state of calcium carbonate shells and on the incorporation of trace elements has been investigated extensively since the pioneering work of Clarke and Wheeler (1922). In the best-documented instances, Mg concentration in calcitic shells increased regularly with temperature in many groups of invertebrates (Dodd, 1967). This change does not, however, appear to be a simple physical-chemical reaction. Mg concentration is mediated by physiological factors which alter the rate of magnesium uptake and which depend upon the taxon (Weber, 1973b).

Many more studies have been undertaken of Mg than of Sr, the other important minor element in calcareous shells. Apparently, there is a positive correlation of Sr with temperature in calcite, but a negative correlation in aragonite (Dodd and Schopf, 1972).

(3) Taxonomic and morphologic methods are traditional in paleotemperature work. There are several aspects of floras and faunas which are correlated with water temperature, but these form no consistent pattern. Briefly mentioned, they include reference to: (a) animal size, (b) animal calcification, (c) density of planktonic shells, and (d) direction of foraminiferal coiling.

(a) Animal size is correlated with temperature (Lewis, 1968). Within benthic marine species, higher-latitude populations tend to breed at a later age and less often than lower-latitude populations, and this difference can have a genetic basis (Frank, 1975). Accordingly, for any adult age, individuals of a higher-latitude population tend to be larger than individuals of a lower-latitude population, a size trend referred to as Bergmann's Rule. Strauch (1971) applied the relationship of increasing adult size and decreasing temperature to interpret the environmental history of beds with the marine bivalve *Hiatella*. For marine pelecypods, the *largest* species of a genus (or of a family) is said to live in the coldest water (Nicol, 1967).

Larger animals have smaller surface area per unit volume than do smaller forms, and thus may expend less energy to maintain body heat. This relationship (known as Allen's Rule) is useful only to animals that maintain a temperature above the environmental temperature.

Specimen size of benthic foraminifera increases as water temperature decreases (and depth increases) (Bandy, 1963). The mean size of the planktonic foraminiferan *Globigerina bulloides* becomes larger during cooler climate, and an equation has even been derived to express this: annual average surface temperature in $°C = -0.1184$ (mean width of *G. bulloides* in μm) $+ 45.06$ (Malmgren and Kennett, 1978).

(b) The extent of calcification of shells appears to be temperature related. Tropical marine animals commonly have thick and heavy shells, whereas, at least for pelecypods, shells of cold-water species generally are thin (Nicol, 1967).

(c) Several morphologic adaptations in phytoplankton (Lewis,

1976) and in planktonic foraminifera appear related to adjustments
in sea-water density (and through that to temperature and salinity)
as a function of latitude. Porosity of skeletons decreases with denser
water, as shown in Figure 4-7. Lower porosity in cooler (heavier)
water would be advantageous in order to sustain the same buoyancy
typical of warmer (lighter) water (Frerichs et al., 1972). Perhaps for
the same reason, the volume of the planktonic foraminiferan *Globigerinoides trilobus* decreases with decreasing temperature (Hecht,
1973b). As another example, *Globorotalia truncatulinoides* changes

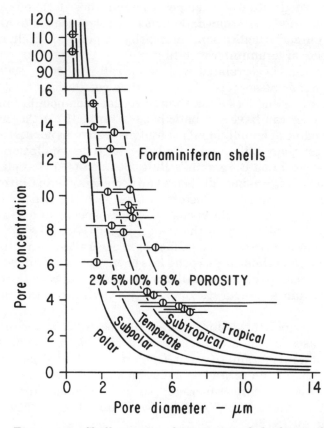

Figure 4-7 Shell porosity of 22 species of planktonic foraminifera
compared with lines of 2, 5, 10, and 18 percent porosity and with suggested climatic interpretation. Horizontal lines are observed ranges
of pore diameters. Pore concentration is number of pores per $25\mu^2$.
Note the increase in pore concentration in cooler, denser water. After
Bé, 1968:883.

from highly conical forms in warm tropical areas to more compressed forms in cooler water, as if the denser shape were adapted to the denser water (Kennett, 1968a).

The direction of coiling in several species of planktonic foraminifera appears to be correlated with summer temperature (Bandy, 1972, and earlier), but this is disputed by some (e.g., Cifelli, 1971; Olsson, 1974), and in any event the causal basis for the relationship remains unknown. Dextral coiled specimens are those in which the chambers are added in clockwise direction when foraminifera are viewed from above (with the dorsal side up), and are characteristic of warmer water; the converse applies to sinistral specimens. Lines connecting points of equal percentage of coiling have been considered to be approximately lines of equal temperature. For example, living populations of *Globogerina pachyderma* with over 90 percent sinistral forms occur in water of less than 6°C. A gradational change in coiling with change in temperature is illustrated by *G. pachyderma* (Kennett, 1968b) and by *G. bulloides* (see Figure 4-8) from the southwest Atlantic (see excellent review of Kennett, 1976).

(4) Faunal and floral gradients in (a) diversity, (b) age, and (c) assemblages are correlated with temperature gradients.

(a) Many groups of marine animals and plants display approximately symmetrical patterns of taxonomic diversity with respect to the equator, with high diversity at low latitudes. Data for several

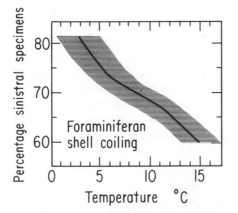

Figure 4-8   The relationship between sinistral and dextral specimens of the foraminiferan *Globigerina bulloides* d'Orbigny and the surface water temperature (southwestern Atlantic and the Drake Passage). Accuracy of temperature from percentage cooling is on the order of ± 2°C. After Boltovskoy, 1973:153.

marine groups have been reported, including foraminiferans (Berger and Parker, 1970), corals (Stehli and Wells, 1971), bryozoa and bivalves (Schopf, 1970), gastropods, copepods, tunicates, and epifaunal and infaunal assemblages (Fischer, 1960). Open ocean zooplankton have much higher diversities in the large subtropical gyres than in the equatorial or polar currents (Reid et al., 1978). Fossil zooplankton groups (such as probably the conodonts of the Paleozoic and Triassic) might also follow this pattern.

Use of worldwide data is necessary in plots of taxonomic diversity in order to minimize the fact that Pacific diversity is approximately twice that of the Atlantic (data for corals, bivalves, bryozoans). In addition, diversity of the Eastern Atlantic is higher than that of the Western Atlantic (for bivalves and bryozoans, but not for corals), but diversity of the Western Pacific is substantially higher than that of the Eastern Pacific (for bivalves, bryozoans, and corals). The latitudinal diversity gradient may simply reflect larger faunal provinces in tropical latitudes because faunal diversity is a function of the area of a faunal province (Schopf, Fisher, and Smith, 1978; see also p. 250).

The ratio of cosmopolitan to endemic families also changes along modern latitudinal gradients, with endemic families much more common in the tropics. Similar patterns exist for Cretaceous foraminifera, and Permian Brachiopoda (Stehli, Douglas, and Newell, 1969), but do not obtain for mammals (Van Valen, 1969).

(b) The average age of genera appears lowest in equatorial regions and increases by a factor of 2 to 3 at high latitudes for benthic foraminifera (Durazzi and Stehli, 1972) and hermatypic corals (Stehli and Wells, 1971). This trend is strongly dependent on the whole ocean being considered (just as for diversity gradients) in that the average age of genera in the Pacific is about half that of the Atlantic, normalized for latitude. In contrast to latitudinal trends in the age of *genera,* the average age of *species* for planktonic foraminifera is independent of latitude (Figure 4-9). Reasons for this difference between genera and species are not understood.

(c) Several techniques for estimating temperature have been derived from assemblages of species. Species of temperate latitudes have their maximum temperature range set by their approach to the equator at the warmest time of the year, and their minimum temperature range determined by their approach to the pole at the coolest time of the year (Hutchins, 1947). Accordingly, if the temperature range is known for several species of a fauna, the temperature range of the fauna as a whole can be approximated. The simplest method is merely to plot the temperature range of each

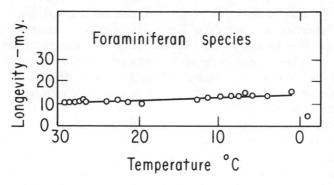

Figure 4-9   Average age (longevity) for species of planktonic foraminifera plotted against mean annual sea surface temperatures (°C).
Note that species longevity is independent of temperature (i.e., latitude). After Stehli, Douglas, and Kafescioglu, 1972:125.

individual species; the zone of greatest overlap is probably the best
guess of temperature, just as with bathymetry (see Figure 2-13).

Approximate temperature also can be estimated by examining
the ratio of cool-water to warm-water genera (as has been done for
Pliocene silicoflagellates). This ratio first must be determined in
both recent sediments and in the overlying water mass (Ciesielski
and Weaver, 1974) before the method can be applied to geologic
problems.

The extent of overlap in temperature for all species within an
assemblage can be evaluated by using appropriate statistical techniques (Imbrie and Kipp, 1971; Hecht, 1973a; Kipp, 1976). To the
extent that fossil foraminifera respond to temperature as do Recent
species, Hecht used the following straightforward procedure:

1. Determine the percentage abundance of species of planktonic foraminifera in a suite of Recent core top samples for some portion of the ocean.

2. Construct a similarity matrix comparing each sample with
every other sample based on the abundance of the species (Hecht
used Parks' Distance Coefficient).

3. Plot the observed surface temperature for each station
against the similarity values for each station (relative to the station
with *maximum* surface temperature).

4. Fit a least squares regression line to the above data.

5. Compare the similarity of fossil samples to Recent samples
by using the Distance Coefficient. By using the equation derived in
Step 4, a paleotemperature may be determined based on the degree
of similarity of the two assemblages.

Results obtained from this method (or the similar but more precise "transfer function" method of Kipp, 1976) provide an independent check for Pleistocene isotopic data. Different "temperature assemblages" can be compared by using only the species in common. Hecht (1974b) reported that "paleotemperature equations based on varying species diversity are similar and give paleotemperatures which differ by only 1–2°C."

If the patterns of abundance of different species reflect temperature differences, than so might the patterns of abundance of morphologic variation within a single species. However, the temperature range within which one is seeking differences for a single species is much less than that for species within an assemblage. It is therefore difficult to obtain meaningful data on temperature by using trends in degree of morphologic variation, although mapable patterns correlated with temperature and salinity do exist for some planktonic foraminifera (Hecht, 1974a; Srinivasan and Kennett, 1974).

Changes in the Sun and the Atmosphere

Air temperatures of the earth's surface may have been significantly lower or higher in the geologic past because of a change in one or more of the following: (1) incoming radiation, (2) atmospheric composition, (3) albedo, and (4) cloud cover. Any proposed change in the earth's surface temperature must, however, fall within very narrow limits. For the time period from about 3.8 b.y. ago (the age of the oldest rocks composed of particles transported by running water) to about 0.7 b.y. ago (when multicellular plants and animals become widespread), the earth's temperature was probably always within the range of 0° to 100°C. From 0.7 b.y. ago to the present, ocean temperatures of most of the earth are likely to have been in the 2° to 30°C range typical of today. Although bounded by these two constraints, the earth's temperature could have changed markedly.

(1) The solar constant is defined as the amount of solar energy of all wavelengths received per unit time per unit area at the top of the earth's atmosphere, for the mean sun-earth distance (SI units as Watts/m^2, but more familiarly as cal/cm^2/min; 1 cal/cm^2/min = 698 watts/m^2). A good estimate of the solar constant is $1,373 \pm 20$ W/m^2 (Frölich, 1977).

During the lifetime of the sun, an increase in solar radiation on the order of 25 percent is the expected pattern of "almost all solar models" (Newman and Rood, 1977), perhaps as great as 1 percent per 50 m.y. (Smith and Gottlieb, 1974). Such a change would have

immense implications for the earth's temperature history because
changes in solar radiation of a very *minor* extent have been postulated to cause significant effects. Budyko (1969) observed that over
the past century, the ratio of temperature change to the change of
direct solar radiation with cloudless sky "turns out to be equal to
1.1°C per 1 percent of radiation change." Theoretical calculations by
the same author also showed that as solar radiation changes by 1
percent (for cloudiness = 0.50, constant albedo, and constant relative air humidity), mean temperature at the earth's surface changes
by 1.2°C. Budyko therefore considered the present thermal regime
to be "characterized by high instability," with a reduction in radiation of 1.0 to 1.5 percent sufficient to cause the glaciers again to
reach temperate latitudes. In addition to increasing temperature,
an increase in incoming radiation may cause a very much increased
precipitation and evaporation cycle (Wetherald and Manabe, 1975).
Others have indicated that an increase in the solar constant on the
order of 2 to 5 percent would be sufficient to cause glaciers to be
melted (Sellers, 1969). In fact, a surface temperature difference of
only 4° to 6°C accompanied the change from full glacial conditions to
the present (Bryson, 1974).

All of this talk about changes in the solar luminosity is not
without a reasonable basis because significant changes in the solar
constant have occurred in recorded history. The Maunder Minimum
is the name given to the period between 1645 and 1715 (Eddy, 1976)
in which there was a prolonged absence of sunspots (named for E. W.
Maunder, superintendent of the Solar Department, Greenwich Ob-
servatory, who documented it in the 1890s). This period also was
marked by ^{14}C values 10 percent higher than normal. The time of
onset of sunspots corresponds to a time of increasing equatorial velocity of the sun (Eddy, Gilman, and Trotter, 1977). Cooling of the
sun's photosphere may also be important in controlling sunspots
(Livingston, 1978). In addition to the evidence from sunspots,
changes in the brightness of Uranus and Neptune from 1956 to 1966
indicate solar variability (Lockwood, 1975). The solar constant was
estimated to have increased at a rate of 0.1 percent per decade for
the years 1921 to 1952 (Öpik, 1968), although some claim uncertainty in measurement of the solar constant is on the order of 1 percent (Lockwood, 1975). If the solar constant has increased as Öpik
suggested, then its present value would be about 1.4 percent higher
than its value during the Maunder Minimum.

These proposed changes in the solar constant correspond to observed temperature changes of about 0.6° between 1880 and 1940,
and of a few degrees since the time of "the Little Ice Age" in 1700

(Wang et al., 1976). Eddy (1977) plotted historical changes in the
level of solar activity versus observed climatic shifts. He found that
"in every case when long-term solar activity falls, mid-latitude glaciers advance and climate cools; at times of high solar activity glaciers recede and climate warms." Eddy therefore proposed that
"changes in the level of solar activity and in climate may have a
common cause; slow changes in the solar constant, of about 1 percent amplitude." Fluctuations in the solar constant of this extent
easily could be caused by variations in the random changes which
accompany solar convection (Dearborn and Newman, 1978).

The spread of luminosities in stars similar to the sun but in the
Praesepe cluster "is consistent with 10 percent variations in the
solar luminosity" (Ulrich, 1975). Eddy (1976) concluded his work on
the inconstancy of the solar constant with the statement which is
the starting quotation for this chapter. One gets the impression that
knowledge of variation in the solar "constant" is in a state of flux,
that variation on the order of 1 percent over time scales as short as
10^3 years is quite possible, but that the climatic implications of such
changes (if they occur) are not clearly established.

These past few paragraphs now bring us to a basic question:
why hasn't the earth's surface temperature changed much more dra-
matically than it has? On the one hand, why hasn't the earth frozen
over, and on the other hand, why hasn't all the water boiled away?
Negative feedback mechanisms both in water and in the atmosphere must somehow be responsible for this steady state, for it
seems clearly within the power of the changing solar constant to
have fundamentally altered the observed temperature history of the
earth's surface.

(2) The atmosphere is the filter through which incoming radiation must pass. Without the atmosphere, the earth's surface would
be frozen 35°C below its present value (Fleagle and Businger, 1975).
An approximate accounting for present solar radiation is that "for
every 100 watts of incoming power, 24 are reflected by clouds, 7 are
scattered back to space by the earth's atmosphere, and 4 are reflected back to space by the planet's surface (subtotal 35) (= the
planetary albedo); 22.5 watts reach the surface directly, 14.5 more
after diffuse scattering from clouds, and 10.5 more after diffuse molecular scattering (subtotal 47.5); of the residue (subtotal 17.5),
about 6 are absorbed by the atmosphere in the ultraviolet and 11.5
in the infrared " (Hobbs and Harrison, 1974). Thus of incoming radiation, about 35 percent is reflected back, 50 percent reaches the
earth's surface, and 15 percent is absorbed.

This section is concerned with the percentage of incoming radiation which is absorbed. Merriam (1974) stated that almost the entire reduction of solar energy (from 2.00 to 1.40 $cal/cm^2/sec$) in its passage from the top of the atmosphere to sea level is due to atmospheric absorption by H_2O, CO_2, and O_2. Of these three components, H_2O is "by far the most important."

Other things being equal, an increase in CO_2 and/or H_2O in the atmosphere will lead to greater absorbance of infrared radiation emitted by the earth's surface, thus causing an increase in global temperature. The magnitude of this "greenhouse" effect can be considered to be the difference between the upward infrared radiation from the earth's surface and the upward infrared radiation from the top of the atmosphere. This difference is now on the order of 0.024 $cal/cm^2/min$ (Bryson, 1974). The greenhouse effect is responsible for the fact that although the average planetary radiative equilibrium temperature is about $-20°C$, the average surface temperature is about $14°C$ (Schneider, 1972a). A 1 percent change in the greenhouse effect is thought to change the surface temperature about $0.3°C$. If relative humidity is held constant, an increase in lower atmosphere temperature of 2.0 to $3.0°C$ is considered probable if there is a 100 percent increase in CO_2 (Baes et al., 1977, and many others earlier). Budyko (1972:873) suggested that even a 50 percent increase in the present concentration of CO_2 would raise high-latitude temperatures enough "for complete melting of polar ice" to occur. Conversely, a reduction in the CO_2 level to one-half its present value "can lead to a complete glaciation of the earth."

If we wanted to be extreme about things, we could let CO_2 pressure vary to 10 times its present value. Sillén (1966) considered that this would result in an increase in average temperature of "perhaps 20° to 30°C." For Precambrian times 4.6 to 2.6 b.y. ago (see Chapter 5), high CO_2 pressures (in the absence of free oxygen) are quite possible. If so, the earth's *average* sea-level temperature is predicted to have risen from its present value of $14°C$ to between 34° and $44°C$. Maximum values would have come closer to $50°C$ instead of the present values of near $30°C$. In addition, the presence of significant NH_3 in a Precambrian atmosphere (Chapter 5) has been claimed to increase further the infrared absorption and to raise temperatures (Sagan and Mullen, 1972), although a change of only $0.1°K$ is predicted from doubling its present value (Wang et al., 1976).

The difficulty with the views of the past two paragraphs relating increasing CO_2 to increasing temperature is that the relationship is not linear. Once the main band for absorption by CO_2 is saturated, increasing CO_2 "does not substantially increase the infrared capacity of the atmosphere" (Rasool and Schneider, 1971). Indeed, these authors suggested that even a tenfold increase in atmospheric CO_2 would result in less than a 2.5°C temperature increase.

Several other gases (O_3, N_2O, CH_4, HNO_3, C_2H_4, SO_2, CCl_2F_2, CCl_3F, CH_3Cl, and CCl_4) occur in trace amounts in the atmosphere and absorb in the atmospheric window, which transmits most of the earth's thermal radiation. H_2O effectively absorbs except for between $\simeq$ 750 and 1,300 μm; this range is further reduced by CO_2, which strongly absorbs for $\simeq$ 600 to $\simeq$ 750 μm and weakly absorbs (with present amounts) at 900 to 1000 μm. Therefore, major changes in transmission of thermal radiation can be caused by any compound which absorbs between $\simeq$ 750 and 1,300 μm. This window is influenced most strongly by H_2O, O_3, CO_2, and NH_3. From doubling N_2O and CH_4 concentrations, Wang et al. (1976) predicted an increase in surface temperatures of 0.7° and 0.3°K, respectively.

The chief difficulty with placing much confidence in any of these postulated minor (or major) temperature changes is that each change in atmospheric composition brings with it its own negative feedback system. For example, the effect on temperature of changes of order 10 percent in CO_2 may be counterbalanced by opposite changes of order 3 percent in water-vapor content (which absorbs in the same 15 μm bands as CO_2) and of 1 percent in cloudiness (Möller, 1963). For this reason Möller concluded that "the theory that climatic variations are effected by variations in the CO_2 content becomes very questionable"

(3) The third factor of interest in changing global surface temperature is the albedo of the earth's surface (i.e., the ratio of radiation reflected from a surface to that originally falling upon it). Removal of the ice at both poles naturally decreases albedo. Indeed, so dramatic is the effect as seen in computerized models of surface temperature, that Sellers (1969:397) predicted "a temperature increase of 7–10°C in the Arctic and 13–17°C in the Antarctic." In the tropics, warming is predicted not to exceed present values by 2°C. Prior to the mid-Tertiary, deep water of the world oceans was 14°C (see below), instead of the present 2°C. Because deep water reflects polar surface temperatures, the increase in temperature postulated by Sellers is certainly approximately correct for an ice-free world. In addition, the paleogeography prior to the mid-Tertiary did not favor the development of Arctic snow and ice. Accordingly, Donn and Shaw (1977) regarded the development of glaciers as the natural consequence of the drastically increased albedo once snow and ice could persist throughout the year as land areas drifted poleward. A larger icecap means increased albedo and less solar radiation absorbed by the earth, thus leading to cooler temperatures and more

ice growth, unless other atmospheric conditions lead to a melting of the ice (as has obviously occurred).

Reflectivity of incident radiation also can be caused by suspended atmospheric particles (dust or aerosols). These particles come from arid lands (Prospero and Nees, 1977) and from volcanic eruptions. As with other climatic factors, there seems to be no simple relationship between amount of dust and the temperature.

Dust which is low in the earth's atmosphere is correlated with a temperature increase, seemingly because the dust decreases the loss of infrared radiation to space (Idso and Brazel, 1978, and earlier). As the amount and elevation of dust increases, however, this relationship reverses and temperature decreases, presumably owing to increased reflectivity. Rasool and Schneider (1972:96) suggested that an increase of a factor of 4 in dust would increase "the albedo of the cloudless earth from 10 to 15 percent, resulting in a global albedo increase from 31 to 33.5 percent." Global temperatures should therefore decrease by as much as 3.5°C.

In the stratosphere, the introduction of dust to heights exceeding 60 km in the Mt. Agung, Bali (8°S, 115°E), volcanic eruption of March 17, 1963, resulted in the largest climatic change ever observed by modern humans (Newell, 1971b). The dust was correlated with a loss in atmospheric transmittance of approximately 1.5 percent (Ellis and Pueschel, 1971) and a decrease of about 0.5°C in the mean tropical troposphere (Newell and Weare, 1976). The change in temperature "is probably caused by absorption of solar radiation by aerosols" (Newell, 1971a). Normal transmittance was not reached again until 7 years after the eruption. Thus it appears that dust may have opposite temperature effects in the atmosphere (warming) and the troposphere (cooling).

The question has long been asked as to whether some periods of geologic time have more volcanism than other periods, and, if so, whether this would be reflected in a changing climate. Broadly speaking, the longer the geologic period, the larger the amount of volcanic rock erupted (Veizer and Compston, 1974:Figure 6), as though the amount of volcanism is time dependent. Within that broad guideline, the variance in extent of volcanism over any short time interval is quite high, but significant differences may exist. For example, through the Cenozoic, the number of volcanic ash horizons in deep-sea cores record two large "pulses": the Plio-Pleistocene (at $t \simeq 2$ m.y. to present), and in the Middle Miocene (at $t = 16$ to 14 m.y. ago) (Kennett, McBirney, and Thunell, 1977; Kennett and Thunell, 1977, disputed by Ninkovich and Donn, 1976). Volcanic activity also was claimed to have increased markedly every 2.5 m.y.

for the past 10 m.y., and every 5 m.y. for the time span 10 to 20 m.y. ago (Hein, Scholl, and Miller, 1978).

There is some scientific basis for the desire to attribute major climatic changes to "supernatural events" like volcanism, principally from the observed effects of the Mount Agung eruption. The primary cause of climate change would be "the radiative effects of volcanic aerosols on the global heat budget" (Pollack et al., 1976). These authors reported that the historical record shows that volcanic eruptions occur often enough to account for global coolings on the order of 1°C "during epochs of intense volcanic activity."

The idea of a relationship between changes in volcanicity and changes in climate can be tested. Geologically, "three conspicuous volcanic ash horizons occur in the late Pleistocene of the western Gulf of Mexico . . . All three volcanic ash horizons coincide with severe climatic cooling episodes" (Kennett, Huddleston, and Clark, 1974). In a more extensive analysis, Bray (1977) examined climatic and volcanicity records over the past 2 m.y., and concluded that massive eruptions were "closely associated with the glacial stages and the cooling episodes." A major difficulty with this and similar single-factor conclusions is that only one aspect of climate is examined, and there is no evidence that another factor (e.g., the solar constant) may not have been equally important.

(4) The fourth factor which has been especially considered in models predicting surface temperature is cloud cover. Cloud cover has a strong latitudinal dependence, as is shown in Figure 4-10. Cloud cover may change either in areal coverage or in cloud thickness, and the effects on surface temperature may be quite different. For the global average temperature, an increase in areal coverage

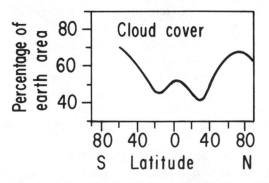

Figure 4-10   The latitudinal distribution of cloud cover. After Arking, 1964:571.

increases both the albedo (thus lowering temperature) and the trapped infrared radiation (thus increasing temperature). However, the rate of *decrease* of temperature because of solar reflectance is faster than the rate of increase of temperature caused by infrared trapping (Schneider, 1972b:1417). The net effect of an increase in areal coverage is a lowering of surface temperatures provided that cloud top height and cloud albedo remain unchanged. In contrast, an increase in cloud thickness has only one effect—to decrease the loss of infrared radiation to space. This leads to an increase in surface temperature. If both cloud coverage *and* cloud thickness increase, they tend to cancel each other out so that a constant temperature could be maintained.

Predicting changes in temperature from changes in cloud cover can become yet more complex. At high latitudes, an increase in cloud coverage may not increase albedo because the albedo of ice and snow is already very high. Therefore, an increase in cloud coverage at high latitudes would lead to an *increased* temperature because more infrared radiation would be trapped. Obviously, modeling of climate is very difficult with the same "cause" having opposite effects depending upon the starting conditions.

What is not yet available, but is developing (Manabe, Bryan, and Spelman, 1975; Gates, 1976), is a model which integrates all of the realistic negative feedback loops for changes in incoming radiation, atmospheric composition, albedo, and cloud cover for geologically interesting extremes. One general result from present models is that *multiple* stable points may exist for the solar constant, CO_2, particulate matter, cloud cover, etc. (Sellers, 1973). It is also possible that a causal relationship exists between a decrease in world temperature and an increase in the intensity of magnetic field, and/or a change in the earth's orbital parameters ("Milankovitch curve").

Temperature over Geologic Time

Estimates of changes in ocean temperature over geologic time are usually derived by calibrating supposed changes against the extremes of glacial and nonglacial conditions. Figure 4-11 summarizes present information on surface and bottom temperature of the ocean for the past 1 b.y.

The conventional view is that even Precambrian ocean temperatures fall within the range of 0° to 100°C, and probably less than 30°C, because Precambrian blue-green algae seem widespread, just as they are today in shallow seas. Oxygen isotopic data from marine

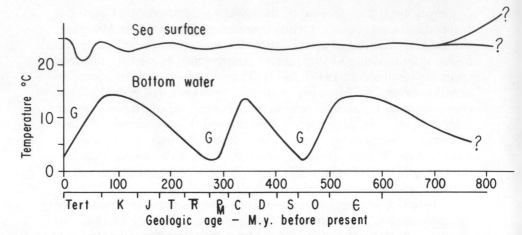

Figure 4-11 Estimated change in mean summer sea surface temperature and in bottom water temperature for tropical latitudes over the past 800 m.y. Note that, as depicted, cold bottom water is limited to glacial periods (G).

cherts of up to 3.8 b.y. old have been interpreted to mean that ocean temperatures may have been 100°C (3.8 b.y.), 70°C (3.4 b.y.), and 22°C (2 b.y.) (Perry, Ahmad, and Swulius, 1978; Knauth and Lowe, 1978). The isotopic changes on which the high temperatures are based may reflect (1) a systematic change in the ^{18}O composition of ocean water through time, and/or (2) isotopic mixing with fresh waters subsequent to initial chert formation (with a greater chance of mixing the older the chert), and/or (3) a temperature effect. The first two alternatives are discussed in Chapter 5 (p. 186), and only the temperature interpretation is presented here.

A hot early ocean is consistent with four other lines of evidence, each of which, however, is independently of questionable significance. First, Hoyle (1972) argued for a decreasing gravitational constant over the history of the earth. In accord with this, he favored early (3 b.y.) temperatures being 100°C, cooling to 70°C by 2 b.y., and to 40°C by 1 b.y., then to 10°C at the present time. However, if the gravitational constant has not decreased through geologic time (McElhinny, Taylor, and Stevenson, 1978; Stewart, 1978), a significant change in surface temperature is not required.

Second, discussed above and in Chapter 5, the inferred atmospheric composition of the Precambrian probably included much more CO_2 and NH_3 than the present atmosphere, and both of these compounds are strong infrared absorbers. Thus more radiation would be confined to the earth's surface, and either surface temperatures would be higher, or a far greater amount of evaporation and precipitation would ensue. This latter course is favored if open ocean temperatures are indeed held to be close to 33°C, as was developed in the introductory paragraphs to this chapter.

Third, δD_{H_2O} values inferred to have been characteristic of ground water which 3 b.y. ago altered gneisses are typical of modern tropical meteoric waters (Taylor 1977:549). Taylor concluded that early Precambrian ocean temperatures may well have been warmer than present-day temperatures.

In addition to these three lines of reasoning, a fourth can be considered. The present geothermal gradient at a depth of 10 km is approximately 15° per km. During the Precambrian (as recently as 1.3 b.y. ago), the geothermal gradient at a depth of 10 km was estimated to have been 70° to 100°C/km (Saxena, 1977). P-T determinations on Archean granulites suggest a surface heat flow at $\simeq 3$ b.y. "between two and three times modern values" (Drury, 1978). If so, the Precambrian ocean bottom would have been heated more than is now the case. Ocean convection would have been aided. Possibly this also would have resulted in an ocean with a higher maximum temperature.

The major argument against a hot early earth is the occurrence of glacial deposits (tillites with striated pebbles) on the order of 2.3 b.y. old in the Gowganda Formation of the Canadian Shield (Symons, 1975; Harland and Herod, 1975). Bona fide later periods of glaciation include the late Precambrian ($\simeq 0.8$ b.y.).

In summary of Precambrian ocean temperatures, some parts of the earth seem to have been cold, as indicated by glaciation, but other evidence is consistent with a surface temperature of 50° to 70°C or higher. By the end of the Precambrian, all evidence is consistent with ocean temperatures that were no higher than about 30°C.

For the Paleozoic, ocean surface temperatures are considered to have ranged no higher than about 30° to 40°C. One independent line of evidence comes from chert deposits. Sediment temperatures at the site of chert crystallization averaged over the time necessary to form a chert nodule indicate values of 20° to 35°C (Knauth and Epstein, 1976). A second independent line of evidence is glaciation. Two times of continental glaciation are well documented: the late Ordovician–early Silurian and the late Carboniferous–early Permian (see Wright and Moseley, 1975).

Beginning in the Mesozoic, reliable oxygen isotopic evidence is available for marine carbonates. Emiliani (1954) discovered, and

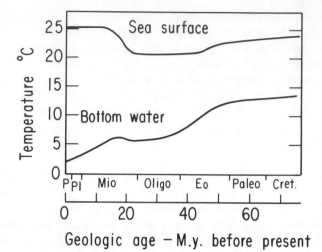

Figure 4-12 Cretaceous to Recent temperatures of the tropical Pacific based on δ^{18}O determinations. Note warm (14°C) bottom water of the late Cretaceous and the onset of glacial cooling in the mid-Miocene. Note also that the temperature difference between surface and deep water has increased from approximately 10°C to 20°C. Based on data from Douglas and Savin, 1973; and Savin, Douglas, and Stehli, 1975.

Douglas and Savin (1975, and earlier) fully documented, that the δ^{18}O values of deep-sea benthic foraminifera show that oceanic bottom temperatures were ≃14° from the early Cretaceous to the late Cretaceous, and then declined to modern values of ≃2°C (Figure 4-12). For the late Cretaceous to the present (approximately 70 m.y.), this pattern was substantiated independently by Shackleton and Kennett (1975), also using benthic foraminifera, and by Kolodny and Epstein (1976), using deep-sea cherts. Sampling from the late Jurassic to late Cretaceous, however, is not extensive, and the possibility remains open that deep-sea temperatures during this interval may have been even warmer.

A change from 14° to 2°C for the vast bulk of the ocean water seems very large. A simple calculation shows, however, that, spread over 70 m.y., the change would be barely noticeable—and could not be measured—if it were going on today!

To lower the temperature of the ocean 1°C requires the loss of approximately 1.4×10^{24} cal since the specific heat of sea water is approximately 0.9 cal/gm, the density of sea water is about 1.04, and the volume of the ocean 1.4×10^{24} cc. A 12°C change over 70 m.y. requires 1.7×10^{25} cal, or a net outward flux of 2.4×10^{17} cal/year. The area of the ocean is approximately 3.6×10^8 km², but heat is lost only over an area north and south of 40°, approximately half of the ocean, or 1.8×10^8 km². For cooling the ocean, the net flux outward would therefore be 1.3×10^9 cal/km²/year.

At the earth's surface, the average amount of radiation now lost at high latitudes (and gained at low latitudes, if we are in a steady state) is 3×10^{14} cal/km²/year. Accordingly, the rate of change necessary to cause a 12° change in temperature of the world ocean since the late Cretaceous is $(1.3 \times 10^9)/(3 \times 10^{14})$, or approximately 4×10^{-4} of the present radiation values/km²/year. This is an extremely small amount of radiation. Thus the gain or loss of radiation necessary to cause geologically significant changes in ocean temperature appears quite easy to account for. Indeed, this suggests that less pronounced, random, variations in ocean temperature could easily occur over time scales of 10 to 30 m.y.

For the Tertiary, the temperature history of the open ocean is much better known than it is for older times. The temperature of oceanic deep water can be determined either from fossils found in deep-sea sediments or from marine deposits of the outer shelf and slope at high latitudes where bottom water forms. The available data, summarized in Figure 4-12, are consistent with the view that a permanent thermocline in the tropics has been a permanent fixture of the world ocean, but that its magnitude has varied from 5° to 20°C. Savin, Douglas, and Stehli (1975) found that during times of more equitable climate (as during the late Cretaceous), the permanent thermocline may have been only 5 to 6°C instead of the present 20°C (Figure 4-12). One point which is not understood is how deep water of the ocean might warm up. Warm surface waters are too light to displace cold deep water, and heat flow from the bottom (or adiabatic

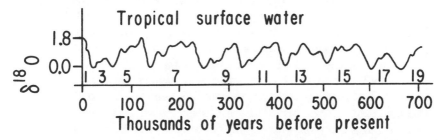

Figure 4-13 Generalized temperature curve for surface waters of the tropics for the past. The numbers along the horizontal axis refer to deep-sea core stages. After Emiliani and Shackelton, 1974:513.

heating) is probably of negligible importance for ocean depths shallower than 5 km, which include most of the ocean.

The temperature history of seas of the continental shelf reflects, of course, more local conditions. Foraminifera are less readily available for shelf studies than are other benthic organisms, and diagenesis and salinity effects also make shelf studies more difficult than those on deeper-water samples. Even with the ease of collecting shelf sediments, only in 1978 did the first isotopic paleotemperature curve for the Tertiary of northwestern Europe appear (Buchardt, 1978). The middle Paleocene to middle Eocene rise in temperature (from 14°C to about 28°C) is followed by a significant reduction in the middle Oligocene (to about 4°C) and subsequent rather uniform conditions (about 5°C to 10°C). These temperature changes reflect different sources of shelf waters as geography and circulation patterns changed, as well as climatic changes, and the exact causal history is complex.

Paleotemperature data of *all* types have been accumulated most extensively for the Pleistocene. Paleotemperature variations in tropical surface waters display a quasi-periodic variation with a wavelength of about 10^5 years over the past 700,000 years (Figure 4-13). This may be due to "astronomical motions of the earth" because the wavelength is of the right order of magnitude for such a cause (Emiliani and Shackleton, 1974).

During glacial maxima tropical sea surface temperatures have been estimated by different authors to have been from 2° to 7°C cooler than present values. Emiliani (1971, and elsewhere) considered from $\delta^{18}O$ data that glacial conditions led to a cooling of tropical Caribbean waters 7° below the present value of 28°C. Hecht (1974b) found the amplitude of glacial to interglacial temperatures to be 4.5° to 5.8°C ($\pm 1.5°$), using the faunal similarity method outlined above. However, these estimated Caribbean changes in temperature are approximately twice the 2 to 3°C estimated by others (Shackleton and Opdyke, 1973, using oxygen isotopic methods, and Imbrie and Kipp, 1971, using their "transfer function.") The differences in temperature from oxygen isotope data are derived from different assumptions of the isotopic composition of sea water during glacial periods, which demands a correction somewhere between 0.5 and 1.7 per mil. The racemization technique has been used to estimate that the coastal temperature of the Mediterranean was lowered by 4°C ($\pm 1°$) during the past glaciation (Schroeder and Bada, 1973).

The outstanding examples of the "transfer function" comparative approach as pioneered by Imbrie and Kipp (1971) are charts of sea surface temperatures during the Pleistocene, when 3 km ice

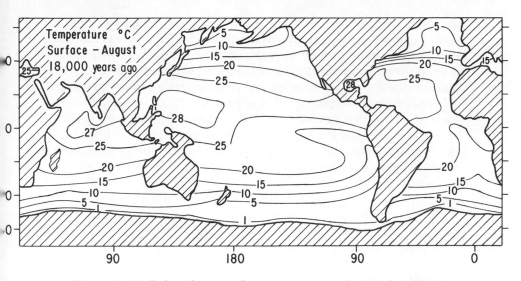

Figure 4-14   Inferred sea surface temperatures for Northern Hemisphere (August) summer, 18,000 years ago, at the time of near maximum glacial advance. In comparison with nonglacial times (Figure 4-2), note the similar wide tropical belt, but the more compressed isotherms at high latitudes. After CLIMAP, Project Members 1976.

sheets covered higher latitudes (CLIMAP, 1976; several chapters in Cline and Hays, 1976). Shown in Figure 4-14 is the summer (August) sea surface temperature for 18,000 years ago. In comparison with modern August temperatures shown in Figure 4-2, note that the wide equatorial band is the same, but that higher-latitude temperature zonation is strongly compressed.

Summary

Negative feedback mechanisms currently appear to set a natural limit on the ocean surface temperature at 33°C. The most important of these mechanisms is the cooling effect of the release of heat on evaporation; this counteracts the warming effect of heating the sea surface. It may be no historical accident that tropical marine species have as their lethal limit the narrow range of 34 to 37°C. That this is *not* an evolutionary limit for metazoans is shown by the occurrence of reproductive populations of arthropods (ostracods) in fresh water in hot springs at temperatures of 50°C.
    Isotopic and other chemical methods, taxonomic and morphologic methods, and floral and faunal gradients have been used to

infer past temperatures. Every method has its difficulties and no method can stand alone. Any given value of temperature is likely to be good to no more than 2° to 3°C. Climatic modeling that incorporates changes in atmospheric composition, albedo, cloud cover, and solar luminosity should provide a very good guide to major changes in temperature over geologic time, but this is yet to be achieved. Ocean temperature has been claimed to have been as high as 70°C 3.4 b.y. ago, but this is uncertain. All evidence suggests it has been less than 30°C from the early Paleozoic to the present. Over the past 70 m.y., the deep water of the world ocean has cooled from 14°C to 2°C today.

5. CHEMISTRY

*It is concluded that climatically induced changes in
the terrestrial biosphere have been of sufficient
magnitude and rate to give rise to the observed
carbonate dissolution cycles . . . The response of
the oceans to Pleistocene changes will never be
understood if only steady-state conditions are
modeled.*

N. J. Shackleton, 1977

This chapter (1) outlines the present chemistry of the world ocean,
and (2) discusses various methods for determining paleosalinity;
these topics lead to (3) a section on ocean chemistry 4.6 to 2.0 b.y.
ago; and the chapter ends (4) with a summary of chemical changes
from 2.0 b.y. ago to the present. As with the development of continental crust, changes in ocean chemistry can be divided into long-
term ($\simeq 10^9$ years) systematic trends, and short-term ($\simeq 10^8$ years)
recycling trends. If an observer had been looking at ocean chemistry
2.0 b.y. ago, would that person have seen all the things which we
identify today as trends in a changing ocean chemistry? Has it taken
incremental changes over very long periods of time to yield "evolutionary" changes? Therefore, an additional focus of this chapter is on
how this stability has been achieved and maintained.

Present Ocean Chemistry

The oceans act as a mixing bowl, as one step in the continuous sequence of transforming igneous rocks to sediments and sedimentary
rocks. In that mixing process, ocean chemistry is regulated, on the
one hand, by the introduction of ions from rivers, and on the other
hand, by removal of ions to sediment and ultimately to rocks.

"Salinity" indicates the amount of dissolved solid material in a
fixed volume of sea water, and the methods for measuring that
amount have changed considerably through the years. The traditional definition dating from the turn of the century is that salinity
is the weight in grams of the dissolved inorganic matter in 1 kg of
sea water, after all bromide and iodide have been replaced by the
equivalent amount of chloride, and all carbonate converted to oxide.
This definition is very difficult to apply with precision for large
numbers of samples. However, chlorinity (Cl) is relatively easy to

determine (by titration with a standard silver nitrate solution), and chlorine is a major component of sea water. Chlorinity was initially related to salinity (S) by the expression

$$S‰ = 0.030 + 1.8050\ Cl‰, \tag{5.1}$$

which was later (1962) abbreviated to

$$S‰ = 1.80655\ Cl‰ \tag{5.2}$$

Within the past 20 years conductivity measurements of sea water have allowed a routine precision of 1 part in 10^4 in total dissolved solids, and the method is easily used (Wilson, 1975). Accordingly, the conductivity of 135 natural sea-water samples from all oceans and four seas (Baltic, Black, Mediterranean, and Red) was determined, and these results were calibrated gravimetrically with previous measurements of salinity. The present recommended method of determining salinity is based on conductivity measurements standardized to water samples whose exact salinity is determined by the silver nitrate titration. In the conductivity method of determining salinity,

$$S‰ = -0.08996 + 28.29720\ R_{15} + 12.80832\ R_{15}^2 - 10.67869$$
$$R_{15}^3 + 5.98624\ R_{15}^4 - 1.32311\ R_{15}^5 \tag{5.3}$$

where R_{15} is the ratio of the conductivity of a water sample to that of water having a salinity of exactly 35‰, both samples being at the same temperature (15°C for R_{15}) and under a pressure of 1 standard atmosphere (Wooster, Lee, and Dietrich, 1969). Tables for computing salinity from conductivity are used for implementing this definition.

Generally speaking, open ocean salinities range from 33 to 37‰. Salinities of large regions of nearshore continental shelf may range down to 29‰ (as in the immense region of the Straits of Juan de Fuca, Washington). The faunas of the region seem to show *no* changes in diversity and to have no physiological problems which are any different from those encountered by forms called "normal marine." Over large areas, fully marine salinities can therefore reasonably be spoken of as 29 to 37‰. Indeed, the specific salinity values could probably be somewhat lower or higher. In contrast to the slow rate of change of waters of the shelf and open ocean, estuarine conditions are characterized by rapid rates of change of salinity over *short* intervals: vertically, horizontally, and temporally. The important contrast between ocean and estuary is the spatial and temporal scale over which changes occur, and not the specific values of salinity or of any other variable.

Salinity in the open ocean changes over several degrees of latitude by 2 to 4‰ because of changes in evaporation and precipitation
(Figures 5-1, 5-2) caused by changing patterns of air movements
(p. 92). At equatorial zones, water-laden warm air rises and is
cooled, thus causing precipitation (and reducing salinity). This
newly formed cool, dry air is transported northward and southward
and descends at about 30° latitude, where it is warmed, and takes up
water, causing an increase in evaporation (and salinity). This air
moves poleward and at 50° to 60° it again rises, is cooled, and precipitation increases (and salinity is lowered).

In local geographic areas, over distances of a few kilometers, salinity can increase by several parts per thousand if evaporation is
intense. To cite three examples, in the lee of Andros Island, Bahamas, salinity increases from 32 to 39‰ in 15 km (Newell et al.,
1959), in the Persian Gulf from 50 to 100‰ over 25 km (Purser,
1973), and in Hamelin Basin of Shark Bay, Australia, from 45 to
65‰ over 75 km (Hagan and Logan, 1974). These distances are not
the hundreds of kilometers envisaged in the epeiric sea model (see
p. 66).

Only 12 elements exist as dissolved solids in sea water in concentrations greater than 1 ppm (Cl, Na, Mg, S, Ca, K, Br, C, Sr, B,
Si, and F; Table 5-1). These 12 elements (and most others) occur in
approximately equal proportions one to the other in sea water; this

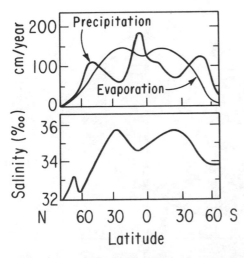

Figure 5-1   Relationship between evaporation and precipitation at
various latitudes, and the salinity of surface waters, for the open
ocean. After Wüst, Borgmus and Noodt, 1954.

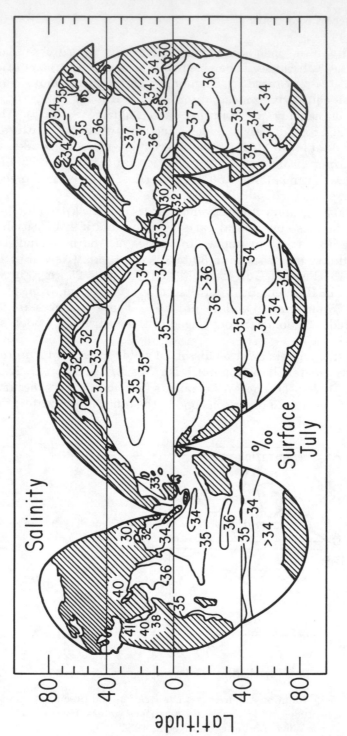

Figure 5-2 Salinity of surface water of the ocean in northern summertime. After Sverdrup, Johnson, and Fleming, 1942.

Table 5-1 Abundances of elements in sea water and residence times. Data chiefly from Goldberg, 1965; Brewer, 1975; Holland, 1978.

Element	Concentration in mg/l	Residence time (yrs)	Element	Concentration in mg/l	Residence time (yrs)
H	108,000		Nb	1×10^{-5}	300
He	2×10^{-6}		Mo	0.1	2×10^5
Li	0.18	2.3×10^6	Ag	1×10^{-5}	4×10^4
Be	6×10^{-6}		Cd	0.0001	
B	4.5	10^7	In	0.0001	
C	28		Sn	1×10^{-5}	
N	150		Sb	0.0002	7×10^3
O	880,000		I	0.06	4×10^5
F	1.3	5×10^5	Xe	5×10^{-5}	
Ne	0.001		Cs	0.0004	6×10^5
Na	10,700	4.8×10^7	Ba	0.02	4×10^4
Mg	1,290	10^7	La	3×10^{-6}	440
Al	0.001	10^2	Ce	1×10^{-6}	80
Si	3.0	1.8×10^4	Pr	6×10^{-4}	320
P	0.07	1.8×10^5	Nd	3×10^{-5}	270
S	905		Sm	5×10^{-7}	180
Cl	19,000	10^8	Eu	1×10^{-7}	300
Ar	.04		Gd	7×10^{-7}	260
K	380	6×10^6	Tb	1×10^{-7}	
Ca	412	8.5×10^5	Dy	9×10^{-7}	460
Sc	6×10^{-7}	10^4	Ho	2×10^{-7}	530
Ti	0.001	1.3×10^4	Er	8×10^{-7}	690
V	0.002	8×10^4	Tm	2×10^{-7}	1,800
Cr	0.0001	10^3	Yb	8×10^{-7}	530
Mn	0.0002	10^4	Lu	2×10^{-7}	450
Fe	0.001	10^2	Hf	7×10^{-6}	
Co	0.00005	10^4	Ta	2×10^{-5}	
Ni	0.0005	10^4	W	0.0001	1.2×10^5
Cu	4×10^5	10^4	Re	4×10^{-7}	
Zn	0.0001	10^4	Au	4×10^{-7}	2×10^5
Ga	0.00003	10^4	Hg	3×10^{-5}	8×10^4
Ge	5×10^{-5}		Tl	1×10^{-5}	
As	0.002	5×10^5	Pb	3×10^{-5}	4×10^2
Se	0.0001	2×10^4	Bi	2×10^{-5}	4.5×10^4
Br	67	10^8	Rn	6×10^{-16}	
Kr	0.0002		Ra	7×10^{-11}	1×10^6
Rb	0.12	4×10^6	Th	1×10^{-5}	2×10^2
Sr	8.1	4×10^6	Pa	5×10^{-11}	
Y	1.3×10^{-6}		U	0.003	3×10^6
Zr	3×10^{-5}				

Table 5-2 Principal ionic constituents of
sea water (34.4‰ salinity). Data from
McLellan, 1965:16.

Ions	G/kg	% salinity
Cations		
Sodium	10.47	30.4
Magnesium	1.28	3.7
Calcium	0.41	1.2
Potassium	0.38	1.1
Strontium	0.013	0.05
Anions		
Chloride	18.97	55.2
Sulfate	2.65	7.7
Bromide	0.065	0.2
Bicarbonate	0.14	0.4
Borate	0.027	0.08

relationship is known as Forchhammer's Principle after the initial discoverer. As Forchhammer (1865:244) thoroughly described more than a century ago, "Thus the quantity of the different elements in sea-water is not proportional to the quantity of elements which river-water pours into the sea, but inversely proportional to the facility with which the elements in sea-water are made insoluble by general chemical or organo-chemical actions in the sea." This was later verified by Dittmar, the chemist on the *Challenger* expedition of 1872 to 1876. Although a few significant variations in Forchhammer's Principle have been subsequently discovered (e.g., ≈0.5 percent increase in Ca/Cl ratio in deep water), the principle remains valid and is the cornerstone of inferences about salinity of the oceans of the geological past.

Although sea water contains some of every element for which a search has been made, salinity is *not* equally distributed among all chemical species. As summarized in Table 5-2, 97 percent of the salinity of the ocean is contributed by four ionic species: chloride (55.2 percent), sodium (30.4 percent), sulfate (7.7 percent), and magnesium (3.7 percent).

Geologic Recognition

Many methods have been used to try to determine ancient salinity. These include (1) the boron method; (2) the sedimentary phosphate method; (3) isotopic methods using ^{18}O and ^{13}C; (4) trace-element distributions; (5) other chemical methods, including examination of interstitial fluids, liquid inclusions, porosity, siderite, and C/N ratios; and (6) floral and faunal evidence.

Use of fossil floras and faunas is the easiest and most decisive method for distinguishing fresh-water from marine conditions. Each of the other methods is (variously) susceptible to diagenetic change and requires specific types of examples. The advantage of some of the chemical methods is that, if all of the conditions are met, their resolution may allow salinity to be estimated to within a few parts per thousand. Thus, for example, a regional reduction in salinity from 35‰ to 28‰ should be able to be documented; or changes in salinity in coal swamp deposits could be deduced.

(1) The concentration of boron in sea water varies as a function of salinity, weakly in estuarine conditions from 2 to 15‰ (Boon and MacIntyre, 1968), more strongly at higher salinities. The uptake of boron by clay minerals is proportional to salinity (Figure 5-3). Properly chosen and prepared argillaceous sediments have been found to contain boron roughly in proportion to observed or inferred total salinity (e.g., Landergren and Manheim, 1963). Marine sediments on the average contain 25 to 50 ppm more boron than fresh-water sediments, normalized for clay content (Shimp et al., 1969). Other normalizations discussed below may enhance the difference.

Boron is also incorporated into carbonate shells. It increases by a factor of 3 in both calcite and aragonite over a salinity range of 5 to 35‰ (Furst, Lowenstam, and Burnett, 1976). Presumably, boron in original aragonite in fossil shells would be suitable for determining

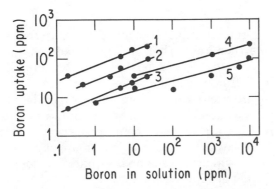

Figure 5-3 Increase in boron in solution and increase in boron uptake for 5 clay samples, 1, 2, 4, and 5 are illite; 3 is montmorillonite. 1, 2, and 3 represent boron adsorption; 4 and 5 represent boron fixation. After Couch, 1971:1831.

ancient salinity, although leaching can be a serious problem (Cook, 1977). The mechanism of incorporation of boron into shell material requires investigation.

Boron occurs in sea water chiefly as boric acid, about 10 percent of which dissociates (Kemp, 1956:57) according to the reaction:

$$H_3BO_3 + 2H_2O \rightleftharpoons H_3O^+ + B(OH)_4^- \qquad (5.4)$$

Boron is incorporated into clay minerals in a three-step process: (1) adsorption, (2) fixation by hydrostatic bonding, and (3) incorporation into the tetrahedral sheets of clay minerals (Couch, 1971).

Many authors have pointed out constraints which must be taken into account in order to obtain accurate results from the boron method. These include the following.

(a) Boron is not accepted to the same extent in all clay minerals. Illite accepts twice as much boron as do montmorillonite and chlorite, and four times as much boron as does kaolinite (Couch, 1971). In addition, the proportion of illite in mixed layered illite/montmorillonite clays increases diagenetically with depth of burial by 30 to 80 percent. If boron is available, it will also increase in the illite so that it remains proportional to the ratio of illite/montmorillonite (Perry, 1972). Quite clearly, mineralogical composition and diagenetic history must be considered in order to interpret boron concentrations in terms of ancient salinity.

The concepts of "adjusted boron" and "equivalent boron" were developed in order to permit the comparison of boron with respect to illite (Walker, 1968).

$$\text{Adjusted boron} = \frac{\text{observed boron} \times 8.5}{\text{percentage } K_2O} \qquad (5.5)$$

The factor 8.5 is the theoretical concentration of K_2O in pure illite. The "adjusted boron" is then converted to "equivalent boron" in order to determine the boron that would exist at equilibrium in illite containing 5 percent K_2O from the same salinity medium.

(b) Sediments that are accumulated from clay minerals which formed during a previous weathering regime may contain boron in equilibration with previous rather than contemporary conditions (Perry, 1972). The incorporation of "detrital" boron appears to produce a systematic error for which there is no adequate control other than the careful evaluation of initial samples.

(c) Incorporation of boron is a function of surface area, and more boron will, on the average, be incorporated into smaller grains. Thus samples must be controlled for grain size (Walker, 1963), with 1 μm a reasonable upper limit for most purposes.

(d) The degree of crystallinity of clay minerals partly determines the extent of boron incorporation. More boron occurs in less well crystallized materials (Porrenga, 1967b).

(e) To an unknown extent, the existence of organic matter may "coat" clay grains, thus preventing normal boron incorporation. This is particularly important in organic-rich mudstones (Hirst, 1968).

Because of these constraints, no absolute values of boron concentrations can be given which automatically differentiate salinities. Indeed, even in the analysis of deep-sea sediments, "it would appear to be more advisable to use the boron content of sediments only as complementary to other criteria for differentiation of environment" (Thompson, 1968:474). However, as shown in Figure 5-4, sediments of marine and fresh-water origin may differ by a factor of 2 in their boron content for constant grain size, crystallinity, and K_2O content of illite.

Boron in conjunction with other trace elements (especially gallium and rubidium) was used to distinguish Pennsylvanian marine and fresh-water sediments (Degens, Williams, and Keith, 1958). In transects going offshore, clay usually increases in abundance, and

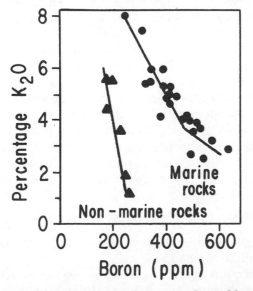

Figure 5-4 Relation between adjusted boron and percentage K_2O for illites of shales and limestones from marine and nonmarine environments. Note that marine rocks have more boron than nonmarine rocks. After Walker and Price, 1963:835.

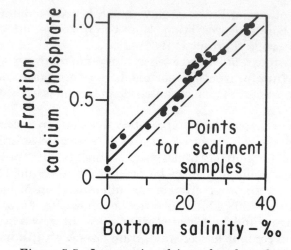

Figure 5-5 Increase in calcium phosphate fraction in sediments and increase in bottom water salinity for recent sediments. The dashed lines represent 95 percent confidence limits, approximately $\pm 4‰$. After Nelson, 1967.

this is paralleled by an increase in boron in sediments (Shimp et al., 1969). In this instance, the boron content of shales may be useful to indicate proximity to strand line, but not salinity per se.

(2) The sedimentary phosphate method of determining ancient salinity is based on the observation that the ratio of calcium phosphate to iron phosphate in estuarine sediments increases with increasing salinity, as shown in Figure 5-5. The molar concentration of $Ca/(Ca + Fe)$ from phosphates increases by a factor of 5 over a salinity range of 0 to 40‰. In soils, the aluminum phosphate variscite ($AlPO_4 \cdot 2H_2O$) and the iron phosphate strengite ($FePO_4 \cdot 2H_2O$) predominate, whereas in the ocean, the calcium phosphate apatite is nearly universal. Nelson (1967) found that salinity was related to calcium phosphate by the expression:

Calcium phosphate fraction $= 0.9 + 0.026$ salinity (‰) (5.6)

As with the boron method, several constraints apply to the use of the sedimentary phosphate method.

(a) Incorporation of detrital apatite in river and estuarine sediments may mask any local precipitation effects (Müller, 1969). It is necessary to evaluate the paleogeographic setting of samples.

(b) Under reducing conditions (which are often present in muds of brackish waters) iron phosphate is reduced (Patrick, Gotoh, and

Williams, 1973), thus leading to the formation of FeS and releasing phosphate. This will preferentially leave calcium phosphate in lower than normal marine salinities.

(c) Mineralized organic phosphate is calcium phosphate. If this phosphate of organic origin (as, for example, from the disintegration of fish bones) is added to "normal" amounts of calcium phosphate in brackish-water sediments, then the calcium phosphate content of the sediments will be higher than that in equilibrium with the true salinity.

Granted these caveats, the sedimentary phosphate method is useful. Marine incursions with rising sea level of the past 3,000 years through a Holocene marsh were charted by changes in pollen abundance and type and by the sedimentary phosphate method (Meyerson, 1972). The two methods yielded similar conclusions.

(3) Both oxygen and carbon isotopes have been used to estimate ancient salinity.

(a) Rain water is approximately 7 per mil lighter in $^{18}O/^{16}O$ than sea water because of preferential evaporation of the lighter isotope (Epstein and Mayeda, 1953). Thus if temperature is uniform in a region, changes in $\delta^{18}O$ can be attributed to changes in salinity. Figure 5-6 illustrates changes in $\delta^{18}O$ with changes in salinity for a Netherlands estuary. This method of paleosalinity determination was used to map variations in an estuarine bay of the Pliocene at Kettleman Hills, California (Dodd and Stanton, 1975), and in charting Pleistocene salinity changes in the eastern Mediterranean (Williams, Thunell, and Kennett, 1978).

In general, the ^{18}O paritioning caused by evaporation is reflected in $\delta^{18}O$ values from marine and fresh-water limestones (Fig-

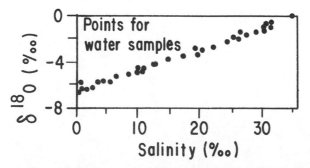

Figure 5-6 Variation of $\delta^{18}O$ values with salinity for water samples from the western Scheldt estuary, the Netherlands (data of Mook). After Dodd and Stanton, 1975.

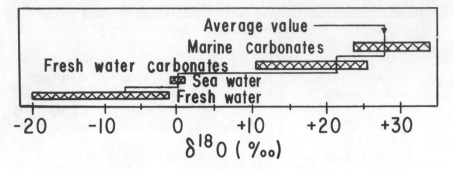

Figure 5-7 Distribution of oxygen isotopes in marine and fresh-water carbonates and in sea water and fresh water. After Degens, 1966:200.

ure 5-7), diatomites (Degens and Epstein, 1962), and brackish-water shells. In addition, precipitation at very high latitudes occurs from water which itself was evaporated from high-latitude sources (and thus is already depleted in ^{18}O relative to mean oceanic values). Such very high-latitude precipitation is depleted in ^{18}O to the extent of -20 per mil or more (Dansgaard, 1964).

(b) Carbon isotopes are sometimes used for paleosalinity work. Because of the small amount of CO_2 in the atmosphere (0.03 percent by volume), most of the dissolved CO_2 in fresh water is derived from soil and humus CO_2. As shown in Table 5-3, these CO_2 sources typi-

Table 5-3 Approximate isotopic composition of carbon from various sources. Data from Broecker, 1964:596.

Source	δC^{13} (per mil)
Atmospheric CO_2	0^a
Soil CO_2	-25
Soil humus	-25
Limestone (marine)	0
Typical river	-12
Typical lake	-5
Surface ocean	0

a. Composition of dissolved CO_2 at equilibrium with atmospheric CO_2.

cally are greatly depleted in ^{13}C, a situation which leads to low $\delta^{13}C$ values in lakes and rivers. Accordingly, fresh-water carbonate deposits are chiefly in the range of -5 to -15 per mil. In contrast, marine limestones are in the range of -5 to $+5$ (Figure 5-8) because carbonate in shells of marine organisms reflects sea-water values. There is further enrichment in ^{13}C in shallow sea water because phytoplankton preferentially take up the lighter isotope (^{12}C), and because water which has been evaporated from the ocean is enriched in the lighter isotope.

Brackish water reflects the mixing of fresh-water and marine ^{13}C sources, and ranges from -10 to $+1$ per mil (Sackett and Moore, 1966). An approximately 3 per mil decrease in ^{13}C occurred going across the 25 to 70 km-wide Flordia Bay from the reef toward the land (Lloyd, 1964). Presumably, lower nearshore values are caused by incorporation of CO_2 from fresh waters and from plant debris. The contribution of organic debris from land to marine sediments is generally very small. [13]C/^{12}C ratios and lignin oxidation products in the Gulf of Mexico indicated that the proportion of land-derived organic matter was less than 50 percent anywhere on the shelf, and less than 10 percent on the outer shelf (Hedges and Parker, 1976). Fresh water sediments and coal are very similar ($\simeq -10$ per mil). Marine shelf and deep-sea sediments are also nearly identical to each other ($\simeq +2$ per mil).

The degree of incorporation of terrestrial organic matter into

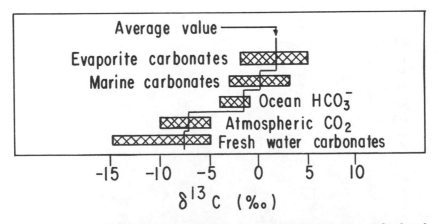

Figure 5-8 Distribution of carbon isotopes in carbonates and related materials. The tails of the distribution for fresh-water carbonates were shortened to conform with more usual occurrences. After Degens, 1966:195.

marine sediments can be checked further with information on the stable isotopes of nitrogen (Peters, Sweeney, and Kaplan, 1978). Whereas marine $\delta^{15}N = +7$ to $+10‰$, terrestrial $\delta^{15}N = 0‰$. Values for 55 coastal marine sediments of the northeast Pacific indicated very little addition of terrestrial organic matter to sediments.

(4) The principle behind the use of trace elements as an indication of marine versus fresh-water conditions is that the abundance of many trace elements is significantly higher in sea water than in fresh water (Table 5-4), and that this will be reflected in marine ver-

Table 5-4 Comparison of concentrations of elements in sea water (S) and fresh water (F). Data from Wedepohl, 1971:310.

Item	S/F
Cl	2.8×10^3
Na	1.8×10^3
SO_4^{--}	2.4×10^2
Mg	3.4×10^2
Ca	2.3×10
K	1.8×10
HCO_3^-	3
Br	$7 \ \times 10^3$
Sr	1.4×10^2
B	2.4×10^2
Si	1.8×10^{-1}
F	6.5
PO_4^{---}	3.5(?)
Li	1.3×10^2
Rb	8.6×10
I	2.5×10
Mo	2.6×10
Ba	3.2×10^{-1}
Zn	1
Fe	$3 \ \times 10^{-2}$
Ni	1.2
Mn	$3 \ \times 10^{-1}$
As	1.5
U	≥ 3
Cu	$3 \ \times 10^{-1}$
V	2
Ti	$2 \ \times 10^{-1}$(?)
Al	2.8×10^{-3}

sus nonmarine sediments. The ideal would be to have a chemically
inert tracer which varied with total salinity just as Forchhammer's
Principle would predict. For example, Broecker (1971) suggested
that the U content of corals may be particularly unlikely to be diagenetically altered, and thus that the ratio of U to Ca and other elements could be determined in order to yield ancient ocean salinities.

The problem with using trace elements is that each place can be
and usually is different from every other place, depending upon historical factors of element supply in the source area as well as upon
grain size, chemistry, and diagenetic history of the enclosing sediment. This variation can be demonstrated by considering how trace
metals are transported by rivers. Elements are transported as crystalline solids (Cu, Fe, Co, Ni, Cr, and Mn), as metallic coatings (Fe,
Ni, Co, and Mn), as organic complexes (Mn), as organic solids (Ni,
Co, and Cr), as well as in solution (Gibbs, 1973). Depending upon the
element, considerable variation may exist from river to river, as is
shown in Figure 5-9. Final accumulation of trace metals may in
turn be related to such a simple thing as the degree of turbulence
and thus to the size of the particles which settle in the region of deposition, as well as to the amount of organic carbon available to act as
a "scavenger" (Tourtelot, 1964).

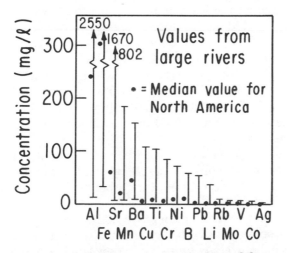

Figure 5-9 Range in observed values of the concentration of several
elements in several of the world's large rivers. Among these elements only B(4 mg/l) and Sr(8 mg/l) have a concentration higher
than 1 mg/l in sea water. Note also the large difference between median North American values and large-river values. After Durum
and Haffty, 1963:3.

Within this general framework, the myriad of studies on trace and minor elements can be grouped into two kinds.

The first type of study reports on analyses of a large number of elements, often using statistical methods to sort out what is correlated with what. This first group includes excellent work on argillaceous sediments (Potter, Shimp, and Witters, 1963), on comparisons of fine- and coarse-grained carbonate sediments (Till, 1970), and on general comparisons of marine and fresh-water sediments (e.g., Dewis, Levinson, and Bayliss, 1972; Shimp et al., 1969), and Spencer's (1966) outstanding paper on Silurian carbonaceous sediments.

The second type of study of trace and minor elements focuses in detail on a single element or pair of elements. Work on Na, Sr, and the Fe/Mn ratio appears to be of widest general application.

Na has been identified as a trace constituent of carbonates of marine and hypersaline origin (Veizer et al., 1977), in which rocks it may occur in the carbonate lattice and/or as fluid inclusions. The Na content used as a complementary characteristic verified the assignment of some parts of a lower Paleozoic limestone and dolomite sequence to normal marine conditions, and some parts to hypersaline conditions (Veizer et al., 1978).

Because of the approximately tenfold decrease in the Sr/Ca ratio in fresh water relative to the ocean, shells of fresh-water mollusks are strongly depleted in Sr (most recently verified by Buchardt and Fritz, 1978). Comparable results also obtain for Sr in carbonate rocks (Veizer and Demovič, 1974).

Fe increases from fresh-water to marine conditions much more than does Mn (Table 5-4). Thus the Fe/Mn ratio increases strikingly in marine sediments (Keith and Degens, 1959) as well as in manganese nodules (data in Manheim, 1965) in comparison with equivalent fresh-water deposits.

(5) Other chemical and geological criteria for estimating ancient salinity involve the use of (a) interstitial fluids and porosity, (b) siderite, (c) carbon/nitrogen ratios, (d) carbonate cements, and (e) sedimentologic criteria.

(a) Pore waters of marine sediments will be initially of the same ionic composition as the overlying sea water, but this can be maintained only in pelagic clays and slowly deposited (<1 cm/10^3 years) biogenic sediments. Samples of pore waters in most ancient rocks provide a sample of ancient sea water modified by subsequent diagenetic history. Notable diagenetic effects are seen in calcareous and calcareous-siliceous sediments deposited at rates of a few cm/10^3 years. Loss of Mg (and less often K) from pore waters is balanced by gain of Ca and Sr, and sulfate reduction by bacteria, with corresponding increase in bicarbonate alkalinity (Manheim and Sayles, 1970). The Ca-Mg carbonate formed in pore waters "is most likely a dolomitic phase" (Sayles and Manheim, 1975).

In sediments removed from proximity to the continental margin, little change in interstitial water composition occurs in rocks as old as upper Cretaceous (Chan and Manheim, 1970). Along the continental margin, fresh-water flushing from continental areas, especially during times of lowered sea level when the aquifer head was enhanced, accounts for markedly fresher waters of the upper kilometer of marine sediments from off New York to off Florida (Manheim and Horn, 1968). Pore waters of marine sediments deposited above salt deposits show an increasing salinity chiefly due to diffusion of ions from the salt. Over geologic time, diffusional migration of salt can proceed through several kilometers of clay sediments and influence the composition of their pore fluids. Because of these postdepositional influences dictated by historical and paleogeographic considerations, the use of pore waters to reveal previous water salinity is only sometimes (Manheim and Chan, 1974) and not generally (Ericsson, 1972) applicable.

A method for determining ancient salinity related to analysis of pore waters is to determine the composition of water which would be in equilibrium with the cation complexes that are adsorbed on argillaceous rocks. Gramberg and Spiro (1964:515) reported that "the change in the composition of the adsorption complex during diagenesis is relatively slight in comparison with the rather significant change in the composition of the readily soluble salts." At the least, this method should distinguish marine from nonmarine shales. Among the intriguing conclusions reached by those authors, which bear further investigation, was that "marine deposits of the Permian type were found to be most similar to Recent marine deposits, while Triassic marine deposits differ appreciably from Recent ones."

Additional information on original pore water chemistry can be obtained from liquid inclusions trapped in calcite grains. The chloride content of such inclusions is approximately ten times greater in marine than in fresh-water limestones (Weber, 1964).

Initial porosities of argillaceous sediments can be estimated from extrapolation of curves of sediment compaction. When this is done, and sediments are normalized for differences in grain size, marine rocks are found to have considerably higher porosities than fresh-water rocks (Heling, 1969). Perhaps this is caused by the greater formation of aggregates in the marine electrolyte solution, and thus a more random orientation of grains (and higher porosity), compared with a more purely physically controlled fabric in the absence of electrolyte solutions (Meade, 1964).

(b) Siderite (Fe, Mg)CO_3 formation requires high iron relative to calcium (greater than 5 percent of the calcium concentration), neutral or basic conditions (acidic conditions dissolve carbonates), a low amount of sulfate (since otherwise sulfides form), and low Eh (from anaerobic bacterial decomposition) (Berner, 1971). This constellation of requirements commonly occurs in fresh-water ponds, and is rare in marine conditions (but siderite does occur in deep-sea sediments; Berger and von Rad, 1972). Geologically, siderite is known chiefly from coal swamps, and even seeming marine occurrences (e.g., Eocene of Arkansas; Weeks, 1938) are often associated with lignite. Accordingly, siderite which is not of deep-sea origin is indicative of a chemistry most often found in fresh-water conditions.

(c) C/N ratios of marine particulate organic matter of the continental shelf are initially at about 4.3 to 7.1/1, reflecting the predominance of marine plankton (zooplankton, $\sim 5/1$; phytoplankton $\sim 6/1$; Müller, 1977). These low ratios reflect the major composition of proteins (C/N $\simeq 3/1$) rather than lipids (C/N $\simeq 113/1$) or carbohydrates (C/N $\simeq 1000/1$). In contrast to marine sediments, terrestrial plant accumulations have much higher average C/N ratios, of 30 to 40/1, reflecting a larger input of terrestrial detritus (Bordovskiy, 1965:48).

(d) In an open, chemical system, the mineralogy and morphology of calcium carbonate cements may be used to distinguish freshwater, brackish-water, and normal marine conditions. Marine deposition is indicated by an aragonitic or high-Mg calcite cement (see p. 43), together with pseudohexagonal (needlelike) crystals. However, aragonite usually recrystallizes within a few tens of millions of years, and unless the original fabric can be identified, or high-Sr remains determined, the signal is lost. Fresh-water conditions are typified by low-Sr and low-Mg calcite rhombs. Many intermediate cements occur in nature, indicating that mixed salinity for cementation is common (Folk, 1974; Alexandersson, 1978; Badiozamani, Mackenzie, and Thorstenson, 1977).

(e) Various sedimentologic criteria suggest near-evaporitic conditions, and taken together indicate a higher than normal salinity. These criteria include the occurrence of collapse-breccias (which may indicate evaporitic solution) and of finely fibrous quartz whose crystallographic c-axis is parallel to the long dimension—so-called length-slow chalcedony. This quartz may form by replacement or precipitation in a sulphate-rich or high pH solution, presumably in the vicinity of nearby evaporites (Folk and Siedlecka, 1974, who cite previous literature).

(6) Fossils of plants and animals provide the easiest and most decisive evidence of marine versus fresh-water conditions. The passage, for example, from a limestone with a typical marine fauna to an underclay with plant roots of terrestrial species, followed by a coal, is decisive evidence for changing salinity. In addition, many groups of marine plants and animals lack mechanisms to control internal ionic composition in very low-salinity environments and therefore are limited in their life cycle to normal marine conditions. These groups include brachiopods, echinoderms, hemichordates, and calcified bryozoans. Even within a phylum, major taxa, such as the cephalopods within the Mollusca, may lack the ability to regulate the ionic concentrations of cells at very low salinities. For foraminifera, agglutinated forms are most evident in regions of low salinity, hyaline calcareous forms in regions of intermediate salinity, and porcelaneous forms where the salinity is greatest (Murray, 1973). At the familial, generic, and species level, very few taxa have wide salinity tolerances. Thus faunal diversity is reduced along steep salinity gradients. However, the restricted fauna which can live under such conditions may exist in tremendous numbers, so that total biomass may not be greatly altered, even though (as is commonly observed) the size of organisms decreases as salinity decreases (Alexander, 1974:639–640).

Because of the latitudinal change in salinity caused by the systematic variation in evaporation-precipitation, open ocean species have "salinity assemblages" in a manner similar to the way in which they have "temperature assemblages" (p. 131). By using multivariate statistical techniques on distributional data of planktonic foraminifera, faunal assemblages can be calibrated to measure salinity in a manner analogous to measuring temperature (Imbrie, van Donk, and Kipp, 1973), to an accuracy of a few parts per thousand.

Ocean and Atmosphere Chemistry: 4.6 to 2.0 B.Y. Ago

This section is divided into three parts: (1) earliest ocean; (2) earliest atmosphere—(a) gases other than oxygen, (b) reducing atmosphere? and (c) origins of oxygen; and (3) banded iron formation. The emphasis in these topics is on an evolving oceanic and atmospheric composition. In the subsequent section on the ocean and atmosphere 2.0 b.y. ago to the present, attention will be given to steady-state recycling.

(1) One approach to determining the chemistry of the earliest ocean is to solve the equations governing chemical reactions during weathering of the important rock-forming minerals in order to de-

termine the ultimate composition and mass of sedimentary rocks, oceans, and the atmosphere. Comparison of these results with the geologic record provides a null hypothesis for the extent to which the present ocean is in geochemical equilibrium. Deviations from this null hypothesis will then permit a focus on the nonequilibrium situations where historical factors should be closely examined.

The initial condition is taken as the assemblage of rock-forming minerals of the average igneous rock, and an acid ocean modeled after volcanic emanations. Lafon and Mackenzie (1974) performed the calculations in which chemical equilibrium was maintained among the aqueous phases and the products of the reaction (i.e., the sediments); these are summarized in Table 5-5.

Oceanic pH initially would be controlled by acid volatiles: (CO, CO_2, sulphur phases (H_2S), and halogen compounds (HCl). These would react with basic igneous rocks, and a higher pH would subsequently develop as silicate mineral phases weathered and formed precipitates. A gradual transition to present oceanic pH values of 7.5 to 8.5 would occur. In addition, *if* ammonia were present in the early atmosphere (see below), it would rapidly dissolve in water, also increasing the pH. Once silicate reactions established a basic pH above 7.5, carbonates could precipitate. The carbonate equilibrium could be effective in controlling pH (as it is today) over a time scale of 10^3 years. Over longer time scales, aluminum-silicate equilibria "have buffer capacities equal to and perhaps greater than" the CO_2–$CaCO_3$ equilibria in sea water (Mackenzie, 1969). Thus both the composition and pH of the oceans may be in long-term dynamic

Table 5-5 Chemical and mineralogical composition of sediment at onset of stable conditions (simulation) compared with the composition of "average sedimentary rock." From Lafon and Mackenzie, 1974:214.

Constituents	Simulated sediment (wt %)	Average sedimentary rock (wt %)	Oxide	Simulated sediment (wt %)	Average sedimentary rock (wt %)
Chert	29	37	SiO_2	65.1	66.0
Clays	32	39	Al_2O_3	15.0	16.1
Feldspars	25	13	MgO	2.6	2.9
Carbonates	14	11	CaO	5.0	5.3
			Na_2O	0.9	1.0
			K_2O	4.2	3.5
			CO_2	7.2	5.2

equilibrium between solutions and solids entering it from land and the removal of solid and liquid phases on the ocean bottom.

As the weathering reactions proceeded, cations, would increase in concentrations in proportion to their abundance in the rock-forming minerals until limited by precipitation of sedimentary phases. The computer simulation of Lafon and Mackenzie was stopped when Na reached a high concentration (and presumably would come to be controlled by evaporite formation). In general, a close agreement is observed in Table 5-5 between the predicted sedimentary phases of the simulation experiment and the observed sedimentary record. In summarizing this whole process, Mackenzie (1974) wrote:

A crude simulation of what might have occurred can be imagined by emptying the Pacific Basin, throwing in great masses of broken up basaltic material, then filling it up with HCl, letting the acid be neutralized, and finally carbonating the solution by bubbling CO_2 through it. Oxygen would not be permitted into the system. The HCl would leach the rocks, resulting in the release and precipitation of silica, and producing a chloride ocean containing Na, K, Ca, Mg, Al, Fe, and reduced sulfur species in the proportions present in the rocks. As complete neutralization was approached, Al could begin to precipitate as hydroxides, then combine with precipitated silica to form cation-deficient aluminosilicates. The aluminosilicates, as the end of the neutralization process was reached, would combine with more silica and with cations to form minerals like chlorite, and ferrous iron would combine with silica and sulfur to make greenalite and pyrite. In the final solution, Cl would be balanced by Na and Ca in roughly equal proportions, with subordinate K and Mg.

The primary minerals of igneous rocks are all mildly "basic" compounds. When they react in excess with acids such as HCl and CO_2 they produce neutral or mildly alkaline solutions plus a set of altered aluminosilicate and carbonate reaction products. It is improbable that ocean water can have changed through time from a solution approximately in equilibrium with these reaction products—clay minerals and carbonates.

Given the rapidity with which chemical equilibrium would seem to be able to be achieved, the greatest changes in ocean chemistry would probably have occurred within the first several turnovers in the sedimentary mass. The oldest sediment now known is the 3.8 b.y. old water-deposited iron-rich chert and carbonate of the Isua Iron Formation of West Greenland. For most sedimentary basins the half-life for the formation and erosion of the earth's sedimentary rocks is now on the order of 100 m.y. (Veizer and Garrett, 1978), and most of the recycling and interaction between land and sea takes place between these downwarped continental margins, which are subsequently uplifted. If this same rate of formation and erosion of basins applied during the earliest intervals of geologic time, then the first 1.0 to 1.5 b.y. of the earth's history would encompass the most striking changes in sediment and ocean chemistry, as equilibrium was attained.

(2) Discussions about the earliest atmosphere are unusually difficult to evaluate because they are couched in terms of one or a few constituents, without much concern for mutually incompatible regimes. It appears that of the many possible (and even more proposed) constituents of the early atmosphere, one might focus on CH_4, NH_3, and H_2, on the one hand, and CO_2 and N_2, on the other. In neither case is free O_2 a likely possibility. Hence this section on the earliest atmosphere is divided into (a) gases other than oxygen, (b) reducing atmosphere, and (c) origin of oxygen. A summary of changes in atmospheric composition is given in Table 5-6 and serves as the basis for discussion.

The modern distribution of atmospheric constituents is dependent upon atmospheric height. Wofsy and McElroy (1973) reported that "height profiles for a number of important atmospheric gases (H_2O, N_2O, CH_4, HNO_3, and NO_2) depend critically on the strength of vertical mixing between 20 and 80 km." This fact cannot be treated here, and I will discuss only the most general attributes of the lower atmosphere, the part which is acting most directly on the earth's surface.

(a) Parallel with the initial neutralization of ocean acid-volatiles, the coexisting atmosphere was developing with gases other than oxygen. As for present atmospheric gases, Junge (1972) considered them according to the way their cycles of occurrence are controlled. Nitrogen is the dominant gas in the present atmosphere (Table 5-6), and it exists as an abundant constituent of present volcanic gases. There are no important mineral reactions which involve nitrogen, and presumably it has simply accumulated in the atmosphere through geologic time. This would apply also to Ar, Ne, Kr, and Xe. (In contrast, O_2, CO_2, CH_4, H_2, N_2O, and CO are considered to have a cycle which is predominantly biological. H_2S/SO_2, NH_3, NO/NO_3, and various hydrocarbons, but not CH_4, have sources which are predominantly microbiological, but sinks which are predominantly physical-chemical. Finally, H_2O, O_3, He, and Rn have cycles which are largely physical-chemical.)

The most important requisite for an NH_3-CH_4 atmosphere is the existence of a continued source of H_2 which can fuel the formation of NH_3 and CH_4 as various processes (reviewed below) break

Table 5-6 Possible changes in composition of dry atmosphere.

Constituents	Time (b.y. ago)					Present atmosphere	Volume (%)
	4.6	4.0	3.0	2.0	1.0		
pO_2		$<10^{-70}$	$10^{>-13}$	$\sim 10^{-3}$	$\sim 10^{-2}$	2×10^{-1}	
Major constituents: $>10^{-2}$	$\begin{bmatrix} N_2 \\ CO_2 \\ CO \end{bmatrix}$ or $\begin{bmatrix} NH_3 \\ CH_4 \\ H_2 \end{bmatrix}$	N_2 CO_2	N_2 CO_2	N_2 CO_2	N_2 O_2	N_2 O_2	78.09 20.95
Minor constituents: 10^{-6} to 10^{-2}	HCl? H₂S?	CO NH₃ H₂ H₂S HCN?	CO Ar	O₂ Ar	CO₂ Ar	Ar CO₂ Ne He CH₄ Kr	0.93 0.0315 0.0018 0.00052 0.00011 0.00011
Trace constituents: $<10^{-6}$						N₂O H₂ CO O₃ Xe $\left. \begin{matrix} NO/NO_2 \\ NH_3 \\ H_2S/SO_2 \end{matrix} \right\}$	0.000025 0.00004 0.00001 0.00004 0.000009 variable, in the ppb range

them down (McGovern, 1969). If H_2 in the lower atmosphere can be maintained at 10 to 15 percent of the atmospheric composition, NH_3 and CH_4 would be renewed. At present, H_2 is a very minor constituent of volcanic emanations, and most of the hydrogen is released as H_2O.

Atmospheric ammonia is very susceptible to degradation by ultraviolet radiation. In the absence of an ozone layer, this process is thought to deplete ammonia rapidly so that "ammonia on the primitive earth would have quickly disappeared" (Abelson, 1966:1365). A possible continuous source of ammonia is suggested by the reserve of 20 times the present atmospheric elemental nitrogen which exists as ammonium in the lattice structure of silicate minerals of igneous and sedimentary rocks (Stevenson, 1959; Eugster and Munoz, 1966). This ammonium presumably can be released on weathering. However, the weathering process is very slow and, in the absence of much free H_2, NH_3 is (and would have been) rapidly broken down once it rises to the upper atmosphere (Ferris and Nicodem, 1972).

Even if NH_3 was not a major constituent of the early atmosphere, it is believed to be necessary in minor amounts for early life. Ammonia at sea-water concentrations of $10^{-4}M$ and atmospheric concentrations of 10^{-6} at 25°C appears to be required for the abiotic synthesis of amino acids (Bada and Miller, 1968). In addition, atmospheric ammonia at concentrations of 10^{-6} would result in a greenhouse effect during the Proterozoic and would maintain a surface temperature between 0° and 100°C (Sagan and Mullen, 1972). The conclusion supported here is that ammonia in low concentrations of 10^{-5} to 10^{-6} may well have existed in the early atmosphere.

A large amount of CH_4 has also been proposed for the early atmosphere. However, between 4.6 and, say, 4.0 b.y. ago, when core formation was proceeding and hot conditions of several hundred to a few thousand degrees C may have developed, "methane would have been converted to water and CO_2 almost simultaneously with the formation of the earth" (French, 1966:245). Carbon isotopic data of Precambrian carbonates (including Archean samples) also led Galimov, Kuznetsova, and Prokhorov (1968:1126) to conclude that there was a "complete predominance of CO_2 in the primitive atmosphere," rather than methane. Nevertheless, methane is an important product of anaerobic metabolism and would be expected to be continuously renewed to ascend in the early atmosphere. As with NH_3, the existence of a CH_4 atmosphere is dependent upon highly reducing conditions in the form of excess H_2. In the present atmosphere, the average lifetime of methane is only a few years, and most of it is destroyed by reaction with OH^- (Walker, 1977:69).

The only other compound of carbon often suggested to have been present in a highly reducing atmosphere is HCN. This is the major product produced in laboratory experiments by irradiation of mixtures of CO, N_2, and H_2. HCN rapidly decomposes, however, and its chief importance may be that it leads to the formation of a variety of amino acids and their derivatives in sea water (Ferris and Joshi, 1978).

A highly reducing atmosphere with prominent amounts of CH_4 and NH_3 seemingly would have been destroyed quite rapidly (in the absence of a continued source of H_2), and the atmosphere would have changed to one in which CO_2, N_2, and CO were prominent. With decreasing NH_3, water vapor would not have been shielded from photodissociation. As H_2O was broken down and O_2 released, CH_4 would have been further converted to CO_2 and CO. The time scale over which this conversion in atmospheric composition occurred has been estimated at less than 10^9 years (Hunten, 1973).

During the earliest Precambrian, CO probably would have been a relatively reactive molecule, and several sinks are available in the presence of OH^- and H_2. The present mean atmospheric residence time for CO is less than a year (Wofsy, McConnell, and McElroy, 1972).

During the later Precambrian, as oxygen became available (see below), the oxidation of $FeCO_3$ to Fe_2O_3 and CO_2 may have become especially important (Garrels and Perry, 1974). This reaction yields 4 moles of CO_2 for each mole of O_2. Accordingly, the Precambrian atmosphere (with little photosynthetic removal of CO_2) may well have contained a much higher concentration of CO_2 than now exists. This is the chemical justification for the accumulation of CO_2, which, if present in large amounts, may have led to high late Precambrian surface temperatures (see Chapter 4).

(b) There are at least five independent geochemical observations which are considered indications (weak or strong, depending on the author) that the atmosphere was reducing (and devoid of any free oxygen) until approximately 2.0 b.y. ago. These observations from Precambrian rocks include (i) detrital uraninite and pyrite, (ii) a high FeO/Fe_2O_3 ratio, (iii) a high greenalite/glauconite ratio, (iv) a high Mn/Fe ratio, and (v) the low oxidation state of europeum and cerium.

(i) "Abundant" *detrital* uraninite (UO_2) and pyrite (FeS_2) occur in conglomerates older than 2.1 to 2.3 b.y. (Cloud, 1974b). The uraninite has also been attributed to precipitation from supergene solutions, but the chemistry of the conglomerates is typical of pegmatitic and granitic composition, and thus the conglomerates are probably of detrital origin (Simpson and Bowles, 1977). Roscoe (1968:129) cited small-scale cross bedding which occurred within "multiple layers of uraninite grains" as evidence of sedimentary origin. On continuous exposure to a partial pressure of oxygen of $\simeq 10^{-2}$ to 10^{-6}, uraninite should convert to U_3O_8 or other oxides (Grandstaff, 1974). This process may set an upper limit to the amount of atmospheric oxygen at about 2.2 b.y. ago.

One difficulty with the view that detrital uraninite indicates anoxic conditions is that the present-day Indus River contains a "stable assemblage" of "uraninite, gold, pyrite and other sulphides" over several hundred kilometers of its length (Simpson and Bowles, 1977). To form the Precambrian Witwatersrand and Dominion Reef deposits, these authors suggested that the uraninite was transported in an oxygenated state and then was attached to decaying organic matter; this presumably was followed by its precipitation at the same time that the organic matter further decayed and sulfate-reducing bacteria were forming H_2S, which in turn resulted in pyrite formation. If this is a general pattern, detrital uraninite should be associated with pyrite in carbonaceous rocks, and it often is.

(ii) A reducing atmosphere is consistent with the ferrous/ferric ratio in iron minerals in the Precambrian. The conglomeratic weathering products and the cement of $\simeq 2.0$ b.y. old diorite have a predominance of FeO over Fe_2O_3 (Rankama, 1955, who also cites other examples of reduced iron in the Precambrian). In general, the amount of reduced iron in sedimentary rocks increases as one goes back into geologic time (Veizer, 1976a:Figure 12).

(iii) The oxidized iron silicate glauconite typically has a Fe^{+2}/Fe^{+3} ratio of 1/7. Glauconite is rare in Precambrian rocks older than 1.0 b.y. However, the related iron silicate greenalite occurs abundantly *only* in pre-Phanerozoic rocks, and it is "the most abundant and widespread iron silicate in the Gunflint Iron-Formation" (Floran and Papike, 1975). In contrast to glauconite, greenalite has more ferrous iron than ferric iron. The relative abundance of these two iron-bearing, early diagenetic minerals may, in a general way, be indicative of changes in the availability of free oxygen during the Precambrian (Bentor and Kastner, 1965).

(iv) Evidence for a reducing Precambrian atmosphere also occurs in the abundance of Mn and Fe. (a) The high Mn/Fe ratio (0.025) of Precambrian iron formations does not show the degree of differentiation expected in an oxidizing environment (0.009), and which characterizes Phanerozoic ironstones (Veizer, 1976a). (b) Intertidal carbonates of the lower Proterozoic of the Transvaal (South

Africa) have unusually high Fe and Mn contents, and both metals are in a low oxidation state (Eriksson, McCarthy, and Truswell, 1975).

(v) Eu occurs in a variety of chemical sediments of Archean and early Proterozoic age (1,900 to 2,300 m.y.) and is in normal chondritic abundances. In the later Proterozoic ($\simeq 800$ m.y.) and younger chemical sediments, Eu shows a large depletion relative to chondrites (Nance and Taylor, 1977). The change between 1,900 and 800 m.y. represents a transition from incorporation of Eu^{+2} to Eu^{+3} (Fryer, 1977). The oxidation state of cerium also shows a time dependence, with oxidation to the $+4$ state not occurring until the early Proterozoic. The more reduced states occur in earliest Precambrian time, and the more oxidized states in later Precambrian time.

The initial state of the atmosphere as oxygen became available can be outlined by following the calculations of Garrels and Perry (1974). At the time that the P_{O_2} pressure reached 10^{-70} atmospheres, any initial CO, H_2, NH_3, and CH_4 would have begun to have been converted to CO_2, H_2O, $N_2 + H_2O$ and $CO_2 + H_2O$. These reactions would have been completed before any free oxygen became available. In the absence of free oxygen, the earliest ocean would have been free of sulfate. The absence of a sulfate reservoir in earliest times is indicated by sulphur isotopic values from $\simeq 3.4$ b.y. old barium sulfates that are only slightly different from the values for contemporaneous sulfides (Perry, Monster, and Reimer, 1971). A $\delta^{34}S$ of $+3.5$ is recorded for this ancient barium sulfate instead of the $+10$ to $+30$ values which are typical of Paleozoic to Recent sulfate deposits. Apparently, sulfate-reducing bacteria had not yet fractionated oceanic sulfate (see p. 184).

The sulphur sink for oxygen (chiefly as sulfate) was, however, in effect prior to the middle Precambrian. Sedimentary barite ($BaSO_4$) is reported from the Archean Fig Tree Group of Swaziland (3.2 b.y. old; Heinrichs and Reimer, 1977). Occurrences of calcium sulfate in the Proterozoic are not common, but do include deposits interbedded with marbles (and some sodium chloride) in the $\simeq 1.2$ b.y. old Grenville Series (Whelan and Rye, 1977), as well as older Proterozoic evaporite pseudomorphs from occurrences in Australia (e.g., Walker et al., 1977) and Canada (Badham and Stanworth, 1977).

(c) Sometime approaching 2 b.y. ago, oxygen probably became a chemically important component of the atmosphere. There are three possible sources of oxygen for the atmosphere and oceans.

(i) H_2O is dissociated by ultraviolet radiation, and some of the H^+ escapes from the earth's atmosphere. $O^=$, which is 16 times heavier than H^+, remains. At the present time, for each 10^6 moles of O_2 produced by photosynthesis, photodissociation yields only 1 mole of O_2 (Walker, 1974). Most of the H^+ released by photodissociation simply recombines with $O^=$. However, the escape of any hydrogen atoms "renders *permanent* the liberation of their original oxygen partners" (Bates and Nicolet, 1950). The amount of oxygen produced is partly dependent upon the amount of water which is available for dissociation. In our present oxygenated atmosphere, 40 to 80 percent of the water vapor in the stratosphere is said to come directly from methane oxidation (Lovelock and Lodge, 1972). One molecule of methane yields two molecules of water vapor. It would appear that H^+ has been leaking from the atmosphere for a very long period of time, and thus that the amount O_2 "rendered permanent" has been gradually increasing from this source. The rate of photodissociation is self-limiting because an increase in O_2 leads to ultraviolet adsorption and less photodissociation. R. H. Becker (personal communication, 1976) calculated from the data of Brinkman (1969) that no photodissociation would occur if the atmosphere had two to three times its present amount of O_2.

(ii) During plant photosynthesis, oxygen is produced by the chlorophyll reaction called Photosystem II in which $CO_2 + 2H_2O \rightarrow O_2 + H_2O + CH_2O$ (i.e., organic matter). The production of oxygen by this pathway requires oxygen in the environment because the synthesis of chlorophyll requires molecular oxygen in one of the steps. Nevertheless, this is the scientifically traditional way for oxygen to have been introduced into the atmosphere (Dole, 1949). This process currently fixes approximately 10^{16} mols C/year, and releases an equivalent amount of O_2. There are 3.8×10^{19} mols of O_2 in the atmosphere, so the atmosphere is renewed in $(3.8 \times 10^{19})/10^{16} = 3.8 \times 10^3$ years. The oxygen in the atmosphere, however, is only a small part of the total amount of oxygen needed to oxidize organic compounds; this requires on the order of 10^{21} mols. At present rates of oxygen production, 10^{21} mols would be produced in 10^5 years, a geologically short period of time.

The occurrence of Precambrian laminated rocks called stromatolites is not in itself an indication of oxygen-producing blue-green algae. Photosynthetic filamentous bacteria (e.g., *Chloroflexus*) might be the sole producer of early Precambrian stromatolites under anoxic conditions when sulfide is present (Garlick, Oren, and Padan, 1977). These organisms use Photosystem I, in which $CO_2 + 2 \ H_2S \rightarrow CH_2O + H_2O + 2S$. The blue-green algae are mainly oriented toward this type of photosynthesis. Photosystem I might especially be favored if the Precambrian was hot because, at the present time, "at temperatures above 60° to 62°C filamentous blue-green algae are never found anywhere in the world" (Doemel and Brock, 1974), and thus do not participate in algal mats formed at these or higher temperatures. At temperatures of 55° to 70°C, flat stromatolite mats are formed by the photosynthetic bacteria. Conical "Conophyton" type mats form only "at temperatures below about 55° to 59°C, and are most common at temperatures around 35° to 45°C" (Doemel and Brock, 1974).

On the other hand, if stromatolites do signify *algal* photosynthesis, then photosynthetic organisms may occur in the Bulawayan stromatolitic complex of $\simeq 2.7$ to 2.8 b.y., and in the somewhat older Pongola Group ($\simeq 3.0$ b.y.) of South Africa and Pilbara Block ($\simeq 3.5$ b.y.) of Australia. One or another of these may mark the first production of free oxygen from plants (J. W. Schopf, personal communication, 1978). Even older (3.8 b.y.) banded iron formation deposits in Greenland suggest at least local oxygen sources, and *may* indicate organic activity.

On balance, it appears that blue-green algae from the earliest Archean to perhaps the late Archean utilized sulfide and Photosystem I. Oxygen continued to be added to the atmosphere by photodissociation. As trace amounts of free oxygen entered the oceans, the evolution of chlorophyll took place, with the subsequent evolution of Photosystem II and the major addition of oxygen to the atmosphere (Towe, 1978).

(iii) Oxygen is released by the chemical attack of highly reduced gases on silicates or carbonates from igneous sources. The magnitude of this process is unknown, but it is generally considered to be small in comparison to the other sources of O_2.

(3) The deposition of banded iron formation in the Proterozoic has been suggested by many authors to indicate that oxygen was entering the ocean. An ocean devoid of oxygen at a pH of $\simeq 6.5$ would have been able to contain a significant amount of soluble iron, and much dissolved HCO_3^-. Under such conditions, the polymerization of silica is favored, especially for pH in the range of 6.5 to 7.0. Thus at this pH, and with oxygen newly available, silica would precipitate as SiO_2, and iron would precipitate as an oxide. This may have been the mode of formation of the enormous deposits of banded iron formation, many of which are 1.9 to 2.2 b.y. old. As Cloud (1974a) stated: "The very widespread dispersal of chemically deposited iron oxides in layers a few microns to a few millimeters thick is hard to explain unless it was dispersed in solution in the ferrous state in an oxygen-poor aquatic system, while its precipitation as ferric or ferro-ferric oxides requires combination with free oxygen." How-

ever, the origin of banded iron formation may not have been purely physical-chemical in nature. Carbon isotopic data of carbonates in the banded iron formation of the extremely well preserved Hammersley Group of Australia suggested to Becker and Clayton (1972:587) that organic activity may have been significant in the "derivation, transport, or deposition of iron and silica." The banded iron formation of the Isua rocks of Greenland is 3.8 b.y. old. Owing to the present implausibility of sufficient nonbiological O_2 production by photolysis of water vapor (Walker, 1978), it is possible that oxygen-releasing photosynthesis was present even at that time.

Banded iron formation extends over hundreds to thousands of square kilometers, and is virtually free of terrigenous material. Several possibilities have been suggested for the mode of transport of the iron to the site of deposition. (a) Upwelling along a continental margin resulted in iron being brought to the oxygenated surface waters from which it was deposited (Drever, 1974). However, for this to be true the paleogeographic constraints on upwelling must be met, and these have not yet been examined. (b) Iron was transported by rivers—a "lateritic weathering model" Lepp and Goldich, 1964). (c) "The restriction of banded iron formation to the Precambrian may well be connected with the abundance of large, stable, alkaline lakes" (Eugster and Chou, 1973) in which high pH ($\sim$9.5) led to the concentration of silica. (d) Volcanism introduced iron (Goodwin, 1973). One or more of these very different ideas may apply to the origin of iron in different deposits. This wide divergence of views by experienced workers indicates how much further attention this topic needs.

In summary, the period 4.6 to 2.0 b.y. ago begins with the origin of the earth and closes with an atmosphere and ocean chemistry very much like that which exists today (Table 5-6), except for very much lower amounts of free oxygen (Berkner and Marshall, 1964). During this time, oceanic waters went from a very low pH to approximately neutrality ($\simeq$7.0), and possibly to slightly basic values typical of those found today ($\simeq$8.0). The gaseous composition of the atmosphere and ocean changed from reducing gases (with $P_{O_2} < 10^{-70}$) to the point of saturating the iron and sulfate sinks. The Proterozoic iron hydroxides (now oxidized to hematite) formed in an atmosphere with $P_{O_2} > 10^{-13}$ (Klein and Bricker, 1977). These changes in ocean and atmospheric chemistry are mirrored in the transition from reduced oxidation states in minerals and rocks to their oxidized equivalents. The culminating geological transition at $\simeq$2.0 b.y. is the deposition of the iron and chert deposits known as banded iron formation. As shown in Figure 5-10, most of the oxygen

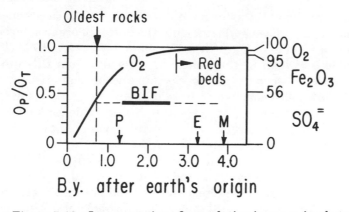

Figure 5-10 Interpretation of cumulative increase in photosynthetic oxygen during the earth's history, expressed as a ratio of the amount of O_2 released prior to a given time (O_p) relative to the total now existing (O_T). Also indicated is b.y. after the earth's origin for banded iron formation (BIF), procaryotes (P), (probable) eucaryotes (E), and Metazoa and Metaphyta (M). At present, of the total photosynthetically produced oxygen 5 percent is free oxygen in the atmosphere and ocean, 56 percent is in $SO_4^=$, and 39 percent (by difference) is in Fe_2O_3. After Schidlowski, Eichmann, and Junge, 1975:50.

which was produced up to this time was taken by the requirements of sulfate and iron.

Ocean and Atmosphere Chemistry: 2.0 B.Y. Ago to Present

The emphasis thus far in this chapter has been on unidirectional changes in atmospheric and oceanic composition. Now it is time to consider more fully a steady-state development in which changes are merely perturbations in a subsequent long-term equilibrium. The material is organized according to (1) oxygen and carbon dioxide, (2) various constraints on ocean chemistry, and (3) possible perturbations in ocean chemistry.

(1) The average content of organic carbon in sedimentary rocks ranges between 0.5 percent and 1.0 percent (Chapter 7). This buried carbon is only about 0.05 percent of the total carbon which is fixed annually. Stated differently, the organic carbon which is buried is the one-hundredth of one percent (0.01 percent) difference between net photosynthesis on the one hand and respiration and decay on the other. An increase in atmospheric oxygen does not result simply from an increase in photosynthesis because for each mole of O_2 re-

sulting from photosynthesis, a mole of organic C is also produced which must be reoxidized at death. Only the *burial* of organic carbon would leave a residue of O_2 in the atmosphere (Walker, 1977). The amount of free oxygen in the atmosphere results from this miniscule difference between oxygen production and utilization. Garrels and Perry (1974:308) concluded that "it seems to us that this remarkably delicate balance [for oxygen] is strong evidence for atmosphere-biosphere-ocean carbon exchange relations similar to those of today for much of the last 600 million years."

Chemical stability of the atmosphere and oceans for the Phanerozoic is also indicated by an abundant fossil record of plants and animals whose general characteristics are indistinguishable from their modern descendants in all important morphological, chemical, and physiological ways. In addition, the chemical composition of the body fluids of both terrestrial and marine species is rather similar to that of sea water, as though physiological processes had evolved in a marine medium (Henderson, 1913:187; Banin and Navrot, 1975). Terrestrial species appear to be carrying a historical tracer of their earlier marine origin.

Metazoa of the tidal flat and of land are highly sensitive to ultraviolet radiation. An effective ozone screen against this radiation occurred when atmospheric levels of free oxygen reached 10^{-3} of the present amounts (Ratner and Walker, 1972). The first metazoans now known are from rocks about 680 m.y. old. At least 1 percent of the present atmospheric level of oxygen presumably has existed since that time or longer. The present amount of oxygen in the atmosphere perhaps could have varied by, say, 10 to 25 percent of its present value over the Phanerozoic, but this is not yet demonstrated, and I know of no way to determine if changes of this magnitude have taken place.

Even though O_2 levels seem delicately balanced, CO_2 levels may even be more so. The steady-state flux of CO_2 is five times faster than that of O_2, and its residence time in the atmosphere is only 2,000 years (Garrels and Perry, 1974). These constraints on CO_2 are important for two reasons. First, land plants would not photosynthesize at a CO_2 level which was as low as one-third that of the present atmosphere, and land plants now contribute 80 percent of the *new* oxygen produced each year (Walker, 1974; Bowen, 1966). Second CO_2 ultimately controls the rate of weathering. By controlling carbonate and phosphate solubilities, CO_2 influences the rate of deposition of cations. Changes in CO_2, however, are very strongly buffered by both magnesium carbonate and magnesium silicate systems (and to a lesser extent by calcium carbonate and organic carbon). Therefore, Holland (1968) believed that "atmospheric CO_2 pressure is not particularly sensitive to changes in the CO_2 input rate." He suggested that since the appearance of land plants in the mid-Paleozoic, "the CO_2 pressure has probably rarely been more than four times the present value."

The net effect of deposition of calcium carbonate is to transform $CO_2 + H_2O + Ca$ into $CaCO_3$ plus H^+ by the series of reactions:

$$CO_{2(g)} + H_2O \rightleftharpoons H_2CO_3 \rightleftharpoons H^+ + HCO_3 \rightleftharpoons XCO_3 \downarrow + 2H^+ \quad (5.7)$$

Similarly, the release of H^+ results from the deposition of SiO_2 in clay minerals or in chert and from the deposition of evaporites. The process opposite to deposition is weathering, and this is character-ized by the fixation of H^+ as the reactions work backward. Thus the two end products (CO_2 and H^+ in the case of $CaCO_3$) are inversely related but linked together through the weathering cycle (Siever, 1968). Times of extraordinarily large amounts of deposition of car-bonate plus silicate plus evaporite should be times of higher H^+ re-lease and lower soil CO_2, and vice versa.

The basic process behind the several ways in which atmospheric oxygen is maintained involves a negative feedback loop so that as O_2 increases, the amount of material undergoing oxidation also in-creases, thus lowering the O_2; and if atmosphere O_2 decreases, the amount of material being oxidized should decrease, thus allowing O_2 to increase again. This concept was first developed by Alfred C. Red-field (1958a), and many distinguished investigators subsequently have elaborated on it.

The negative feedback loop which controls a change in oxygen has been attributed to at least five different causes: (1) the sulfide-sulfate cycle; (2) the pyrite-iron oxide cycle; (3) the organic carbon-carbonate-CO_2 cycle; (4) the organic phosphorus-phosphate cycle; and (5) the organic nitrogen-nitrate cycle. At the present time, nearly all of the "old" (i.e., recycled) carbon which is not buried and should be available to be oxidized is already combined with oxygen as CO_2. Nearly all of the carbon being deposited is "new" carbon (i.e., newly combined in organic matter); (for dissenting evidence that old carbon is being recycled see Sackett, Poag, and Eadie, 1974). If "old" carbon is being oxidized about as rapidly as it comes to the surface, then this "weathering sink for present day oxygen is independent of the oxygen content of the atmosphere" (Walker, 1974:206). The control over present atmospheric oxygen must reside in some factor controlling the production of new oxygen, rather than in rock weathering releasing ancient carbon.

The important component of buried carbon seems to be the

marine component—the phytoplankton (Walker, 1977:97). Both in
the plankton and in the benthos, biomass increases significantly
going toward shore, especially in places of upwelling (Bogorov et al.,
1969). Evidently, the terrestrial system (since the Silurian) has
been a self-contained cycle which has little influenced marine organic carbon (see p. 236). Probably the expansion and contraction of
shallow water basins, such as occurred during the Cretaceous, has
had a much larger influence on reservoirs of organic carbon than
have changes in the terrestrial biosphere. Since carbon fixed by land
plants is oxidized on death, it does not matter that most of the new
oxygen produced each year is from land plants, or that land may
outproduce the sea by 2:1 (173×10^9 tons of dry matter a year of
which 55×10^9 is marine, and 118×10^9 is terrestrial; Odum,
1976).

Redfield (1942) figured out, and it has been well confirmed (e.g.,
by Alvarez-Borrego et al., 1975), that for marine plankton there is a
stoichiometric balance between, on the one hand, oxygen production
combined with use of nitrate and phosphate and, on the other hand,
oxygen utilization combined with the release of nitrate and phosphate at death. The "Redfield ratio" reflects the fact that the theoretical consumption of 276 oxygen atoms (measured value 263) results in the production of 106 carbon atoms, 16 nitrogen atoms, and
1 phosphorus atom. Carbon is in great excess in the ocean. Accordingly, oxygen production should obtain to the extent that phosphorus and nitrogen are available for phytoplankton growth.

The main reservoir of nitrogen is the atmosphere, but most nitrate in sea water is believed to be resupplied by river water bring-
ing nitrate fixed by leguminous plants (Emery, Orr, and Rittenberg,
1955). In fact, the problem has been to find the route of reflux of free
nitrogen from the ocean back into the atmosphere (Deuser, Ross,
and Mlodzinska, 1978). In the sea, nitrogen fixation has been conclusively demonstrated for blue-green algae and for bacteria, but the
quantitative importance is yet to be evaluated (Stewart, 1971).
However, in intensive work on Pacific coral reefs far from land, Johannes et al. (1972) discovered that "atmospheric nitrogen fixation
must have been occurring at rates of the same high order as those
reported for alfalfa fields in order to balance the observed export of
fixed nitrogen." Therefore, it appears that nitrogen will not usually
set an upper limit on plankton growth.

Phosphate is the other major element which might be limiting
to phytoplankton growth. The main reservoir of phosphate is igneous rocks. In addition, phosphorus is limited by its solubility in sea
water (Redfield, 1958a). In lakes (which are sufficiently small to be

manipulated experimentally), "evidence has mounted for an as-
toundingly precise relationship between the concentration of total
phosphorus and the standing crop of phytoplankton in a wide variety of lakes, including many in which low nitrogen-to-phosphorus
ratios should favor limitation by nitrogen" (Schindler, 1977). It appears that phosphorus would usually set an upper limit for plankton
growth.

Why should the amount of oxygen in the atmosphere be 21 percent instead of 2 percent or 50 percent? The answer might be a true
historical accident based on the limiting solubility of a single element, phosphorus. If so, the amount of oxygen in the atmosphere
might have been always about the same once the solubility of phosphorus was attained, and for as long as the major oxygen sinks had
been saturated. The oldest known sedimentary phosphates are approximately 1 b.y. (Veizer, 1976a). Therefore, "excess" phosphorus
has been available for at least that long.

According to ideas favored here, variation in the amount of oxygen in the atmosphere should result from variation in the amount of
organic carbon being buried (see also p. 231). Periods during which
much organic carbon was buried (e.g., the middle Devonian—see
Figure 5-11) might have had a higher atmospheric oxygen concentration than periods in which there was less organic carbon buried
(e.g., the Silurian, Carboniferous, and Permian). Burial of phytoplankton and zooplankton results in the joint burial of organic car-

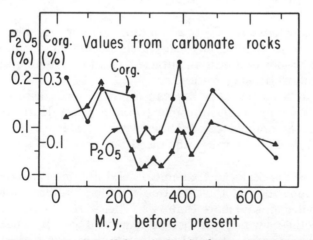

Figure 5-11 Parallel variation in the average percentage of P_2O_5
and organic carbon in carbonate rocks of the Russian platform. After
Ronov and Korzina, 1960:Figure 5.

bon and P_2O_5; and as shown in Figure 5-11, the organic carbon and P_2O_5 content of rocks is positively correlated.

Oxygen solubility is temperature dependent, with higher concentrations at lower temperatures. The thermal structure of the ocean through time therefore may exert a control over the degree of development of anoxic basins (see Chapter 3) and the burial of organic carbon. For example, from the late Cretaceous to the Recent, ocean water has cooled approximately 12°C (see Chapter 4), and this should have resulted in a progressively more oxygenated deep ocean and therefore more oxidation of organic matter. If the total amount of oxygen in the Cretaceous and the Recent were the same, there is relatively less oxygen in the present atmosphere than there was during the Cretaceous, although there would be more oxygen in the present ocean than there was in the Cretaceous seas.

(2) Constraints on changes in ocean salinity have been considered from several points of view.

(a) Broecker (1971) summarized the response time of major components of sea water. These are:

1. Residence time of water in the surface ocean ($\simeq 10^2$ years)
2. Residence time of water in the deep ocean ($\simeq 10^3$ years)
3. Response time of $CO_3^=$ ion content of the sea ($\simeq 10^4$ years)
4. Response time of C, P, and Si contents of sea water ($\simeq 10^5$ years)
5. Response time of Ca content and O_2 content of the atmosphere and of sea water ($\simeq 10^6$ years)
6. Response time of K and Mg contents ($\simeq 10^7$ years)
7. Response time of Na, S, and Cl content of sea water ($\simeq 10^8$ years)

As Broecker stated, events of any given duration. for example, glacial cycles lasting 10^4 years, may cause a significant change in any event of equal or shorter duration, but are likely to leave unaffected a cycle of significantly longer duration.

Changes in residence time are caused by biologic factors (reviewed below), and by ocean-lithosphere mixing rates. Hart (1973) suggested that the abundance of elements with residence times in the range of 10^6 years may be influenced by the rate of sea-floor spreading.

(b) Because the ocean is a chemical solution, one can predict changes in composition of that solution if individual components varied by a factor of, say, more than two. These predicted changes form null hypotheses which can then be compared with the chemistry of ocean precipitates (especially evaporite deposits), and inferences can then be drawn concerning original sea-water composition. However, evaporites become increasingly rare as older and older rocks are examined [and may have a half-life of only 200 m.y. (Garrels and Mackenzie, 1969)]. Thus these comparisons apply chiefly to the past 1.5 b.y. Holland (1972) compared hypothetical perturbations with known evaporite chemistry, and concluded the following:

1. A threefold increase in the present calcium concentration would produce an ocean saturated with respect to gypsum ($CaSO_4$). Gypsum has never been known to have been an open ocean deposit, and is not a constant companion of limestone deposits, and thus calcium has not exceeded this limit.

2. A thirtyfold decrease in the calcium concentration would make gypsum and halite equally saturated during evaporative concentration. Gypsum invariably precedes halite deposition in such sequences, and thus calcium has never been below this lower limit.

3. A reduction in the calcium to bicarbonate ratio from its present value of 4 to 1 to 1 to 2 would leave excess bicarbonate after calcium carbonate deposition, and later evaporite deposits would have carbonates of other cations. Although reported from freshwater lake sediments, such carbonates are not known in marine deposits.

4. At a pH of 9.0 to 9.5, a change in the ratio of Mg to Ca in sea water from its present 3 to 1 to less than 1 to 1 would result in saturation with brucite, $Mg(OH)_2$. And if pH were above 8.6, sepiolite ($H_4Mg_2Si_3O_{10}$) should be a common authigenic mineral. These minerals are not common, and authigenic brucite is not known from marine sediments, and thus an upper limit on oceanic pH is suggested.

5. An increase in the ratio of concentration of calcium to (sulfate plus half of the bicarbonate) would result in late bitterns depleted in sulfate and enriched in calcium. However, known deposits indicate depletion of calcium and enrichment in sulfate. These processes would continue to occur to the point of reduction in sulfate from its present value of 0.028 mol/kg to one-tenth of this value, at which time no combination of calcium and bicarbonate concentrations can yield the mineral sequence of marine evaporites. On the other hand, if sulfate were three times its present concentration, gypsum and halite would reach saturation simultaneously, and this has apparently never occurred.

6. If we take into account the considerations cited above for calcium, bicarbonate, and sulfate, and omit the combinations of these variables which call for precisely fixed values, it appears likely that these constituents have varied by less than a factor of 2 from their present oceanic values.

7. The total amount of NaCl now known as salt deposits would,

if dissolved, double or triple ocean salinity. Since salt deposits are known from all of the Phanerozoic, this limit has never been reached.

(3) The view favored in this book is one of long-term stasis in ocean chemistry punctuated by short-term perturbations. This section considers many of the lines of evidence for the proposed perturbations: (a) salt deposits and sulphur isotopes, (b) oxygen isotopes, (c) strontium isotopes, (d) Mg/Ca ratio, (e) organism influences on sea water, and (f) a variety of other changes.

(a) The greatest known change in ocean salinity through the Phanerozoic appears to have been the Permian deposition of salt. Stevens (1977) thoroughly reviewed the data and concluded that those deposits still contain an amount of salt equivalent to 10 percent of the amount of salt now in the ocean (or about 3‰). Therefore, ocean salinity prior to that unusual Permian deposition *may* have been on the order of 10 percent higher than present values, say, between 32 and 41‰, instead of present values of between 29 and 37‰.

On the shorter time scale of the Pleistocene, oceanic salinity increased owing to removal of fresh water to glaciers, probably about 1 to 1.5‰ (Worthington, 1968). Conversely, during rapid glacial melting, a less saline "melt-water lid" may have developed on the ocean surface if mixing of the fresher water was not immediate (Worthington, 1968). Berger (1978b) presented evidence for a fresher upper lens based on $\delta^{18}O$ data from planktonic foraminifera, 12,000 to 10,000 years ago; this supported a decrease in salinity of approximately 10 percent in surface water for the Gulf of Mexico, as previously suggested by Kennett and Shackleton (1975). If more of these records of a fresher surface layer are presented, then they will have implications for mixing processes in the ocean for time scales of 10^3 years.

Sulphur isotopic values have changed markedly through the Phanerozoic (Figure 5-12). From the Cambrian to the Permian, values from evaporites of $\delta^{34}S$ decreased from $+30$ to $+10$. They then increased to $+20$ to the Recent.

$$\delta^{34}S = \left[\frac{{}^{34}S/{}^{32}S \text{ of sample} - {}^{34}S/{}^{34}S \text{ of standard}}{{}^{34}S/{}^{32}S \text{ of standard}} \right] \times 10^3; \quad (5.8)$$

(the standard is troilite from the Canyon Diablo meterorite.) The major means of sulfate isotope fractionation is by the bacterium *Desulphovibrio* during the reduction from sulfate to sulfide (Berner, 1972), a fractionation of about $-29‰$ (Schidlowski, Junge, and Pietrek, 1977:2561). This enriches the remaining oceanic sulfate in ^{34}S, making sulphate deposits (presently) about $+20$ and reduced

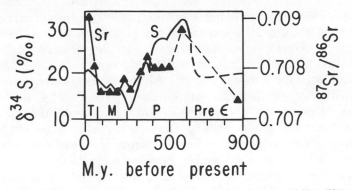

Figure 5-12   Parallel variation in $\delta^{34}S$ and oceanic $^{87}Sr/^{86}Sr$ over geologic time. $\delta^{34}S$ after Schidlowski, Junge, and Pietrek, 1977; $^{87}Sr/^{86}Sr$ after Faure, Assereto, and Tremba, 1978.

sulphur as sulfide (presently) about -9 (review in Kellogg et al., 1972). The low Permian sulfate values of $+10$ could result from a larger amount of reduced sulphur in the reservoir of oceanic sulfate. This might follow from the exceptional weathering of previously deposited Pennsylvanian (coal measures) sulfides during Permian low stands of sea level. Really severe perturbations in $\delta^{34}S$ are very localized in space and time (Holser, 1977).

The large amount of sulfate which was deposited during the Permian buried considerable oxygen. However, oxygen is also partitioned into a carbon reservoir (about twice as large as that of the sulphur reservoir), and an iron reservoir (about a third as large as that of the sulphur reservoir) (Holland, 1973). Data on only a single sink, in this case sulphur, are insufficient to evaluate overall changes in atmospheric oxygen. For example, during the Permian an increase in oceanic phosphate and CO_2 (from erosion of carbonates) should have stimulated photosynthesis. However, during the Permian, little new organic matter was being buried. Both organic carbon and P_2O_5 in limestones reached record lows (Figure 5-11), but the implications of this observation can be argued in different ways. On the one hand, this might be indicative of less phytoplankton biomass. On the other hand, if carbon and phosphorus were continually recycled back into the oceans, they together would have formed a feedback loop which would have stimulated photosynthesis and countered any short-term lowering of atmospheric oxygen. It would be very desirable to have knowledge of the number and size of anoxic marine basins through geologic time as an indication of long-term carbon burial.

(b) A consistent, time-dependent trend in $\delta^{18}O$ values of carbonate and chert (Figure 5-13) exists for the past 3.2 b.y. The "facts" are, however, subject to more than the ordinary degree of conflicting interpretation for such an important line of evidence.

(i) The data have been interpreted to represent progressive change in the oxygen isotopic composition of the oceans through time (Perry, 1967). If the Archean represents the time of initial mixing of sea water and mafic rocks (p. 6), then it makes sense for the ^{18}O values of sea water to be very low. Subsequent values would be a mixture of low basaltic values and high granitic crustal values. This is the general picture outlined below for $^{87}Sr/^{86}Sr$, and if it holds for that isotope pair, then in principle it could also apply for oxygen isotopes. Possibly the exchange of oxygen isotopes between sea water and the oceanic crust is so extensive that this process controls the ^{18}O value over geologic time (Muehlenbachs and Clayton, 1976).

(ii) The change in $\delta^{18}O$ may be indicative of a change in temperature of the ocean. To summarize briefly, cherts which are $\simeq 3.2$ b.y. old have isotopic values which *could* result by precipitation in equilibrium with sea water at about 90°C (Perry and Tan, 1972), and rocks 1.3 b.y. old could have formed at about 50°C (Knauth and Epstein, 1976). To follow the $\delta^{18}O$ values, cooling then occurred at a

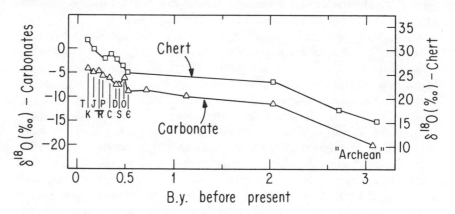

Figure 5-13 Oxygen isotope composition of limestones (PDB $\delta^{18}O$ standard) as a function of geologic age (data from Veizer and Hoefs, 1976, and Degens and Epstein, 1962) and of chert (SMOW $\delta^{18}O$ standard) (data from Perry and Tan, 1972; Degens and Epstein, 1962; and Knauth and Epstein, 1976). Different sources may cite $\delta^{18}O$ values that differ for rocks of the same age. In drawing the figure, I have assumed that changes in $\delta^{18}O$ values should be approximately parallel for marine limestone and marine chert.

constant rate until the Cretaceous but from then to now, temperature has been essentially uniform. (A more conservative interpretation of these data was given above, p. 140).

(iii) The decrease in $\delta^{18}O$ as cherts become older is the natural
result of increased probability of isotopic exchange with fresh water,
richer in ^{16}O (Degens and Epstein, 1962). This view is supported by
data on $\delta^{18}O$ values of petroleum brine waters (which have presumably been protected against dilution), in which no change in $\delta^{18}O$
values was observed from the Cambrian to the present (Degens et
al., 1964). Values for limestone and chert over this same time span
changed appreciably (Figure 5-13).

(iv) Each of these factors may be involved, and that is the interpretation favored here. The gradual decrease in $\delta^{18}O$ over the past
0.5 b.y. seems to stabilize from 0.5 to 2 b.y. (Figure 5-13). Change
over the past 0.5 b.y. may represent mixing with fresh water, which
reaches an asymptote at typical fresh-water values. The markedly
lower Archean values at 3.2 b.y., however, are below fresh-water
values and thus represent initial upper mantle input, plus a further
isotopic change because of somewhat increased temperature.

(c) Granite derives its $^{87}Sr/^{86}Sr$ ratio either from a mantle
source (at $\simeq 0.703$) or from previously differentiated continental
crust (at $\simeq 0.720$) (Faure and Powell, 1972:43). Sea water at any
given time has a $^{87}Sr/^{86}Sr$ ratio which is intermediate between
values for mantle and continental sources, depending upon the relative contribution of each. Archean carbonates (reflecting Archean
oceans) have $^{87}Sr/^{86}Sr$ ratios which are very close to mantle values
(Veizer and Compston, 1976). As time progressed, carbonates came
to have higher $^{87}Sr/^{86}Sr$ ratios, as shown in Figure 5-14. Close sampling of Phanerozoic rocks has revealed secular variation, as shown
in Figure 5-12. Early Paleozoic sea water is characterized by high
values, decreasing in the Permian and Mesozoic and increasing to
the present day.

The time scale of major changes in $^{87}Sr/^{86}Sr$ is hundreds of millions of years. The Phanerozoic trend in the $^{87}Sr/^{86}Sr$ curve roughly
parallels the trend of the $^{34}S/^{32}S$ curve (Figure 5-12). Possibly both
of these Phanerozoic trends are controlled by patterns of erosion,
which in turn are a function of amount of land area, climate, and
paleogeography.

(d) The present mass ratio of Mg to Ca in sea water is approximately 3/1 (= molar ratio of 5.7/1). For three reasons this ratio may
be much higher now than in the past. First, the composition of liquid
inclusions in salt and chert of Silurian age had lower Mg and increased Ca relative to modern oceans (but otherwise very similar

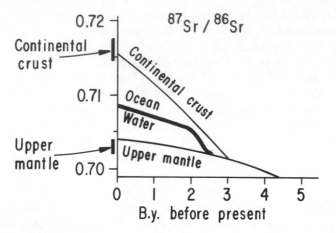

Figure 5-14 Strontium isotopic evolution of sea water showing early similarity to upper mantle values and later approach to continental crust values. After Veizer, 1976b.

SO_4, Na, K, F, and total ionic strength) (Kramer, 1965). Second, with increasing age, Mg is progressively lower in petroleum brines (Degens et al., 1964): (3) Third, as shown in Figure 5-15, as one goes back into the geologic past (especially the Precambrian), Mg (and dolomite) increase markedly in sedimentary rocks relative to Ca (and limestone) (Daly, 1909; Veizer, 1978).

Three interpretations have been offered in order to account for these observations.

(i) Mg is being incorporated into limestone by continental, meteoric water to make dolomite. This explanation is supported by the fact that $\delta^{18}O$ is progressively isotopically lighter with increasing age and it may simply be responding to exchange with fresh water (Mackenzie, 1975:322). A difficulty with this explanation is that Mesozoic and Cenozoic carbonates are still limestone instead of dolomite, and yet there has been plenty of time for diagenetic change to have occurred, especially since the source of older dolomite is generally considered to be penecontemporaneous (Veizer, 1978).

(ii) Mg may be uniform through time, but it is the Ca which is being preferentially transferred from the ocean to carbonate sediments, thus yielding an increasing Mg/Ca ratio in the ocean. The currently high Mg/Ca ratio of sea water would, by this hypothesis, be a reflection of the observation that "the major change in the last 150 million years is that calcite has become a major mineral in pelagic sediments," owing to plankton (foraminiferal and coccolith)

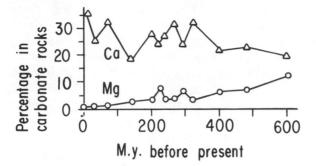

Figure 5-15 Changes in average percentage of Ca and Mg in carbonate rocks of the Russian platform. Note that Mg was more abundant in older carbonates. After Vinogradov, Ronov, and Ratynskii,
1957:273.

deposition (Sibley and Vogel, 1976). Furthermore, Mg inhibits nu-
cleation of low-Mg calcite and hence favors the precipitation of aragonite (and high-Mg calcite) in modern interstitial fluids (Folk and
Land, 1975). If Paleozoic seas had a more equitable Mg/Ca molar
ratio ($\simeq 2/1$), then calcite, rather than aragonite, would be more
likely to have been deposited, and this is in fact observed (Sandberg,
1975). These observations support the idea that Mg is increasing in
the ocean, and calcium is being depleted.

(iii) A third explanation for the historical trend in Mg is that at
high CO_2 pressure (e.g., in the Precambrian), Mg leaves the ocean as
dolomite, and calcium as calcite. As CO_2 pressure decreases through
the Phanerozoic, calcite is preferentially deposited (Veizer, 1973;
Holland, 1976).

(e) Organisms extract elements and thus alter their abundance
and change their residence times in sea water. Previous authors
(e.g., Nanz, 1953) attributed the increase in CaO, CO_2, SO_4, and
P_2O_5 in fine-grained rocks going from Precambrian to the Cenozoic
to the increased scavenging effects of lime-secreting organisms. At
present, the most important biologic control of sea-water composition is plankton, especially for the elements silica, phosphorus, and
nitrogen (and some trace elements). Low values of surface waters for
each of these three major elements are a function of biological extraction. The concentration of each of these three elements increases
with depth because oxidation of sinking organic remains releases
them.

The chief mechanism of transport of Si, P, and C through the
water column is fecal particles (Cherry, Higgo, and Fowler, 1978)

and other aggregates of one type or another, including "marine snow" (Silver, Shanks, and Trent, 1978). Before the development of grazing phytoplankton and zooplankton populations, ocean bottom sediments probably received far less of these elements per unit time than they now do. There would have been no efficient way to transport these elements to the bottom. In general, residence times of these elements in the ocean through the first seven-eighths of earth history may have been much longer then they now appear to be.

For silica, diatoms are considered to extract "at least an order of magnitude more opal per year than radiolaria" (Heath, 1974:84). Siliceous oozes due to skeletal accumulations of both groups occur in open ocean areas of high rates of sinking from surface waters, namely, near the equator and at high latitudes. Owing to biological causes, the silica cycle may have undergone a considerable change in the past 100 m.y., especially because Cretaceous and Tertiary cherts once thought to be volcanic in origin have now been found to be phytoplankton accumulations (Weaver and Wise, 1974).

Another instance of biologic control over sea water compositon is seen in the incorporation and concentration of trace metals in shells of planktonic organisms and in organic detritus. Metals which remain dominantly in solution have their cycles largely controlled by incorporation into organisms and organic matter. The residence times of such metals "probably represent the time scales in which organic scavenging, biogenic particle settling and dissolution, and final association with inorganic particulate surfaces on the deep water are occurring" (Bewers and Yeats, 1977). Thompson and Bowen (1969) reported that the planktonic foraminifera remove much more cobalt, nickel, and zinc than to the phytoplankton coccolithophorids. Aluminum removal from sea water may be linked with siliceous diatoms (Mackenzie, Stoffyn, and Wollast, 1978). Therefore, sea-water chemistry may have varied depending upon the nature of the major carbonate- and silica-secreting members of the plankton.

(f) There are a variety of other observations on both the stability and the change in abundance of elements in the excellently documented changes in the chemistry of rocks through geologic time (Mackenzie, 1975, who cites earlier literature). Whether the variations reflect true secular changes or merely reflect the anticipated results of mixing of the sedimentary package (such that the more resistant parts have a longer lifetime) is uncertain (Garrels and Mackenzie, 1969). Nevertheless, additional patterns have been noted (recently reviewed by Veizer and Garrett, 1978; Veizer, 1978; Schwab, 1978).

(i) The K/Rb ratio in authigenic illites of late Precambrian age (Beltian) is similar to that of the Phanerozoic which indicates that "little change has occurred in the K/Rb ratio in sea water since Lower Beltian time" (Reynolds, 1963:1097). On the other hand, the K/Rb ratio was found to increase with geologic age in samples analyzed by Veizer and Garrett (1978). They attributed the increase chiefly to an increase in K because of more authigenic K-feldspar and illite in older rocks.

(ii) The boron content of marine illite is uniform over at least the Phanerozoic (Landergren, 1945) and probably to 1.9 b.y. ago (Reynolds, 1965:8). Thus, following Forchhammer's Principle, "The salinity of the ocean has been essentially constant for the last two or three thousand million years" (Gregor, 1967:9). However, the candid caption to Reynolds's figure illustrating the data gives one pause: "The boron values are displayed on a logarithmic axis because of the logarithmic nature of the laboratory errors."

(iii) The 1,000 Sr/Ca ratio has considerable variation. Possibly it represents a time-dependent change (El-Hinnawi and Loukina, 1972), but more likely it is diagenetic in origin (Veizer, 1977).

(iv) The SiO_2/Al_2O_3 ratio decreases the older the rocks (Ronov and Migdisov, 1971). Low ratios indicate more of a mafic contribution and less of a granite component. Therefore, this trend "is consistent with the progressive development of an increasingly granitic crust" (Schwab, 1978).

In summary of the period from 2.0 b.y. ago to the present, the major changes have been the addition of oxygen to the atmosphere and the evolution of a geochemically important biosphere. The level of oxygen in the atmosphere may be set by the ability of phytoplankton to utilize phosphorous, which itself is held to a fixed value by its low solubility in sea water. A variety of changes were possible in the composition of sea water and of the atmosphere during the Phanerozoic. However, salinity has certainly not varied by a factor of 2 from its present value, and atmospheric oxygen has never gone to 0. The general picture is one of steady-state interaction among the atmosphere, biosphere, hydrosphere, and lithosphere.

Summary

The chemistry of sea water is a function of the input of elements from the land, partial utilization and mixing by the biosphere in the ocean, and passage of elements onto the sea floor, where they become incorporated into rocks. Thus the composition of sea water is to be understood as part of long-term cyclic exchange among the lithosphere, biosphere, and hydrosphere. Negative feedback mechanisms appear to constrain the system from having wide perturbations over geologic time.

The salinity of the ocean at least since the Cambrian generally has been judged to be the same as that of the modern ocean because highly complex organisms which today occur exclusively in the oceans have existed throughout the Phanerozoic. More precise estimators of ancient salinity include methods based on boron, sedimentary phosphate, and isotopic abundances of ^{18}O and ^{13}C; more general indicators of salinity are based on trace-element distributions, interstitial fluids, siderite, and specific taxa of plants or animals. By far the easiest way to distinguish fresh-water from marine conditions is by analyzing the flora and the fauna.

The view adopted in this book supports the notion of an early, nearly complete degassing of the earth's interior and early equilibration with volatiles which described unidirectional trends that resulted in essentially modern salinity values and sea-water chemistry by about 2.0 b.y. ago. Atmospheric evolution is seen as changing from an early oxygen-deficient environment to a clearly oxidizing environment, possibly by 2.0 b.y. ago, and certainly by 700 m.y. ago. Variation in sea-water chemistry probably has occurred over the past 1 b.y., especially where elements have entered into biological cycles, but overall salinity has not varied by a factor of 2.

6. CLIMATOLOGY

*There is no such thing as moderation in nature—
that's a human invention.*

Saying of F. von Voigtlander, 1977

For present purposes, climatology pertains to when and where it rains, and what the consequences are, and what can be inferred about patterns of atmospheric circulation. This chapter covers (1) latitudinal zonation of climatic zones, (2) patterns of sediment yield, and (3) storms.

Since temperature and salinity are latitudinally zoned, several subjects covered in other chapters are related to climate. Previously considered topics involving climate include the carbonate lysocline and CCD (Chapter 2), Mediterranean versus estuarine type of circulation (Chapter 3), siliceous sediments and cherts (Chapter 3), and atmospheric composition and albedo (Chapter 4). Many aspects of paleoceanography influence other areas, and this (or any other) division of topics is to some extent arbitrary.

Latitudinal Variation

Latitudinal trends are divided into (1) those which affect continental shelves and (2) those which affect the deep sea. At present, each hemisphere is broadly divisible into polar (and subpolar), temperate, subtropical and tropical (or equatorial) climatic zones, as shown in Figure 6-1. These climatic zones correspond to a poleward gradient in decreasing temperature, and to wet belts in the tropics and temperate regions, and to dry belts in the subtropics and polar regions. These climatic zones also correspond to changes in current systems (since both climate and currents are related to atmospheric circulation), and as a result, climate is reflected in faunal and floral provinces.

(1) The sediment types of the continental shelf are related to climate (Keary, 1970; Lisitzin 1972:133), and especially to rainfall and temperature (Figures 6-2 and 6-3). Even within a limited basin, the rate of sedimentation may be chiefly a function of rainfall in the source area (Soutar and Crill, 1977).

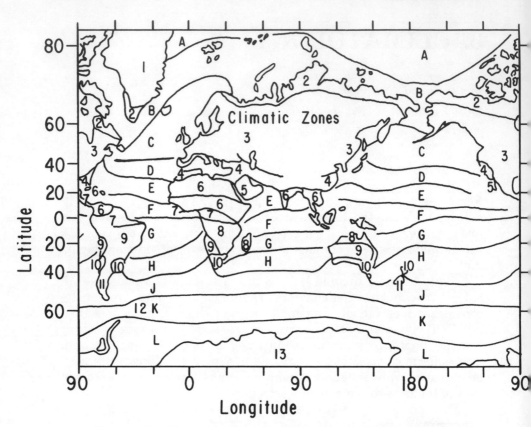

Figure 6-1   Diagrammatic map of the natural climatic regions of the
world ocean compared with the land climatic regions. A (polar, arctic), B (subpolar, subarctic), C (temperate), D (subtropical), E (tropical, trades), F (equatorial), G (tropical, trades), H (subtropical), J
(temperate), K (subpolar, subantarctic), L (polar, antarctic). For
comparison, the land regions are also given: 1 (artic, ice desert), 2
(subarctic, tundra, and forest tundra), 3 (temperate, taiga, deciduous
forest, steppe), 4 (dry, Mediterranean, and humid subtropics and the
northern parts of deserts), 5 (tropical deserts), 10 (dry and humid
subtropics), 11 (temperate, mainly forestless), 12 (subpolar), 13 (glacial zone of Antarctica). After Bogdanov, 1963; Petterssen, 1969.

Patterns of sedimentation related to rainfall and temperature
are in part caused by different efficiencies of erosion. As shown in
Figure 6-4, with increasing rain, chemical weathering becomes
more important and clay is the anticipated most likely end product
(see Chapter 5). The material is transported both in suspension
(Douglas, 1967) and in solution. Clay is most pronounced where

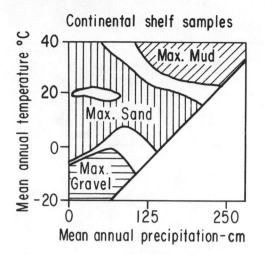

Figure 6-2   Areas of concentration of gravel, sand, and mud on the continental shelf as a function of mean annual temperature and rainfall. After Hayes, 1967b:119.

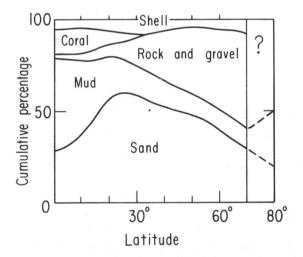

Figure 6-3   Change of sediment type on inner continental shelf with latitude. Note that mud (calcareous) and coral are most abundant in equatorial regions, that sand is at a maximum in the mid-latitudes, and that sediments are coarsest in high latitudes. After Hayes, 1967b:121.

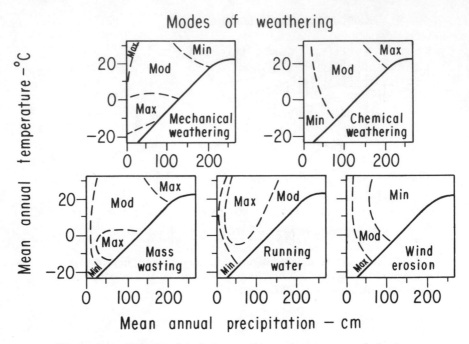

Figure 6-4 Relationship between the various geomorphologic processes and average annual temperature and precipitation. Maximum (Max), moderate (Mod), and minimum (Min) effects are noted. After L. Wilson, 1969:307.

temperatures are highest and where precipitation is always in rain instead of snow. At low latitudes, the ratio of solid detrital suspended sediment to dissolved salts is approximately 2.3/1 (Emery and Milliman, 1978).

In the tropics, clay is predicted to be the dominant product of weathering because of the large amount of rain from rising equatorial air. Deep weathering profiles are the rule because "compared with temperate climates, *the decomposition of the rock* (mainly the result of chemical weathering) *is more rapid than the transport of material on the slopes,* which in turn *is more effective than fluvial erosion*" (Birot, 1968:73, italics his). The tropical sun also insures a luxurious growth of calcareous algae, and the algae yield immense amounts of clay-sized calcium carbonate to the sediments. On a worldwide basis, approximately 50 percent of the tropical, inner-shelf sediments are mud (Figure 6-3), probably mostly algal in origin.

In the subtropics, between 5° and 35°S and 15° to 40°N, the de-

scending dry air is a sink for moisture from the ground, and this
causes the formation of desert and evaporitic conditions, summarized in Figure 6-7. With low precipitation and high temperature, mechanical weathering is of prime importance. Weathering in
semiarid climates should give rise to greater amounts of rock fragments, feldspar, and accessory minerals, and lesser amounts of
quartz, in comparison to weathering in humid climates (Young et
al., 1975; Mack and Suttner, 1977). Inner-shelf sediments are dominantly sand-sized (Figure 6-3). (However, the problem with interpreting outer-shelf sediment patterns with present climate is that
climate has changed much faster than the shelf conditions! Present
sediments of the outer shelf reflect a climate which existed at a
lower stand of sea level, but how much different the conditions were
must be determined for each case; see pp. 36, 201.)

Atmospheric circulation is again reversed at about 50° to 60°
latitude with increased precipitation from rising air. The air, however, is cooler than the tropics to begin with and so has less moisture
to release. Thus in temperate regions, chemical weathering is less
intense than in tropical climates, and mechanical weathering is relatively more important. In shallow seas of the continental shelf, sed-

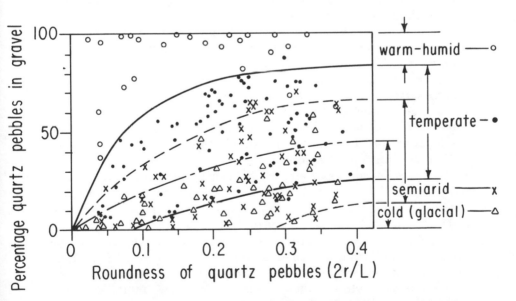

Figure 6-5 Use of gravels to distinguish various climatic zones according to roundness and percentage of quartz pebbles. *L* is pebble
length; *r* is the bending radius of the sharpest protrusion along its
contour. After Cin, 1968:1098.

iment may become coarser, especially under glacial conditions (Figure 6-3). In the finest fraction, silt is dominant over clay (McManus, 1970:77).

The extreme contrast with tropical conditions is near the poles, where low temperatures and reduced precipitation bring mechanical weathering and wind erosion to a maximum (Figure 6-4). There are both shelf and deep-sea indicators of polar climate in addition to results from paleotemperatures and from terrestrial glacial deposits. In a coastal zone with yearly ice formation, beach deposits include ridges of poorly sorted sand and gravel (Greene, 1970). Small outwash flats result from melting snow banks, and kettles result from melting sea ice. Thus beach deposits under Arctic conditions are as likely to be in equilibrium with ice as with water, as the depositional agent.

In the microcosm of gravels, the latitudinal trends outlined in the past few paragraphs are reflected in a general way. As shown in Figure 6-5, temperate and warm-humid regions have the fewest nonquartz remnants in gravels and the highest degree of rounding indicating the most complete weathering (Crook, 1968). Semiarid and cold regions have the highest inclusion of metamorphic and granitic pebbles in gravels and the lowest degree of rounding, indicating the least complete weathering.

In terms of deposits on the shelf, other latitudinal trends involve carbonates, siliceous sediments, evaporites, dolomite, and phosphates. The abundance of carbonate sediment decreases away from the equator (Figure 6-6), and the composition of this sediment changes from an algal-coral dominance in the tropics to a foraminiferal-molluscan (and sometimes bryozoan) dominance at higher latitudes (Lees and Buller, 1972). Siliceous sediments (from diatoms) may also become important on the shelf at mid-latitudes (Niino and Emery, 1966:160).

Evaporites are associated with (a) arid latitudes (chiefly 30 to 40°, Figure 6-7) and (b) favorable geographic settings, especially grabens, which form during the opening of ocean basins, (Burke, 1975; Evans, 1978) in which oceanic circulation may be greatly restricted (see p. 99). However, not all "arid" conditions are equally arid. An equilibrium is reached between the mean relative humidity and the water-vapor pressure of the brine (Kinsman, 1976). Calcium sulfate is precipitated when the mean relative humidity of the atmosphere is between 93 and 76 percent, halite between 76 and 67 percent, and potash facies minerals are less than 67 percent. Thus the local aridity and humidity can be inferred from evaporitic facies (and has been by Schreiber and Schreiber, 1977).

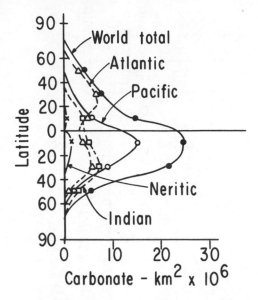

Figure 6-6   Distribution of present-day marine carbonates (pelagic and neritic) in terms of latitude. Note the bimodal peaks of carbonate in the Atlantic caused by relatively high terrigenous inflow of the Amazon and Niger rivers; the effect is absent from the open Pacific and is not noticed in the world total. After Fairbridge, 1964:473.

Low-latitude coastal zones commonly have mean relative humidities of 70 to 80 percent. Lower values occur only where a marine basin is surrounded by large expanses of land. This relationship may account for the fact that calcium sulfate (gypsum and anhydrite) is so tremendously abundant in evaporite deposits (Kinsman, 1966), even though sodium chloride represents about 85 percent of the dissolved salt in the sea. The association of potash deposits with high continentality is also of paleogeographic importance.

Dolomite and phosphate occurrences are related to climate. Both arid and humid conditions may have supratidal dolomite, but each climate leaves quite different imprints on supratidal deposits. Arid regions are marked by gypsum or anhydrite, whereas humid regions have brackish algal marshes with calcereous crusts (Shinn, Lloyd, and Ginsburg, 1969; Hoffman, 1975). The vast bulk of sedimentary phosphates (see p. 102) are attributed to continental margin accumulations of the arid zone (Ronov and Korzina, 1960) at places of upwelling. Under arid conditions, phosphate and organic carbon are strongly correlated; under humid conditions, phosphate and iron covary.

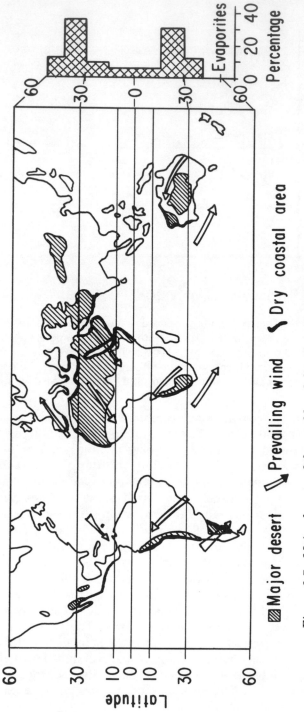

Figure 6-7 Major deserts of the world and distribution of present-day evaporites. Deserts after Glennie, 1970; evaporites after Drewry, Ramsay, and Smith, 1974:547.

Marine faunal and floral provinces are of course latitudinally zoned, probably because water masses reflect the wind-driven circulation. Reefs, with their attendant high diversity, rock-forming capability, and association with algae and other photosynthetic organisms, are the main indicators of tropical conditions.

Figure 6-8 summarizes the anticipated sediment regimes for shelf deposits in relation to climate and other factors. Equatorial sediments are chiefly biogenic, mid-latitude sediments chiefly water contributed, and high-latitude sediments are glacial (at present!). Authigenic minerals (especially phosphorites) occur in tropical to temperate latitudes in regions of intense upwelling.

In contrast to these general equilibrium considerations, historical and geographic factors often prevent immediate interpretations. Coarse outer-shelf sediments reflect conditions of many thousands of years ago when sea level was tens of meters lower (pp. 36, 197). Presumably, historical factors are most important when dealing with intervals of "rapid" sea-level fluctuations; unfortunately, that could be most of geologic time. Even the usual idea that mineral stability of feldspar weathering increases with decreasing difference

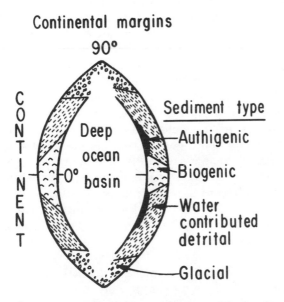

Figure 6-8 Idealized distribution of major classes of sediment on continental shelves bordering an ocean when sediments are in equilibrium with their environment. The width of shelves is exaggerated. After Emery, 1968:446.

between crystallization temperature and surface temperature (Goldich, 1938) is not always reflected in sediments (Todd, 1968). The net effect is that any specific region is bound *not* to be in total equilibrium with the predicted depositional regime, and some regions will be very far from predicted "equilibrium."

Although rainfall is latitudinally distributed, any given terrestrial latitude may have both high *and* low rainfall, depending on how mountainous is the relief. The Tropic of Cancer passes both through the middle of the Sahara and Arabian deserts and through the rain forests of Southeast Asia! The reason for this is that the Himalayan Mountains force air to rise, thus releasing moisture gained from air passing over the Arabian Sea and the Bay of Bengal. The deserts of Asia are therefore displaced *north* of the Himalayas to 40 to 50°N, a latitude which is moist in Europe, adjacent to the Atlantic and Mediterranean. Another example of the orographic effect is that North America is generally dry to the west of a line drawn northward from the western extreme of the Gulf of Mexico. East of this line, however, North America is quite moist because water from the gulf is carried north and released.

Perhaps the most spectacular evidence for the orographic effect on rainfall is along the Andes. South of 35°, the Chilean coast has an immense amount of rain released by the currents coming from the west off the ocean, whereas between 35°S and the equator, air currents are *from* the east, and they drop all of their rain into the Amazon basin. Therefore, with a change in latitude of about 5° along the coast, the region which was forest becomes a desert, and one goes from more than 300 to less than 25 cm annual rainfall.

The net result of these orographic considerations is that just as striking a gradient in wet or dry conditions is encountered when going north to south as when going east to west, in both North and South America and from Europe across Asia. Therefore, the question "Why is a given region a desert?" has three equally correct possible answers: it could be adjacent to mountains, or it could be in the arid latitudinal belt, or both. Because topography controls climate, and climate controls vegetation, which in turn controls soil type (see Fullard, 1971:8, 9), topography is of fundamental climatological importance.

(2) Just as with shelf sediments, deep-ocean deposits are latitudinally zoned (Figure 6-9). This zonation is seen in (a) glacial deposits, (b) carbonate sediments, (c) clay minerals and quartz of eolian origin, and (d) volcanic ash.

(a) Glacial deposits are indicated by ice-rafted detritus, which consists of poorly sorted angular fragments of igneous, sedimentary,

Deep-sea surface sediments

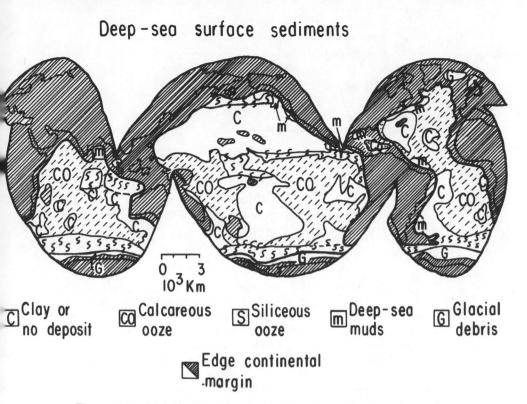

$\boxed{C}$ Clay or no deposit	$\boxed{CO}$ Calcareous ooze	$\boxed{S}$ Siliceous ooze	$\boxed{m}$ Deep-sea muds	$\boxed{G}$ Glacial debris

$\boxed{\blacksquare}$ Edge continental -margin

Figure 6-9   Distribution of recent sediments on the deep-ocean floor. Note the dominance of calcareous ooze in the Atlantic and Indian oceans and of clay in the Pacific. After Berger, 1974:214.

and metamorphic rocks, or mineral grains such as quartz and feldspar from such rocks (Connolly and Ewing, 1965a, 1965b). This material occurs in transitions to and from fine-grained sediments of differing composition, and this is indicative of a rapidly changing ocean climate. As depicted in Figure 6-10, during interglacial times ice-rafted debris is abundant close to highest-latitude continental regions where ice calving is intense. During glacial times, icebergs float equatorward and, on melting, cause ice-rafted debris to be most abundant away from the source of ice. Thus a deposit of ice-rafted debris may be of different ages in different places, just as for zones of carbonate productivity (p. 104). As is usual in interpreting specific records, paleogeographic reconstruction provides essential information.

The surface of grains of quartz sand may be altered by various

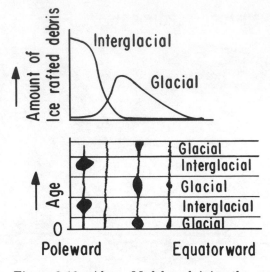

Figure 6-10   Above: Model explaining the spatial and temporal accumulation of ice-rafted debris in deep-sea sediments of the Southern
Ocean during glacial and interglacial periods (mg/cm^2/10^{3}y). During
interglacial periods ice-rafted debris accumulates close to the source
because of increased iceberg calving. During glacial periods, ice-
rafted debris extends much further northward, corresponding to the
movement of the 0°C isotherm. *Below:* Ice-rafted debris will be greatest during interglacials in southernmost latitudes and during gla-
cials in more northern latitudes. After Watkins et al., 1974:534.

chemical and mechanical means, depending upon the agent of erosion. Those grains are subjected to pressure and ice wedging in a
glacial regime and exhibit many textures (Table 6-1), the observation of four or more of which "over large areas of a single grain may
be taken as adequate evidence of a glacial origin" (Krinsley and
Donahue, 1968). The occurrence of such grains from the early and
middle Eocene and Oligocene in oceanic sediments near Antarctica
indicated the existence of Tertiary continental glaciation to Margolis and Kennett (1971).

　　(b) The total carbonate concentration preserved in oceanic sediments varies according to the balance of surface primary productivity and of subsequent dissolution. During warmer, interglacial intervals, dissolution of deep-sea carbonates appears to be enhanced,
relative to glacial periods (Berger, 1973). Accordingly, cycles of
abundance of carbonate in Pleistocene cores can be interpreted to result from broad climatic changes, sometimes related to oceanic circulation (p. 104).

Table 6-1 Summary of surface texture characteristics of sand grains under glacial conditions. Glacio-fluvial patterns are the same, but the grains are rounded. Data from Krinsley and Donahue, 1968).

1. Large variation in size of conchoidal breakage patterns
2. Very high relief (compared with grains from littoral and eolian environments)
3. Semiparallel steps
4. Arc-shaped steps
5. Parallel striations of varying length
6. Imbricated breakage blocks, which look like a series of steeply dipping hogback ridges
7. Irregular small-scale indentations, which are commonly associated with conchoidal breakage patterns
8. Prismatic patterns, consisting of a series of elongated prisms and including a very fine-grained background

(c) Clay minerals and quartz of eolian origin are washed out of atmospheric dust by rain into the ocean. There they may adsorb trace metals and then sink and accumulate in the fine fraction of deep-sea sediments (Chester, 1972). The eolian contribution is on the order of 10 to 30 percent of deep-sea sediments (Windom, 1975). Deposited aerosols follow the latitudinal zonation of winds, and thus they reflect climatic zonation on the continents, sometimes for a considerable distance. A dust storm in the Sahara can result in dust crossing the Atlantic in five to six days (Carlson and Prospero, 1972).

The clay minerals chlorite, illite, and kaolinite characterize different latitudinal weathering regimes, as shown in Figure 6-11 (Griffin, Windom, and Goldberg, 1968). Chlorite is most abundant in highest latitudes, where mechanical degradation is important relative to chemical weathering. In eolian debris, chlorite occurs in low, but consistent amounts, increasing slightly at the highest latitudes sampled (Figure 6-11).

In eolian materials, illite is in lowest relative concentration in equatorial regions but increases with latitude (Figure 6-11). This distribution reflects the terrestrial pattern, in which soils in lower latitudes have less illite. Kaolinite forms under concentrations of strong chemical leaching (Keller, 1956) and is in highest concentration in equatorial latitudes in both eolian materials (Figure 6-11) and deep-sea sediments. These differences in clay minerals are of course mirrored in sediment chemistry. Al is higher and Si, Ti, and K are lower in tropical latitudes than in higher latitudes (Donnelly,

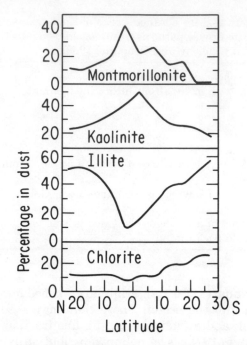

Figure 6-11   Variation of clay minerals with latitude in eolian
dusts. Values from the total $< 2\mu$m fraction are averaged for each 5°
of latitude. After Chester et al., 1972:100.

1977). In all of these cases, clay mineral distribution appears to reflect the weathering characteristics of the source areas on the adjacent continents.

Detrital quartz is also a common constituent of eolian debris,
and thus its distribution in deep-sea sediments reflects zonal wind
patterns. The highest Atlantic concentrations are seaward of the Sahara Desert (Chester et al., 1972). When glacial ice was greatly expanded 18,000 years ago, "this high quartz band moved southward
by 7° to 8°," partly owing to a southern shift in the trade wind belt
(Biscaye, Kolla, and Hanley, 1977). Eolian sand turbidites have also
been reported (Sarnthein and Diester-Haass, 1977). In the Pacific,
detrital quartz in deep-sea deposits is indicated by unusual oxygen
isotopic values ($\delta^{18}O \simeq 18$), which suggest an eolian origin (Churchman et al., 1976). No geographic variation in $\delta^{18}O$ of deep-sea detrital quartz has been observed between 8°S and 43°N, although systematic differences do exist in the size of the particles (going from 1
to 44μ), suggesting different relative contributions of quartz from
igneous versus sedimentary eolian sources. In times of aridity, sili-

ceous fresh-water diatoms and terrestrial plants ("phytoliths") may be transported in dust to the ocean, where their occurrence in sediments provides a climatic marker (Parmenter and Folger, 1974).

(d) The introduction of volcanic ash to the atmosphere may have a strong influence on temperature, as was reviewed in Chapter 4. As the ash settles, it acts as a tracer for atmospheric zonal wind circulation. Eaton (1963) demonstrated a close relationship between the observed east-west circulation of the lower atmosphere and the direction of the long axis of the ash accumulation for both marine and terrestrial deposits. A survey of explosive volcanism covering the past 200 years indicated two primary latitudinal belts of volcanic additions to the stratosphere (Cronin, 1971). One of these is a broad belt centered at about 15°N to 8°S. The other is a narrow belt at 56°N to 65°N. Volcanic ash may be transported over 3,000 km by the wind and is generally composed of particles which are too large to be transported for such distances by surface or deep-sea currents (Huang et al., 1973). In the May 7, 1902, eruption of the Soufrière on St. Vincent in the Lesser Antilles, winds at 7 to 16 km height played the dominant role in distributing ash at 60 km/hour (Carey and Sigurdsson, 1978). Accumulation can be traced for 3,000 km, and rates of deposition were on the order of several $mg/cm^2/10^3$ years. Considerable local variability can exist in thickness of ash fall. Dowding (1977:1144) reported that cores 25 km apart show "great variability in the thickness of the [ash] layers (0–17 cm), and certain layers may be absent," for reasons unknown.

The clay mineral montmorillonite is considered indicative of volcanic sources. Its high equatorial abundance (Fig. 6-11) agrees with the intensity of equatorial volcanic activity reported above. However, sediments from volcanoes are of more importance in providing marker horizons than in providing larger amounts of material to oceanic sediments. "The bulk of sediments in all parts of the ocean is ultimately terrigenous, including many sediments originally identified as vulcanogenic" (Donnelly, 1977).

(3) Increased attention should be given to possible coupling between deep-sea and shelf deposits. The critical factor is that organic production, sedimentation rates, and elemental input all have high gradients across continental shelves, and thus each strongly influences what is recorded in deeper water. Changes in climate, together with eustatic rise and fall of sea level, control both deep-sea and shelf sedimentation patterns. High stands of sea level cause a stoppage of transport of sediments to the deep sea (Damuth, 1977). High deep-sea accumulation rates are strikingly correlated with low stands of sea level (and vice versa) for the past 48 m.y. (Worsley and

Table 6-2 Climatic criteria based upon the sedimentary products produced by climate. Data from McManus, 1970.

Climate	Inner-shelf sediments	Deep-sea sediments
Polar climate	1. Mud, minor clay minerals	1. Quartz grains with glaciated surface texture
	2. Gravel, if present in adjoining land ice	2. Terrigenous grains in pelagic sediments, abundant if present in ice on adjoining land
	3. Quartz grains with glaciated surface texture	3. Chlorite of certain characteristics in silt- and clay-sized fractions
	4. Chlorite of certain characteristics in silt- and clay-sized fractions	
Tropical rainy	1. Mud, abundant clay minerals	1. Kaolinite
	2. Kaolinite, abundant near smaller rivers	
	3. High quartz content in gravels	
Dry	1. Sand, also present in mid-latitudes	1. Charcoal, grass seeds, fresh-water diatoms, opal phytoliths, etc., in pelagic sediments
	2. Quartz grains with eolian surface texture	2. Quartz grains with eolian surface texture
Intermediate	1. Sand, also present adjacent to dry climates	

Davies, 1978). This much seems reasonable; whether or not there is also a causal relationship between changes in sea level and climate, as suggested by some (e.g., Damon, 1968; Berger 1974:228), is open to question.

Table 6-2 summarizes most of the climatic criteria discussed above. Chemical sediments are concentrated between 40°N and S, whereas clastic sediments dominate in higher latitudes. The same low-latitude localization of chemical sediments of the shelf is seen in the middle Devonian (paleolatitude determined by paleomagnetic evidence) and in the lower Permian (Figure 6-12). As stated by Briden (1970:442), "The sharp cut-off in precipitated sediments at paleolatitude 40° is in line with the well-defined climatic zonation which one would expect at a time when a large polar ice cap existed, as it did in the southern hemisphere." As an outstanding climatolog-

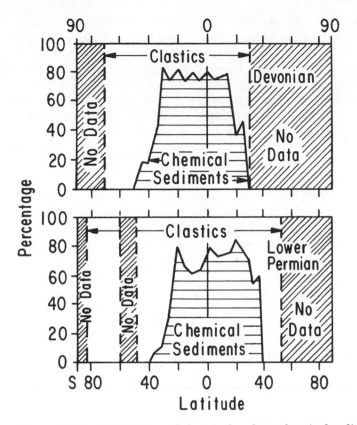

Figure 6-12 Distribution of chemical and nonchemical sediments.
Lithofacies distribution in the middle Devonian, (*above*) and lower
Permian (*below*) plotted against paleolatitude. After Briden,
1970:443.

ical application of geological data, Ziegler, Hansen, et al. (1977)
demonstrated that the latitudinal track of a continent could be determined from the predicted occurrence of climatically influenced
rock types. Kazakhstania traveled from cool wet climates at about
60°S (in the late Precambrian) successively through warm dry subtropics, hot wet tropics, and in the Permian reached the warm dry
subtropics at about 30°N.

Sediment Yield

The rate of introduction of sediment into the ocean seems to be primarily a function of regional geography and secondarily one of cli-

mate. These two factors combine to produce variation in rates of sed-iment yield through geologic time. In a steady-state world, the mean rate of erosion must equal the mean rate of accumulation (or loosely speaking, sedimentation). Hence data on both erosion and sedimen-tation are presented. The general topic of sediment yield is divided into (1) the influence of geography, (2) the influence of climate, and (3) the geologic history. Although over the long term the sedimenta-tion rate is important, over the short term the resuspension of already deposited sediment is of much greater significance for day-to-day and tide-to-tide activities of marine organisms and for sedi-mentary features.

(1) The difference in sediment yield attributable to different ge-ographic regions is so great as to overwhelm average values for whole continents. Approximately 80 percent of all of the suspended sediment carried to the ocean is from Asia, and at least half of this comes only from rivers draining the Himalayas and the loess de-posits of north central China (Holeman, 1968). The five rivers which contribute the largest amount of suspended sediment (Yellow, Ganges, Bramaputra, Yangtze, and Indus) are not the longest (ranking is 6, 40, 25, 5, and 21, respectively). The longest river in the world (Nile) ranks only sixteenth in amount of suspended sedi-ment, and the second longest river (Amazon) contributes even less. The historical accidents of local geography and starting materials have as much to do with sediment yield as the climatic regime per se.

One of the most interesting geographic factors in predicting sediment yield is the influence of the size of the area drained. Smaller basins tend to have steeper slopes and a limited floodplain in comparison with large basins, with the result that relatively more sediment is flushed seaward in small basins. In addition, in large basins the sediment is more likely to be chemically rather than mechanically weathered because particles tend to spend a longer time in large basins. This longer residence time is indicated since a tenfold increase in basin area increases basin yield by only a factor of 2 (Wilson, 1973:344).

There is a general geographic relationship between area drained, number of kilometers of river (Potter, 1978), and number of major rivers. For every 10×10^6 km^2 of land, there are about 30×10^3 km of major rivers on wet continents, and 7×10^3 km on dry continents, as shown in Figure 6-13. If sea level covered 75 per-cent of current land area, the remaining land area would be about 2.9×10^7 km^2, and could be compared with Asia, which is a third larger at 3.7×10^7 km^2. Such a hypothetical land mass would, if all

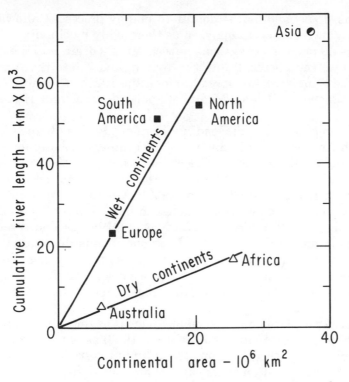

Figure 6-13 Cumulative river length versus continental area. Note the difference between wet and dry continents, with Asia intermediate. B. Rabe, personal communication, 1973.

wet, be predicted to have 67×10^3 km of major rivers, and Asia has 72×10^3 km (between the figures for wet and dry continents, but closer to that for the wet continent, as shown in Figure 6-13).

The number of major river systems can also be estimated. At present there are 60 major river systems. A land area a quarter the size of that presently existing would be predicted to have 15 major rivers; Asia, about a quarter of the total land mass, has 19. Shreve (1974) showed that mainstream length in river networks varies with area to the 0.6 power for basins 1 to 10^3 km² and to the 0.5 power for basins of approximately 10^7 km². Thus for a given paleogeographic setting, the mainstream length and the number of major river systems should be predictable. The occurrence of sedimentary basins and deltas and submarine canyons should also be predictable (Potter, 1978). However, although deltas have been classified by tectonic regime (Inman and Nordstrom, 1971; Audley-Charles, Curray, and Evans, 1977) and by the dominant energy process (Galloway, 1975), they have not been analyzed as thoroughly by climate.

Of the 60 major river systems now extant, 20 drain into the Atlantic, 9 into the Pacific, 9 into the Indian, and 7 into the Arctic Ocean. The remainder are spread among the Black (4), Baltic (3), Aral (2), Caspian (2), and Mediterranean (1) seas, and Hudson Bay (3).

The high relief along the leading (Pacific) margin of both North and South America results in most of the drainage flowing toward the opposite coast. The dominance of the Atlantic as the terminus for draining rivers seems reflected in the large amounts of sediments which it receives (Olausson, 1969). Rates of sedimentation were two and a half to four times higher in the Atlantic than in the Pacific. This general difference is also reflected in suspended sediment (Emery and Milliman, 1978). Both noncombustible matter in surface waters and discharge of solid sediment are larger along western sides of oceans than along eastern sides. Combustible matter, however, is highest in upwelling regions of eastern sides of oceans.

The tectonic placement of ancient basins is as important in predicting the rate of accumulation (Schwab, 1976) as is the comparison of an opening or closing ocean. As shown in Figure 6-14, the rate of sediment accumulation is markedly lower in basins on cratons than in any other type of basin.

(2) The climate of a region influences the runoff volume and sediment concentration of rivers, and thus the sediment yield. Six climatic regimes are broadly defined by temperature and precipitation, three of which are nonseasonal—glacial, desert, and selva (= dominance of year-round tropical air masses); and three of which are seasonal—Mediterranean (wet winter, dry summer), continental (warm moist summer, and cool moist or cold dry winter), and tropical wet-dry (wet summer, dry winter).

The *amount* of sediment yield appears to be very different for these six different regimes. However, well-established sediment yield–precipitation curves are known for only one climatic regime (continental), and are generally known for only two others (Mediterranean and tropical wet-dry) (Wilson, 1973:341). For the continental regime, the maximum sediment yield occurs at about 70 cm mean annual precipitation, and declines with both lesser and greater amounts of rain (peak *B*, Figure 6-15). However, when values from small basins developed on loess in regions not now subject to agricultural use are considered, the maximum yield occurs at about 4 cm mean annual precipitation (peak *A*, Figure 6-15). This

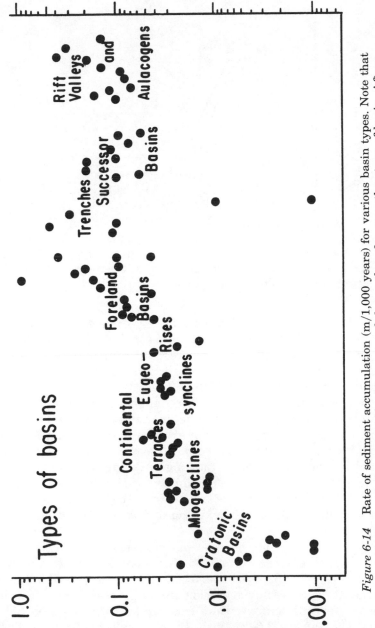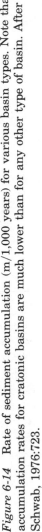

Figure 6-14 Rate of sediment accumulation (m/1,000 years) for various basin types. Note that accumulation rates for cratonic basins are much lower than for any other type of basin. After Schwab, 1976:723.

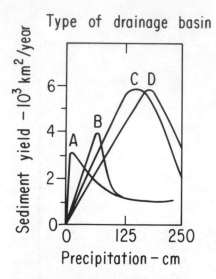

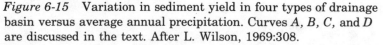

Figure 6-15 Variation in sediment yield in four types of drainage basin versus average annual precipitation. Curves *A, B, C,* and *D* are discussed in the text. After L. Wilson, 1969:308.

smaller figure is presumably more closely correct for preman rates of sediment yield. In both cases the reason for the general shape of the curve is that initially the increasing rain increases sediment load, but that after a point, increased rain merely aides the growth of grasslands and forests, which stabilize the soil, thus decreasing sediment load (this is known as the Langbein-Schumm Rule).

For the Mediterranean regime, maximum sediment yield is at about 130 to 165 cm mean annual precipitation (peak *C,* Figure 6-15). Possibly the high yield from the Mediterranean climate is caused by the wet season coinciding with the nongrowing season (winter), thus decreasing the possibility for plant growth to stabilize the soil. For the tropical wet-dry regime, the peak sediment yield occurs at about 180 to 190 cm mean annual precipitation (peak *D,* Figure 6-15).

Few data exist for nonseasonal climates, but relatively little erosion appears to occur under conditions of continuous rain-forest cover (selva) (p. 196), or no rain at all (deserts and glacial regimes). In general, "rates of fluvial erosion are far greater from areas with a strongly seasonal climate than from areas with non-seasonal climates" (Wilson, 1973:342). In the tropics, the rate of chemical denudation increases approximately as the two-thirds power of runoff, independently of the type of silicate rock being weathered (over the range of 0 to 200 cm mean annual runoff; Dunne, 1978).

During periods of widespread epicontinental seas, a maritime climate would have been more widely developed than it is today. Our present marginal seas enhance continental climatic extremes. The difference between continental and maritime climates is well displayed by the much greater temperature differential between summer and winter for the North American east coast ($\approx 5°$ to $10°C$, continental) and west coast ($\approx 1°$ to $3°C$, maritime) (Fleming and Elliott, 1952). Although much smaller than epicontinental seas, even large irrigated areas may influence climate by recycling water as rainfall (Stidd, 1975; but disputed by Fowler and Helvey, 1975).

The total drainage of a region is also a function of climate. As indicated in Figure 6-13, the dry continents of Africa and Australia have a much lower cumulative river length than the wet continents, and this is reflected in total stream discharge (Garrels and Mackenzie, 1971a). In general, the percentage of semiarid, arid, and extreme arid climate in Australia is 83 percent, in Africa 64 percent, in Asia 39 percent, in South America 17 percent, in North America 16 percent, and in Europe 1 percent (Meigs, 1952). Among the wet continents, the rate of chemical denudation is roughly independent of elevation, but the rate of mechanical denudation increases with increased average gradient (Figure 6-16), perhaps exponentially.

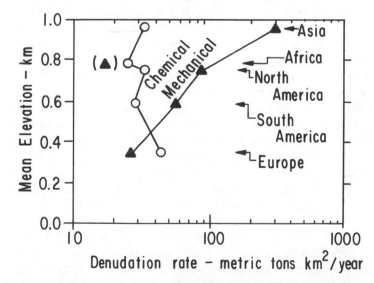

Figure 6-16 Chemical denudation is independent of continental elevation, whereas mechanical denudation generally increases with increasing elevation. Africa shows abnormally low values for both types of denudation, even though Africa stands abnormally high (Figure 1-5). After Garrels and Mackenzie, 1971a:122.

Considering all land areas, the rate of mechanical erosion exceeds the rate of chemical erosion by a factor of approximately 6 (Holland, 1978:145).

Marine coastal erosion is concentrated on cliffed coasts. The annual product of suspended matter from sea-cliff erosion is on the order of 10^6 metric tons, compared with 10^9 metric tons from major rivers (75 percent solids, 25 percent dissolved) and something greatly in excess of 10^7 metric tons windblown from large deserts (Emery and Milliman, 1978). The world average erosion rate is close to 1 cm/10^3 years (Holland, 1978:144). However, there is considerable variation from region to region (Menard, 1961), and the rate reaches 40 cm/10^3 years on volcanic islands (Vogt, 1974).

High sediment yield has been singled out as the most important characteristic leading to the formation of submarine canyons (Il'in and Lisitzin, 1968), and climate is a strong component. Zones of high sedimentation (30 to 100 cm in the Recent) are much wider in humid than in arid regions, and submarine canyon concentration "is greatest in precisely these places."

(3) Rates of erosion during the Precambrian have been suggested to have been (a) much lower than present values, (b) about the same, and (c) much higher.

(a) The view of low rates of erosion during the Precambrian was based on the belief that the earth was much cooler and without any plant cover (and resulting organic acids), and had *lower* amounts of CO_2 in the atmosphere than now (Donn, Donn, and Valentine, 1965). Present views of the Precambrian are that, if anything, it was warmer than the present, and CO_2 was probably a major component of the atmosphere (p. 169).

(b) Most authors consider that rates of erosion have been rather uniform over geologic time. In order to test this, the rate of chemical weathering was measured in Iceland in regions devoid of cover by higher plants. Rates of chemical weathering were two to three times lower than in places with plants. However, a CO_2 pressure of "no more than five times the present value would have been needed to produce present-day mean rates of chemical weathering under comparable climatic and geographic conditions before the development of higher land plants" (Cawley, Burruss, and Holland, 1969:392). In further support of the view of similar intensity of chemical weathering in Precambrian and modern times, Holland (1972:648) wrote that the mineralogy and chemistry of shales 3.2 to 3.4 b.y. old are "virtually indistinguishable" from those of shales 0.5 b.y. old. In addition, the proportion of inorganic carbon to organic carbon in limestones has been constant for 3.3 b.y. (Schidlowski, Eichmann, and

Junge, 1975). As Holland (1976) stated, if the rate of erosion had been higher in the Precambrian, because of markedly higher CO_2, then Mg and Ca demand for CO_2 should have been higher, and the $\delta^{13}C$ of limestones would have changed through time to reflect the changing sources of carbon. A changing $\delta^{13}C$ is not observed, and so a subsequent slackening of Mg and Ca demand for CO_2 is not supported. If the demand for CO_2 has been uniform through time, this suggests that rates of erosion did not change through time.

However, even if the rate of erosion was approximately the same in Precambrian days as now, the absence of land plants prior to the Devonian may have had (at least) two influences. First, in the absence of vegetation, soils as we know them would not develop, and erosion would act more directly on bedrock. Lepp and Goldich (1959) considered that "in Precambrian times weathering akin to laterization retained aluminum, titanium, and phosphorus in the regolith, but unlike recent laterization processes, the Precambrian processes also supplied iron and silica, alkaline earths, and alkalies to the ground waters." This rock may have been partly chemically eroded by fungi (which have been reported from the Precambrian), which attack basic igneous and metamorphic rocks. In laboratory experiments, fungal attack "markedly increased the quantity of Si, Al, Fe, and Mg in solutions when compared with uninoculated controls" (Silverman and Munoz, 1970). Lichens also significantly chemically degrade basalt (Jackson and Keller, 1970). This evidence is in agreement with the view that the mid-Paleozoic development of land plants might not have significantly altered rates of erosion. In contrast, Holland (1978:51) emphasized that decaying organic matter increases soil CO_2 pressure by 10 to 100 times the atmosphere pressure. In the absence of this organic matter, and without a higher atmospheric CO_2 pressure, chemical erosion would probably have been markedly less than modern values.

Second, the chief factor which certainly was different from modern times that might have contributed to a higher sediment load before the Devonian was the lack of a binding agent for soils. Possibly this absence of a binding agent accounts for the nonrecognition of a distinctive shoreline facies in the otherwise typical tidal flat and lagoonal succession ("water depth never becoming more than 8-10 m") of the 2,000 m thick, late Precambrian quartzite-carbonate rocks of central India (Singh, 1976). The influence of tides in redistributing sediments over vast areas which lacked a well-developed beach may have been the rule and not the exception prior to the land plants (see p. 82 for review of the Ranns of Kutch, a possible modern analogue). Sediment yield presumably would have been linearly related to precipitation over a much wider range of values than now is encountered. Possibly a more pronounced effect of floods and braided rivers should be anticipated, resulting in episodes of cyclic sedimentation (Schumm, 1968).

In the absence of land plants, winds also may have been a more potent agent for sediment erosion and transport. Evidence for this resides in extensive late Precambrian rounded quartz *and* feldspar sandstones (Chaudhuri, 1977).

(c) Prior to the widespread acceptance of the concept of sediment recycling, the prevalent view of rates of sediment yield was that "if rates of deposition for past ages are judged by the apparent maximum thickness of sediment per unit time for each of those ages, there appears to be little question that the rate of sedimentation has increased exponentially with time" (Nanz, 1953). There is so much more Tertiary sediment than Mesozoic, and Mesozoic than Paleozoic, etc., that a historical, non-steady-state view could yield no other conclusion than that the rate of erosion was increasing through time. Rates of erosion also are often interpreted in terms of extent of continentality (discussed in Chapter 1). However, in the absence of any clear secular trends in continentality, acceptance of suggestions of covarying trends in erosion must wait.

For shallow-water sediments, rates of deposition follow two smooth curves: one for the late Precambrian through the Devonian and one for the Carboniferous to the present, as shown in Figure 6-17. Taken at face value, these curves indicate that rates of denudation before the Carboniferous were about four times their subsequent value (Gregor, 1968). Possible this reduction in rate of denudation was caused by the development of extensive vegetation from the late Devonian onward. Alternatively, the same disparity in sediment volume could result from a much lower probability of destruction of the group of older sediments, perhaps owing to the initiation of new major tectonic cycles of erosion and deposition (Garrels and Mackenzie, 1971b). Preman "natural" rates of erosion and deposition appear to have been approximately the same from the Carboniferous to the present for the continental regions.

Rates of deposition in the deep sea for the Paleocene and for the late Eocene to early Miocene are only 10 to 25 percent of anticipated steady-state values (Davies et al., 1977; Donnelly, 1977). Certainly, some portion of that deficiency must be part of any natural variance about a mean value. A full explanation for perturbations by a factor of four or five, however, cannot be given, but in part may reside in the fact that much of the time represented in the deep sea is recorded nondeposition (possibly owing to erosion) rather than simply

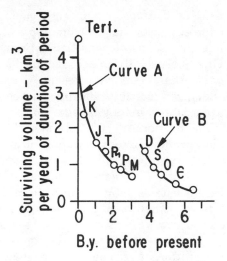

Figure 6-17 Survival rate of sediments as a function of their age. Points and letters are for geologic periods. The surviving volume of each geologic system is expressed in km³ per year of duration of geologic period. For curve A, $s = 2.1 \, (e^{-2.0t} + {}^{-0.37t})$. For curve B, $s = 1.4$ $(e^{1.7 \, (t-3)} + e^{-0.44 \, (t-3)})$. Note that the earliest five points (late Precambrian through Devonian) have a decay rate different from later data, as though there have been two major cycles of Phanerozoic continental denudation. From Gregor, 1970:274.

a very low continuous rate of deposition. In the South Atlantic, a third of the time is represented by hiatuses (Van Andel et al., 1977). From the point of view of deciphering climatic changes of duration as short as 10^3 or 10^4 years, problems caused by vertical mixing of faunas, ash layers, etc. are serious if hiatuses are common. Climatic events probably cannot be resolved over large areas of the Pacific and Indian oceans, where rates of sedimentation are less than 1 cm/10^3 years, but they are optimally resolved in the North Atlantic, where rates of sedimentation are more than 3 cm/10^3 years (Ruddiman, 1977:119).

The present rate of denudation appears to have been approximately doubled to tripled by man, chiefly by turning forested land into crop land, and then into urban area (Meade, 1969; Meade and Trimble, 1974). In the 48 contiguous United States, 4 billion tons of sediments are carried by surface runoff into waterways each year, and 3 billion tons of it are from farmland; an additional billion tons is lost from farmland by wind erosion (Pimentel et al., 1976:150).

In summary, there is no firm evidence for either markedly

higher or markedly lower rates of denudation through geologic time, but circumstantial evidence suggests that rates may possibly have been measurably higher in pre-Carboniferous times, prior to the evolution of land plants. This is especially so because mechanical erosion is much more effective than chemical erosion. At present, rates of sedimentation for deep-sea red clays and siliceous oozes are on the order of a few mm/10^3 years (= m/m.y.), being highest in the tropical Atlantic (summarized by Berger, 1974; Bandy, 1972). Equivalent rates of accumulation occur in Ordovician and Silurian graptolitic shales (of probable deep-basin origin) (Churkin, Carter, and Johnson, 1977). In nearshore basins, like those off southern California, or in deltas, sedimentation rates are on the order of a few m/10^3 years (= 100's m/m.y.). Rates of sedimentation on continental shelves or in deep-sea calcareous oozes are intermediate between these extremes. Vastly different amounts of sediment are deposited during the same time interval in different regions. On continental shelves, lower sedimentation rates may occur during a transgression (as is going on today) and higher values during regressions. Considering overall sedimentation rates of the continents and shelves, Barth (1968) accepted as a "reasonable figure" 3 kg/cm^2/m.y. For present land area, Barth calculated that this would mean erosion of 150 km over 4 b.y., a figure which "simply proves that extensive recycling must have taken place," since erosion has obviously never been that extensive.

Storms

Why be concerned with rare events like hurricanes, severe floods and other "abnormal" happenings which occur once a year, a decade, or a century? The answer is that these unusual events, rather than the much lower energy day-to-day events, may be responsible for imparting the final character to deposits (Wolman and Miller, 1960; Baker, 1977). For example, more than 90 percent of the load of the Mississippi River is transported during less than 10 percent of the time. The major deposits of clay and silt in the Gulf of Mexico are probably the result of conditions prevailing only very rarely. As another example, in a one-week period following Hurricane Agnes (1972), more sediment was discharged from the Susquehanna River into Chesapeake Bay than had been transported during the past several decades (Schubel, 1974). The reason for the importance of these rare storms is that sediment load is the *product* of the streamflow and the concentration (Meade, 1976), and the elevation of *both* of these at the same time rarely occurs. And as a biological example,

85 percent of each year's diatom biomass off the coast of southern California results from three major blooms which coincide with upwelling (Tont, 1976). Thus short-duration climatic fluctuations control the bulk of sediment input.

The geographic location and time of occurrence of severe storms is not random. For tropical storms, both the spatial distribution (Figure 6-18) and summation by latitude of the place of initiation (Figure 6-19) indicate a strong preference for 10° to 20° latitude. These latitudinal zones correspond to the poleward sides of the doldrum equatorial trough, and they alternate north and south of the equator depending upon the season (Figure 6-20).

As these storms migrate away from their place of origin, their major effect also is restricted within latitudinal belts (Figure 6-21). For the period 1900 to 1963, 86 percent of the North Atlantic storms are recorded from 15° to 40°N (mean of 28°N). Similarly, 84 percent of Western North Pacific storms occurred from 10° to 35°N (mean of 23°N).

The lunar tides have been considered to enhance preexisting meteorological disturbances in the atmosphere (Figure 6-22). Data from both hemispheres on the time of development of 1,013 hurricanes and typhoons and 2,481 tropical storms exist over a 78-year period. The time of the new moon (and less clearly the full moon) is

Disturbances which became tropical storms

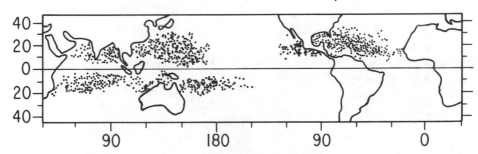

Figure 6-18 Location of first detection of disturbances which later became tropical storms. Tropical storms are defined as warm-core cyclonically rotating wind systems in which the maximum sustained winds are 35 knots (i.e., 64 km/hour) or greater. Hurricanes, typhoons, and cyclones (Southern Hemisphere) are also included in this definition. Note that tropical storms do not occur in the southwest Atlantic or the north central Pacific. Data ranges from 20 to 70 years of observation depending on the region. After Gray, 1968:670.

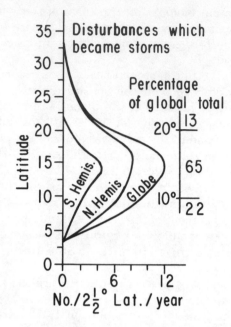

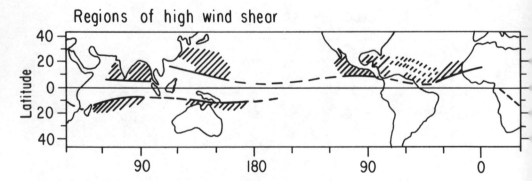

Figure 6-19 Latitude of first detection of disturbances which later became tropical storms. After Gray, 1968:672.

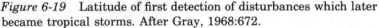

Figure 6-20 Hatched area of high wind shear (between 200 and 800 mb, shear is ±10 kt or less) is on the poleward side of doldrum equatorial troughs during August for the Northern Hemisphere and January for the Southern Hemisphere. Note that the hatched areas agree well with the areas of tropical storm development (Figure 6-18) except for the North Atlantic, where storm development can occur in the region of the dashed lines (and intensify disturbances generated just to the east). After Gray, 1968:681.

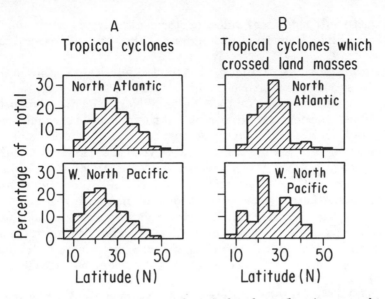

Figure 6-21 A. Occurrence of tropical cyclones (hurricanes and typhoons) arranged by latitude in the North Atlantic for the period 1900–1963 and in the Western North Pacific for the period of 1958–1962. The North Atlantic median is 28°N and the Western North Pacific median is 23°N. B. Latitudes in which tropical cyclones crossed major land masses in the North Atlantic during 1900 to 1963 and the Western North Pacific during 1958 to 1962. After Hayes, 1967a:5.

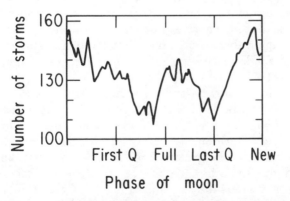

Figure 6-22 Number of storms (1,013 total) with respect to the phase of the moon. The curve is a smoothed, running mean. After Carpenter, Holle, and Fernandez-Partagas, 1972:459.

correlated with a greater number of storms than is any other time in the lunar cycle. Rainfall is also highest at the times of new and full moon.

How rare is rare? In the North Atlantic, hurricanes occur on the average about once a year between 25° and 30°N, their latitude of commonest occurrence. In the Western North Pacific, typhoons occur about four to five times per year at their latitude of commonest occurrence (20° to 25°N). Over the past 30 years, the duration and frequency of storms generating high waves (and the length of the winter storm season) have increased (Hayden, 1975). Ball, Shinn, and Stockman (1967) described the effects of Hurricane Donna (1960) in southern Florida and estimated that such storms come through on the average about 16/century, or approximately 500,000 such storms during the 3 m.y. of the "Pleistocene." Within the general group of items called storms, however, a very severe storm may have much more significance than several moderate storms (Smith and Hopkins, 1972).

Tidal currents that are amplified by storm waves may move sediment at depths and velocities far in excess of normal conditions (Shepard and Marshall, 1973). At 6 m depth off the Gulf Coast, water normally moves over sediment at about 10 cm/sec, but during Hurricane Camile (August 15 to 17, 1969) velocities reached 160 cm/sec (Murray, 1970). For the shelf off southwestern England, the length of time during the year when maximum particle speeds exceed various values has been calculated (Figure 6-23). Values which occur only 1 percent of the time at 90 m occur 40 percent of the time at 30 m. Similar calculations have been made for the coast off southern Texas. The effect of rare, high-current velocities is amplified on a shelf which is incised by canyons that channel the currents by wave refraction. Oscillatory motion apparently related to tidal movements causes ripple formation and other sand movement to depths of 200 m off England (Ewing, 1973) and Oregon (Komar, Neudeck, and Kulm, 1972).

Independent of tides, storm winds are important in causing water to transport sediments. Dott (1974) estimated a water depth in Cambrian seas of about 50 m within 500 m of shore, based on inclinations in sandstone of former islands. Quartzite boulders adjacent to the islands indicated hurricane-force storms with breakers "at least 7 to 8 m high." Even in the absence of extreme hurricane conditions, the effect of winds may be substantial. In water of 3 to 10 m depth, waves with a length of 60 m, and a period of 10 sec can be sustained (King, 1959:13). This is certainly not the "low-energy" nearshore picture created in the usual epeiric sea model (Chapter 2).

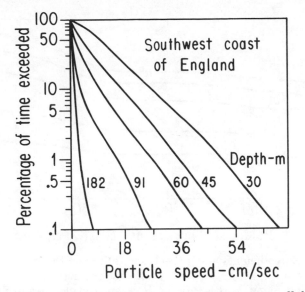

Figure 6-23 Particle speed near Sevenstones, off the southwest coast of England. Particle speed is the highest momentary speed at the sea bed attained by a particle during the passage of each wave equal in height to the significant height, which is the average height of the highest one-third of all waves present at the time. The percentage of the time during which particle speeds exceed any given speed is shown for 30, 45, 60, 91, and 182 m. For example, near Sevenstones at a depth of 60 m, particle speeds of 27 cm/sec should be exceeded for 1 percent of the time, or for about three or four days per year. After Draper, 1967:137.

In shallow water, the effect of the storm surge is the geologically most significant result of major storms. Water levels may be several meters higher than normal high tides (Rossiter, 1954). Surge heights recorded once in 100 years will be three times that recorded, on the average, for any 5-year interval (Figure 6-24). Hayes (1967a) described the effect of the storm surge of Hurricane Carla (1961) on the inner neritic zone: "As the storm moved landward, various bottom materials (rock fragments, macro-invertebrates, coral blocks, etc.) from depths as great as 50 to 80 feet were transported to the beach and scattered over the barrier-island complex and wind-tidal flats. After the storm passed inland and the storm surge began to recede, strong currents spilled out of the numerous hurricane channels that were cut through the barrier island during the storm-surge flood. These currents, which presumably

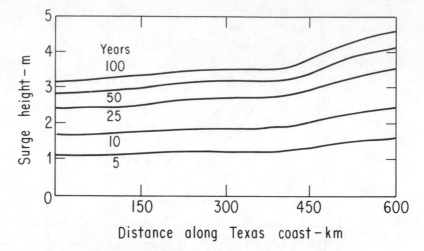

Figure 6-24 Surge heights along the Texas Gulf coast that could be expected every 5, 10, 25, 50, and 100 years. The transect covers Port Isabel, Mustang Island, Port O'Connor, Freeport, Galveston, and Sabina. After Bodine, 1969:22.

evolved into density currents, spread a 'pure' sand layer (1 to 3 centimeters thick) over formerly homogeneous mud bottom in depths of 48–60 feet and spread a graded bed, or turbidite (up to 9.0 centimeters thick) of fine sand, silt, and clay farther out on the shelf." Further inland in the barrier island complex, faunal assemblages were thoroughly mixed, and a pavement of the heaviest shells was left on the surface. Yet another significant effect of storms is to increase concentrations of noncombustible matter in suspension over the inner shelf (Meade et al., 1975).

Several general criteria may be used for geologic recognition of storms. Basically, a storm redistributes materials so that sediments which could be eroded or transported only in very high currents are juxtaposed to background material which has been transported in low currents. Seven examples follow. (1) Coarse blocks or coarse sands are transported to regions with normally fine sediments, as when gravel sheets were added over soils to islets on Jaluit Atoll in the Marshall Islands owing to Typhoon Ophelia (January 7, 1958) (Blumenstock, 1958; McKee, 1959). (2) Large ripples with coarse sediments occur in regions that normally have reduced tidal activity, as off North Carolina at 30 to 50 m (Macintyre and Pilkey, 1969), off Plymouth, England at approximately 50 m, and off Oregon at 200 m (Komar, Neudeck, and Kulm, 1972). (3) Marine faunas are juxtaposed to terrestrial or brackish-water forms, of which nearly all of the shells are broken. (4) Unusual amounts of fine sediment are carried from tidal flats seaward and deposited on sands. Bedding that may result from mud droppings over sand closely resembles (or is identical to?) "flaser" bedding (p. 79). McCave (1970) argued that under normal conditions not enough fine sediment is transported by the retreating tide to account for the significant amount of clay in flaser bedding. In addition, not enough time may exist during slack water for whatever fine sediment that is present to be deposited. Accordingly, McCave suggested that large amounts of fine sediment must result from inundations that occur during storms, and that flaser bedding indicates storm conditions first, and intertidal conditions second. (5) Unusual amounts of sand are carried seaward and deposited over silts and clays. These thin sand layers have been called "rhythmites" in sediments north of Germany (Reineck, Gutmann, and Hertweck, 1967). (6) Supratidal flats are by definition too high to be washed by tides, yet they lack vegetation and terrestrial life, apparently because the life is killed by salt water during the storms. Such areas may not have a chance to reestablish themselves before additional storms occur. Sediment in such areas is therefore deposited during storm conditions, as by Hurricane Donna in the Florida Keys (September 9, 10, 1960) (Ball, Shinn, and Stockman, 1967), or by monsoon conditions. (7) An erosional basal contact is followed by coarse (pebble or shell) debris (Wright, 1974).

Although storms account for considerable sediment transport, the ultimate importance is slight if the effects are rapidly brought back to "normal." In the New England hurricane of 1938, 6 to 12 m of many cliffs were eroded in three to four hours (Brown, 1939), and this cannot be repaired. But inlets cut into lagoons behind beaches by storms may become clogged and filled in time. Beach ridges established by high water during storms will be broken down, and the rubble spread out (Blumenstock, Fosberg, and Johnson, 1961). Immense washover deposits caused by storms (e.g., Hurricane Belle, 1976) are often transported back to the beach six months later (Fisher and Stauble, 1977). Offshore, storm ripples in deep water may be modified even if they are out of equilibrium with local hydrology. Finally, fine sediment transported seaward by storms will not be returned to tidal flats, although large amounts of it are trapped and returned to estuaries (Meade, 1972, 1974; Nichols, 1977).

Ancient deposits attributed to storm conditions include those of the Precambrian (Hobday and Reading, 1972) and of every geologic period of the Phanerozoic. To the extent that storms were concentrated in the lunar new-moon and full-moon periods, a greater frequency of severe storms would be expected, with the stronger tidal forces of earlier geological time. During the Precambrian, tidal forces and storms may have been more pronounced, but subsequent storm deposits of the Phanerozoic do not appear to have been any more extensive than they are now.

Summary

The purpose of this chapter has been to summarize three climatologically important aspects of paleoceanography. (1) Because of latitudinal zonation of rainfall, temperature, and biologic productivity, several types of sediments of both the continental shelf and the deep sea are roughly latitudinally distributed (mud, sand, gravel, tillite, evaporite, dolomite, phosphate, carbonate, and wind-blown quartz and clay minerals). Attempts have been made to couple shelf and deep-sea patterns, and as a general guide, times of high stands of sea level are related to low deep-sea depositional rates (and vice versa). In paleoclimatic interpretations, perhaps the most important single decision is the identification of the tropical humid zone and its flanking subtropical arid zones.

(2) Judged over geologic time, there appear to be large regional differences in sediment yield that are only partly related to precipitation and temperature. The total river runoff and the rate of mechanical denudation may be predicted from the climatic regime and topography. The highest sediment yield is not predictable from climate alone, and at present, the largest amount of sediment is from regions with extensive Pleistocene loess deposits and very high elevations. The mainstream length of rivers and the number of major river systems (on a world basis) should be predictable from climate alone. Modern weathering regimes may have been in existence for about 3 b.y.

(3) Storms may be responsible for most sediment transport, in only a small amount of time, and are of very great geologic significance.

7. BIOLOGY

Biogeography cannot confine itself simply to describing the occurrence of living forms, arranging them regionally, investigating the ecological causes of distribution. It must also proceed historically.

S. Ekman, 1953

Biology in relation to paleoceanography (and as dealt with in previous chapters) is important in recycling of elements by phytoplankton (e.g., diatoms), as a tidal indicator (e.g., blue-green algae in stromatolites), as a marker of oceanic currents (e.g., foraminifera), and as an indicator of bathymetry, salinity, temperature, and climate (each species has its own range). In short, biology has already entered into every previous chapter. In this present chapter I will consider three general topics: productivity, patterns of taxonomic change, and biogeography.

Productivity

I will consider two ways in which productivity through geologic time can be estimated: (1) by comparing the carbon content of shales over time and (2) by examining $\delta^{13}C$ ratios. In both cases, the conclusion is reached that organic carbon production has varied by about a factor of 2 over time periods of 10 to 100 m.y. In addition, from 3.3 b.y. ago to the present, the total production of organic carbon per unit time has increased by about a factor of 10.

(1) The production of organic carbon through geologic time can be related to the amount of organic carbon buried in sediments (R. M. Garrels, personal communication, 1975). The basic relationship is that the (total number of grams/cm²/year produced) times (percentage of that which is produced which is buried) should be equal to (fraction in sediments in grams C/cm³ sediment) times (rate of sediment deposition in cm/year) minus (diagenetic loss of C).

To apply this method of determining organic carbon production in a comparative way, sedimentation rates, loss to oxidation, and diagenetic loss must be approximately equivalent and other systematic biases must be absent. In making comparisons through time one should compare sediments (and rocks) of the same general type.

Data are available on carbon production, the fraction of carbon in sediments and rocks, and the rate of sediment deposition. We can use these values to calculate the percentage of that which is produced which is buried. Table 7-1 cites the percentage C which is buried in environments yielding most of the shales of the geologic record, and exclusive of "special" circumstances, such as upwelling. The highest percentage burial ($\simeq$7.0 percent) is in deltas, next highest in the continental slope ($\simeq$1.0 percent), and the lowest percentages are in the abyssal plain and the continental shelf. The highest percentages of organic carbon are related to places of highest rates of sedimentation, a factor which effectively removes the carbon from an oxygenated environment. Organic carbon content increases by about a factor of 2 for a 16-fold increase in sedimentation rate (Müller and Suess, 1979).

Carbon production in ancient times can be evaluated by examining percentage carbon in shales of different geologic age. The appropriate modern analogue for shales may be either continental slopes or deltas, or both! I will use both comparisons.

Table 7-1 Calculation of percentage burial of modern organic carbon in different environments. Numbers in the third column will be increased by the average density of the sediment ($\simeq$2.7), and decreased by the final porosity ($\simeq$1–0.5), to give figures of the same order as the initial fraction of sediments, and these corrections have not been made here.

Environment	Carbon Production—gross production[a] (g C/cm²/yr)	Fraction in sediments (g C/cm³)	Rate of deposition[b] (cm/yr)	Calculated % burial of organic C
Deltas, estuaries, lagoons	0.0194[c]	1.3[d]	0.100	6.7
Shelf	0.0194	0.12[d]	0.002	0.012
Slope	0.0183[e]	0.7[d]	0.030	1.12
Abyssal plain	0.0033	0.3[f]	0.003	0.27

a. Values from Emery and Uchupi, 1972:305–306.

b. Values from Emery and Uchupi, 1972:417–418.

c. Assumed the same as the shelf.

d. Values of organic carbon from Emery and Uchupi, 1972:Table 18 (A, C–E, and H–I), p. 384.

e. Assumed that 80% of the organic C comes from shelf photosynthesis via slumps, etc., and 20% from the direct contribution of overlying photosynthesis.

f. From Sanders, Hessler, and Hampson, 1965:Table 2, Stations MM, NN, OO.

The organic carbon content of ancient shales is known, and let us assume for the moment that the percentage burial is the same through geologic time. Using the relationship expressed on the previous page, we calculate carbon production for times in the past. The results of this calculation are shown in Table 7-2 and are plotted in Figure 7-1. The general conclusion of Figure 7-1 is that carbon production may have increased about a factor of 4 to 8 *since* the Archean (the Archean is discussed below). If deltas are the best environmental counterpart to ancient shales (curve *B*), then mean values have been approximately 100 g C/m²/year, with excursions as high as 200.

The Precambrian may have been anoxic for a considerable part of its duration (Chapter 5), in which case the percentage burial of organic carbon may have been higher than it is today. Conceivably as much as 50 percent of that which was produced could have been buried. The absolute amount of organic carbon in Archean rocks is higher than in Proterozoic rocks (Table 7-2). If a much larger percentage of organic carbon has been buried in the Archean than was buried subsequently, this would account for the otherwise anomalously high Archean values.

Figure 7-2 shows how organic matter would have increased through time if the percentage burial of organic carbon was much higher in the past than at present. In general, present organic carbon production per unit time would be on the order of 50 times greater than during the early Precambrian if 50 percent burial obtained and if we use the 1.0 percent value for modern sediment burial.

This estimate of a 50-fold increase in organic carbon production from anoxic Precambrian times to the present seems to be an upper limit for two reasons.

(a) If more preserved sediment had its origin with deltas than with any other environmental regime, than a percentage burial closer to 10 percent than 1 percent is more reasonable. Precambrian anoxic environments need then have buried only 5 times present amounts of organic carbon in order for constant amounts to be found in sediments of all ages.

(b) It is not at all certain that one should expect 50 percent burial of carbon under anoxic conditions. No large increase is observed in organic carbon in anoxic water columns (Richards, 1971 and earlier). In the sediments of anoxic basins, values of only 4 to 8 percent organic carbon are characteristic (Richards, 1970); even under exceptional conditions, muds only as high in organic carbon as 14 percent have been reported (Piper and Codispoti, 1975). In the Black

Table 7-2 Calculated production of organic carbon for clays and shales of different ages. For paired values, the upper value refers to the continental slope, and the lower value to deltas.

Age of rocks	Percentage buried[a]	Fraction in rocks[b] Russian Platform	Mean	North America	Rate of deposition[a] (cm/yr)	Calculated production of organic carbon g C/cm²/yr	g C/m²/yr
>2.4 b.y.	1.12 / 6.7	—		0.74	0.030 / 0.10	0.020 / 0.011	200 / 110
2.4–1.7 b.y.	1.12 / 6.7	0.17		—	0.030 / 0.10	0.0045 / 0.0025	45 / 25
1.7–0.8 b.y.	1.12 / 6.7	0.34		—	0.030 / 0.10	0.0091 / 0.0051	91 / 51
0.8–0.225 b.y.	1.12 / 6.7	1.04	0.88	0.72	0.030 / 0.10	0.024 / 0.013	240 / 130
0.225–0.0 b.y.	1.12 / 6.7	0.90	1.21	1.52	0.030 / 0.10	0.032 / 0.018	320 / 180
Modern						0.0188 / 0.0194	188 / 194

a. See Table 7-1.
b. Archean (>2.4) from Cameron and Baumann (1972); other values from Ronov (1968:40).

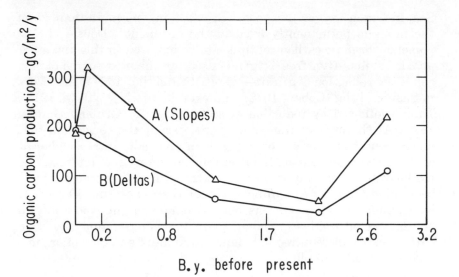

Figure 7-1   Relationship between organic carbon production and geologic time for two contrasting situations. *A*. Ancient shales represent modern continental slope deposits. *B*. Ancient shales represent modern delta deposits.

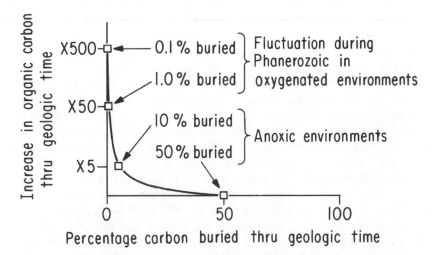

Figure 7-2   The relationship between percentage carbon buried through geologic time and the increase in organic matter through geologic time which is required in order to maintain a "constant" 0.5 to 1.0 percent organic carbon in shales. For the Precambrian, the most probable figure is less than 10 percent buried.

Sea, in which oxygen disappears at $\simeq 125$ m, "no more than 4% of the input is permanently fixed in the sediment" (Deuser, 1971). Somehow organic carbon continues to be oxidized in this and other anoxic basins. After free oxygen is used up, the next source of oxygen is the reduction of nitrate—to nitrite and then to N_2O, as occurs in Saanich Inlet (Cohen, 1978), and even just to N_2 (Wilson, 1978). This is followed by reduction of sulfate to H_2S (Orr and Gaines, 1973). At the present time, one cannot escape the conclusion that anoxic conditions per se do *not* result in markedly increased *burial* of organic carbon. During the Precambrian interval of interest, I do not know if nitrate was available, but the sulfate sink seems to have been filled (p. 173). Accordingly, rather than expecting 50 percent burial under anoxic conditions, one reasonably might expect only 10 percent burial (or less). The increase in biomass from anoxic to "oxic" conditions required to balance equivalent amounts of organic carbon in shales would be correspondingly lowered from a factor of 5 to present values.

Anoxic metabolism has a much lower energy yield than does oxic metabolism. Thus the same average biomass would turn over much less carbon in an anoxic era than in an oxic era. A reduced production in an anoxic environment should cause Precambrian rocks to appear to have very reduced biomass if burial is roughly proportional to production. However, the fact that the difference between anoxic and oxic eras is very small is in agreement with the idea that Precambrian and Phanerozoic carbon production were not very different from each other. Is it possible that much of the "anoxic" Precambrian may have been oxygenated (see also p. 175)?

We can now place a lower limit on the amount of organic production in going from Precambrian anoxic to present oxygenated conditions. With 10 percent burial of organic carbon in sediments in anoxic conditions, and 1 percent burial under oxygenated conditions, one would have to allow for a tenfold increase in biomass in order for there to be a constant organic carbon content of ancient and modern shales. This seems to me to be a very reasonable value.

A scenario for changes in organic carbon production and preservation through geologic time is as follows. The somewhat higher values of organic carbon in Archean rocks could reflect a higher percentage of burial in anoxic Archean times. The subsequent drop in percentage buried in the lower Proterozoic might reflect the beginnings of an oxygenated atmosphere and water column. The gradual increase in organic carbon in rocks from the Proterozoic to the present would mirror a tenfold increase in organic carbon production.

(2) A second estimate of the variation in amount of organic car-

bon produced per unit time can be obtained from measurement of $\delta^{13}C$. The data are consistent with a variation in organic production by a factor of 2 over the Phanerozoic, as is discussed below. Presumably, most of the change in production of organic carbon occurred before the Phanerozoic during the time the atmosphere became oxygenated.

Isotopic values of ^{13}C from organic carbon are uniform over the past 3.0 b.y. at -25 to $-30‰$ (Figure 7-3), and then rapidly change at about 3.5 b.y. to approximately $-15‰$ (Oehler and Smith, 1977). The heavy values of organic carbon found in these 3.5 b.y. old rocks of both Australia and Greenland are characteristic of "primordial organic matter" in carbonaceous chondrites. Their occurrence in very old rocks "may mark the time of origin of biochemical mechanisms capable of fractionating carbon isotopes in a manner similar to that of modern autotrophs" (Oehler, Schopf, and Kvenvolden, 1972:1247). (Alternatively, the anomalously heavy values may be due to metamorphic thermal alteration; McKirdy and Powell, 1974.)

$\delta^{13}C$ values of modern blue-green algal mats are mostly -8 to $-16‰$ (Behrens and Frishman, 1971; Calder and Parker, 1973).

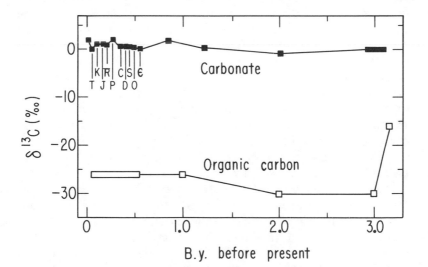

Figure 7-3 Average ^{13}C (‰, PDB standard) for limestones (upper curve) and for organic carbon (lower curve), by geologic age. Note that there is very little change with age except for the oldest organic carbon values. Limestone data from Veizer and Hoefs, 1976; Schidlowski, Eichmann and Junge, 1975; organic carbon data as summarized by Garrels and Perry, 1974.

This indicates that the earliest undisputed biologic carbon at 3.0 b.y. (with $\delta^{13}C = -25$ to -30) could not have been derived solely from blue-green algae in a modern atmosphere. Alternatively, values of $\delta^{13}C$ are further fractionated by -15 to $-19‰$ if atmospheric CO_2 is increased (Epstein, 1968), possibly from 0.3‰ (its modern value) to 3‰ (an ancient value?). The Precambrian $\delta^{13}C$ values from organic carbon are consistent with a higher CO_2 content in the early atmosphere (Garrels and Perry, 1974).

The $\delta^{13}C$ values for marine carbonates must be looked upon differently than the $\delta^{13}C$ values of organic carbon. The $\delta^{13}C$ values of carbonates (Figure 7-3) reflect the relative contributions of both organic carbon and of carbonate carbon in the total sedimentary carbonate reservoir (Broecker, 1970). $\delta^{13}C$ of carbonate is quite uniform for the past 3.3 b.y. (0‰ $\pm$ 2‰, as shown in Figure 7.3), and seemingly even for the past 3.8 b.y. (Schidlowski et al., 1979).

The $\delta^{13}C$ values of modern carbonates result from a fractionation of the original carbonate reservoir (at $\delta^{13}C \simeq -5‰$) by removal of 20 percent of the light ^{12}C and enrichment of the remaining carbonate to a heavier $\delta^{13}C$ value of $\simeq 0‰$. Thus organic carbon makes up approximately 20 percent of the reservoir of sedimentary carbon, and inorganic carbon approximately 80 percent. An excursion from 0‰ to $\pm 2‰$ would reflect a change of 50 percent in the removal of the lighter isotope (the one which is preferentiallly removed by phytoplankton). This leads to two conclusions.

(a) Schidlowski, Eichmann, and Junge (1975:1, 2) concluded that the $\delta^{13}C$ values are "just a measure of the gross amount of photosynthetic oxygen produced . . . As a whole, the carbon isotope data accrued provide evidence of an extremely early origin of life on Earth since the impact of organic carbon on the geochemical carbon cycle can be traced back to almost $3.5 \times 10^9 y$."

(b) The amount of organic carbon production has varied by ± 50 percent over the past 3.3 b.y. (Garrels and Perry, 1974).

The resolution of $\delta^{13}C$ is within tenths of 1 per mil, and therefore the technique is suitable for examining changes in the seawater fractionation for closely spaced carbonate samples through the Pleistocene and Tertiary. The preferential use of the light isotope (^{12}C) by phytoplankton (p. 124) and their (and its) subsequent burial in muds would raise the content of the heavy isotope (^{13}C) in the remaining sea water and in precipitated carbonates. Higher $\delta^{13}C$ values are therefore characteristic of limestone during periods of presumed high phytoplankton burial (and lower $\delta^{13}C$ values, as in the Triassic, represent a time of presumed low phytoplankton burial). Changes in values of carbon isotopes therefore may reflect changes in phytoplankton productivity and/or burial (Broecker, 1973). Alternatively, the value of atmospheric CO_2 input may not be constant, just as the changing Pleistocene ocean volume led to different initial values of $^{18}O/^{16}O$. In the case of CO_2, Shackleton (1977) emphasized the effect on the Pleistocene CO_2 cycle of expanded and contracted tropical rain forests. Over periods of a few thousand years, the arid-pluvial cycles may have resulted in rain forests going from less than a third their present extent to more than half again as large. The transfer of atmospheric CO_2 to organic matter, and back again, is considered by Shackleton to result in measurable changes of tenths of a part per thousand in the Pleistocene isotopic records. This is yet another instance of the interplay of historical interpretations and equilibrium models.

I arrive at the end of this section on productivity with the following conclusions: (1) the most reasonable estimate for a change in biomass over geologic time is a factor of 10 increase from the anoxic Precambrian oceans to the modern oceans; (2) the amount of organic carbon production has varied by ± 50 percent from one geologic period to the next.

Patterns of Taxonomic Change

This section focuses on the early development of life and on the subsequent radiation of marine invertebrate organisms. Attention is given to (1) taxonomic composition, (2) taxonomic diversity, and (3) faunal extinctions.

(1) The oldest widely publicized possible fossils are $\simeq 3.4$ b.y. from the Swaziland System of South Africa (Knoll and Barghoorn, 1977). However, if organisms are involved in the origin of the $\simeq 3.8$ b.y. old banded iron formation of Greenland, as is envisaged for other banded iron formations (Chapter 5), life may be concurrent with the oldest rocks on earth.

The oldest certainly accepted fossils are $\simeq 2$ b.y. from the Gunflint Iron Formation of southern Ontario. These organisms are blue-green algae and bacteria, the two major groups of procaryotes, an informal grouping of organisms which lack a nuclear membrane, lack organelles such as mitochondria, and also lack miosis in reproduction. Such organisms today are able to metabolize a very large variety of chemical substrates and to live and to reproduce in a very wide variety of habitats (Mitchell, 1974). In this regard, procaryote organisms are more versatile than the eucaryote organisms into which they evolved.

Small cells (~ 20 μm) occur throughout Proterozoic time, but large cells (~ 40 to 80 μm) first occur in organisms which are $\simeq 1.4$ b.y. old. In modern microbiotas, cells of this large size are characteristic only of eucaryotes. This indicates the probability that by 1.4 b.y. ago eucaryotes had by then come into existence (J. W. Schopf, 1978). The most diverse flora of eucaryote organisms is $\simeq 800$ m.y. old, from the Bitter Springs deposits of central Australia. Eucaryotes are characterized by having a nuclear membrane, membrane-bound organelles such as mitochondria, a much larger cell size than procaryotes, and both miosis and mitosis in reproduction. The Archean and the early Proterozoic can be viewed as times of major evolutionary innovation with the evolution of basic biochemical and cellular mechanisms.

The first multicellular organisms (both metazoa and metaphyta) occur in rocks $\simeq 0.7$ b.y. old, and were first described from the Ediacara beds of south Australia (Glaesner and Wade, 1966), but are now known from many areas (see Cloud, 1976a). Metazoan organisms diversified rapidly and by the early Paleozoic came to have approximately as many orders as existed in even the latest Tertiary (as shown in Figure 7-7, see below).

Multicellular organisms disguise the packaging of their biochemical and cellular mechanisms in a vast array of external morphological forms. The way in which the organic carbon is repackaged in the Phanerozoic differs from species to species, genus to genus, etc., as it is recycled through time (many "evolutionary patterns" summarized in chapters in Hallam, 1977b). The documentation of this diversity is most thoroughly illustrated in the nearly 30 published volumes of the Treatise on Invertebrate Paleontology.

Figure 7-4 summarizes the major change in taxonomic composition of metazoan fossils through the Phanerozic. There are four major faunas:

(a) Trilobites accounted for close to 75 percent of the marine metazoan species in the early Cambrian, and the percentage remains high through the early Paleozoic. Organisms with phosphatic impregnation of carbonate are especially common relative to later faunas. This long interval of the Cambrian to middle Ordovician (570 to ~ 450 m.y.) was also a time of extraordinary morphologic experimentation of molluscan and echinoderm lineages.

(b) The second major Phanerozoic fauna is marked by the earliest typical biohermal mounds and reefs, together with widespread level-bottom shelly faunas consisting chiefly of brachiopods, as well as regions with higher current velocities with stalked echinoderms (especially crinoids). This second characteristic fauna began with the middle Ordovician and lasted throughout the Permian (450 to

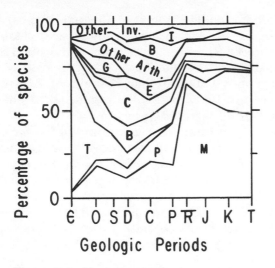

Figure 7-4 Variation in the taxonomic composition of the invertebrate fossil record through the Phanerozoic: M (mollusks), P (proto-
zoans), T (trilobites), B (brachiopods), C (coelenterates), E (echinoderms), G (graptolites), other arthropods, B (bryozoans), I (insects),
and other invertebrates. After Raup, 1976a:287. Note the early Paleozoic demise of trilobites and the Mesozoic expansion of mollusks.

225 m.y.); during the Permian, half of the marine invertebrate
families became extinct (Figure 7-6).

(c) The Mesozoic faunas (225 to 65 m.y.) are noted by the abundance of bivalves (instead of brachiopods) and the fossilization of
enormous numbers of a major group of cephalopods—the ammonites. Molluscan families came to represent more than 50 percent of
the marine metazoan families.

(d) With the Tertiary (65 m.y. to the present), the general aspect of the faunas changed once again. Gone are the ammonite
faunas, and in their place are bivalve, snail, bryozoan, and coral assemblages. These have continued today to characterize the continental shelves of the present world. Thus, the geologic record of me-
gascopic invertebrates is one both of stability (in the orders) and of
change (in the families, genera, and species).

The geologic record of diversity of shelled plankton is not nearly
so "uniform" as is the record of shelled benthos. Prior to the Jurassic, calcareous or siliceous phytoplankton are virtually unrecognized. However, Paleozoic deep-sea cherts and deep-water fine-
grained limestones are most plausibly related to the remains of such
groups, subsequently altered to unrecognizable form by diagenesis.

A large variety of organic-walled plankton, including the acritarchs, chitinozoa, dinoflagellates, and green algae, are certainly abundant in Paleozoic rocks (chapters in Tschudy and Scott, 1969; Tappan and Loeblich, 1973), as are the phosphatic remains of conodonts (which were probably members of the zooplankton). Although conodonts became extinct in the Triassic, the organ-walled phytoplankton remain abundant to the present day.

Beginning in the early Jurassic and continuing to the present day, the chiefly planktonic algal coccolithophores were abundant, and these introduced enormous amounts of carbonate to deep-sea sediments on disintegration of their calcareous plates (chapters in Ramsay, 1977a). Protozoan planktonic foraminifera are also calcitic, but although known from the Jurassic, they do not appear to be the source of major sedimentary deposits until the Cretaceous, when they became the major group of carbonate producers for deep-sea deposits. Thirty-seven modern species are distributed among five biogeographic realms, polar, subpolar, transitional, subtropical, and tropical (Bé, 1977), as is characteristic of all open ocean organisms. The fossil remains from organisms in each of these realms can be followed in cores for the past several hundred thousand years (Ruddiman, 1977:136) and longer.

Among siliceous plankton, the oldest confirmed records (Ordovician) are for a protozoan zooplankton group, radiolaria; by the Cenozoic some 200 radiolarian species lived at any given time (chapters in Ramsay, 1977a). The known diversity of radiolaria from Paleozoic rocks is much lower than expected if these zooplankton were a major source of siliceous material (although diagenesis into cherts could be responsible for the obliteration of their distinguishing features). As with calcareous phytoplankton, the first siliceous phytoplankton (in the form of diatoms) are known from the Jurassic and reached major importance in the Cretaceous.

(2) Changes in the *number* of marine invertebrates through the Phanerozoic have been summarized at the species, family, and ordinal levels (Figures 7-5, 7-6, and 7-7). Both the species-level and family-level plots indicate a reduction of approximately a factor of 2 from the mid-Paleozoic to the lower Triassic, and a subsequent rise to much higher values for the Cenozoic. The number of metazoan orders, however, is essentially uniform through time (see below). This suggests that the number of major units through which carbon is recycled does not change through time.

Three explanations for changes in taxonomic diversity at the species and family levels will be considered: (a) sampling bias, (b) increase in diversity within an adaptive zone, and (c) changes in diversity because of changes in biogeography.

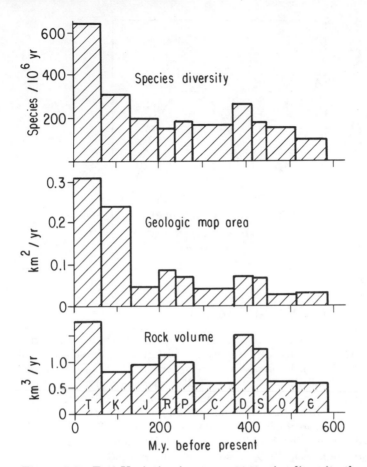

Figure 7-5   Top: Variation in apparent species diversity through the Phanerozoic. *Middle:* geologic map areas covered by sedimentary rocks. *Bottom:* estimated sediment volumes (excluding deep-sea deposits). After Raup, 1976a, 1976b.

(a) The actual record of *species* diversity through time is highly correlated with the volume and area of sedimentary rocks available for study, as is shown in Figure 7-5. Thus the observed patterns of diversity may be nothing more than two centuries of sampling of the available rock record. "True" diversity through the Phanerozoic is, then, this record emended to include any changes attributable to differences between the available and the original record. Any "emendations" in observed diversity are most likely to be required for species and genera (i.e., taxa with the smallest geographic and stratigraphic ranges) and are least likely to be required for orders and classes (i.e., those taxa with the widest distributions). Thus at

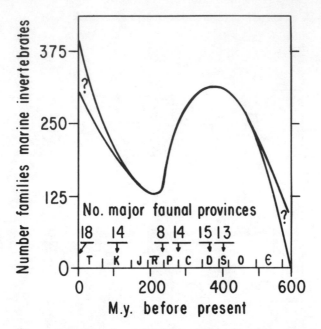

Figure 7-6   Number of families of marine invertebrates through the
Phanerozoic (in m.y. before present). Letters at bottom are symbols
for geologic periods; numbers above letters (18, 14, 8, 14, 15, 13) refer to
number of major faunal provinces. The question mark for 600 million years refers to lack of knowledge of soft-bodied forms; the question mark for the Tertiary refers to the bias of the more adequate
fossil record, which by itself makes it appear that diversity has increased. After Schopf, 1979. Point for 15 Devonian provinces after
Richard Bambach (1979, personal communication).

the species level, diversity is highly correlated with available rock,
but at the ordinal rank diversity is nearly level through geologic
time. The sampling bias model presents the possibility that real
variations have not occurred in the number of taxa of marine invertebrates through most of the Phanerozoic (Raup, 1976b).

(b) A second explanation for the observed patterns of change in
species and family richness (Figures 7-5 and 7-6) is that variations
not only are real but reflect a meaningful biological pattern. One
such pattern could be that increased diversity reflects the exploitation of new habitats. Bambach (1977) explored this possibility by as-
sessing species richness in nearshore and offshore benthic deposits
over geologic time. These deposits are likely to retain a large proportion of the fauna which had mineralized skeletons (T. J. M. Schopf,

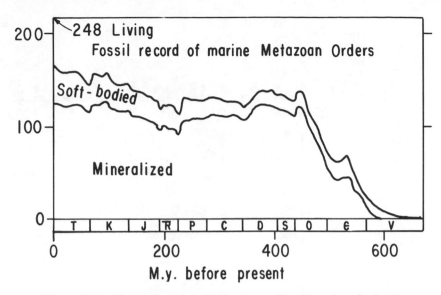

Figure 7-7   Plot of metazoan orders as a function of geologic time. Note uniformity after the middle Ordovician. After Sepkoski, 1978.

1978). As shown in Figure 7-8, species richness in nearshore deposits is the same from the Cambrian to the present. In contrast, species richness in the offshore regime increased significantly from the upper Paleozoic to the Mesozoic and the Tertiary, and this is reflected in the observed increase in overall species diversity (Figure 7-5).

The increase in species richness in offshore deposits is a reflection of the inclusion in the Tertiary sample of mollusks, which live *in* the sediment, whereas most species of the offshore deposits in older times lived *on* sediment. The exploitation and addition of this new mode of living may account for the fact that changes in diversity approximate a step function rather than a continuously increasing curve. Such changes in marine diversity would be analogous to changes in terrestrial diversity as large unoccupied regions (e.g., the air) were colonized owing to new structural modifications (e.g., wings). Bambach's approach led to the conclusion that a true increase in diversity in marine invertebrates has occurred, and that it is due to the occupation and exploitation of new "adaptive zones."

(c) A third explanation for changes in Phanerozoic marine diversity is that these changes mirror the number of biogeographic provinces through geologic time. As continental configurations change, so does endemism, so that periods of continental congealing

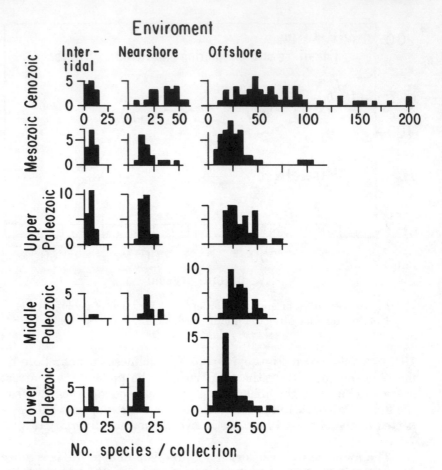

Figure 7-8 Histograms of numbers of species in fossil deposits, grouped by environment and time segment. Vertical dimension is number of deposits; horizontal dimension is number of species in the deposit. After Bambach, 1977.

(Permian) are times of few biogeographic provinces (and low diversity), and periods of widely dispersed continents (such as the present) are times of many biogeographic provinces (and high diversity) (Valentine and Moores, 1972). In Figure 7-6, the numbers at the bottom of the graph (18, 14, 8, 14, 15, and 13) refer to the number of major marine faunal provinces inferred for various intervals of geologic time. High diversity therefore is related empirically to biogeographic differentiation. The ecologic basis for this biogeographic view of faunal diversity will be discussed below.

(3) The third part of this section on biotic patterns concerns faunal extinctions. Changes in faunas have been attributed to two general causes: density-dependent changes and density-independent changes (Valentine, 1972).

(a) Density-dependent changes in diversity are those in which diversity is a function of the number of individuals, or taxa, already in existence. For example, density-dependent changes normally occur where habitable area expands or contracts because habitable area controls population size. The larger the area, the more populations (and taxa) which can be supported by finite resources. Changes in diversity are gradual to the extent that factors which control habitable area (e.g., climate) change gradually.

Density-independent changes in diversity are those in which changes are independent of the number of individuals or taxa. Density-independent changes in diversity may occur because of some change which affects a species as a whole—such as the hypothetical effects of a supernova, of a reversal of the earth's magnetic field, of chemical poisoning of the sea—or because of other "special events" which kill individuals and species independent of their abundance. Additional density-independent mechanisms of extinction are those associated with "innate degeneracy," such as the multitude of variations on the theme of orthogenesis (directed evolution) and racial senescence. Density-independent explanations have in common that a specific cause (sometimes external, sometimes internal) literally preordains individuals or species to extinction.

I will consider three general explanations which have been put forth for faunal extinctions: changes in temperature; changes in ocean chemistry; and, in the next section, "Biogeography," the effects of changes in habitable area. Two recently suggested density-independent explanations of animal extinction which are not considered here are changes in oxygen concentration and genetic racial sensecence. These have been previously discounted (Schopf, Farmanfarmaian, and Gooch, 1971; Schopf and Gooch, 1972). In studying faunal extinctions, what we are concerned with are overall patterns of extinction for whole faunas, rather like an actuary following the statistical properties of a cohort of human beings. We are not concerned with what caused the extinction of any *particular* species (or *individual* person).

(i) Temperature has been suggested to be the reason why metazoan life did not evolve before the late Precambrian. Let us grant, for the moment, the possibility that an extremely high temperature regime (50° to 60°C) existed for the early to intermediate part of the earth's history (p. 140). As pointed out above, metazoans of a high degree of organic complexity (ostracoda of the phylum Arthropoda) reproduce in hot springs with a constant temperature of 50°C, and other eucaryotes (fungi) reproduce in temperatures up to 60°C. Procaryotes live in natural hot springs to temperatures of 100°C as long as liquid water is present (Brock, 1978). Thus, given sufficient availability of an environment, there is no known reason why some eucaryotic organisms will not adapt to high temperature conditions. Because of this counterexample, pervasive high temperature to 50°C *by itself* is probably not the reason why metazoans did not evolve until they did, since metazoans can and do exist in those conditions.

A naturally occurring change in temperature has been suggested to be by and of itself a prominent agent of extinction. One typically meets this argument where, on the one hand, a faunal change has been noted, and, on the other hand, a "sharp" change in temperature is hypothesized or observed. For example, with regard to the Eocene/Oligocene boundary, "The oxygen isotope record indicates a severe and abrupt drop in surface and bottom water temperature near this boundary coincident with a global crisis among oceanic plankton. The very sharp drop in bottom-water temperatures produced a major crisis in deep-sea benthonic faunas affecting both ostracods and benthic foraminifera" (JOIDES, Executive Committee, 1977:70).

Let us consider this claim of a relationship between temperature change and faunal extinction. An assuredly sharp drop in temperature would be a 10°C change over a million years. Yet this is only a 1°C change per 100,000 years, well within the adaptive capabilities for every species ever examined, and probably every species which is now in existence or which has ever been in existence. A time of 100,000 years is 50,000 to 100,000 generations for nearly all the species we are concerned with. However, only about 10 generations are required to adapt to environmental changes *far* more severe (Lewontin, 1974). Even if there had been a 10°C change in only 10,000 years, this is only 1° every 1,000 years—again a trivially small change in temperature as far as natural populations of organisms are concerned. For comparison, we may note that plants and animals of the Cape Cod Canal (north of Woods Hole) undergo in late summer a 10° change during the half hour it takes for 22°C water at one end of the 20-mile canal to be replaced by 12°C water from the other end of the canal—and vice versa $12\frac{1}{2}$ hours later. The fauna and flora of the canal are a luxuriant mixture of both northern and southern species. It is also now well known that species of the deep sea, in as stable an environment as exists on earth, are just as genetically variable as are shallow-water species, and maybe even more so (Schopf and Gooch, 1972; Gooch and Schopf, 1972). Sufficient genetic variability exists for adaptation to any broadly distributed environmental regime which has existed on earth.

All available physiological and genetic evidence—and there is considerable—indicates the immense adaptational capabilities of organisms (see, e.g., Mangum, 1972:113; Dales, Mangum, and Tichy, 1970:375–376). Temperature, like oxygen, salinity, or any density-independent environmental factor merely provides the background within which the important causes of extinction operate—those which are density dependent. Probably the real significance of all of those figure captions which read "there is good agreement between the fluctuations of diversity and of isotopic temperature for the last x-number of millions of years" is that *both* temperature and diversity are tracking changes in water masses with different characteristics. My main conclusion is that temperature per se, within the known limits in which aquatic life occurs, probably has had little to do with *any* pattern of origination or extinction of any group of animal or plant species.

(ii) Many have concluded that a reduction in salinity could be responsible for the extensive Permian extinctions (e.g., Stevens, 1977). Approximately half of the families of marine organisms died out during the latter part of that geologic period (Figure 7-6). This chemical mechanism for species extinction is independent of organism abundance and is based on an innate property (physiological response) in a changing environment.

The comparison is sometimes made between changes in diversity in an estuary and presumed changes in diversity if the whole ocean had as low a salinity as that of the estuary (Lantzy, Dacey, and Mackenzie, 1977). The ocean, however, is not an estuary in terms of temporal instability, which is the most important factor in limiting estuarine diversity (p. 148). Conditions in an estuary are simply not stable enough to permit *detailed* morphologic and physiologic adaptation of the kind routinely found in physiologically stable fresh-water and marine conditions. It is a mistake to compare changes in diversity over the temporally and spatially unstable situation which occurs from one end of an estuary to another with the slowness and vastness of change which must accompany *any* oceanic change. A rapid oceanic change in salinity would be ± 5‰ in a million years. Yet this is only a 1‰ change in 200,000 years, or (approximately) 200,000 generations. More striking adaptations are performed every month in studies of laboratory selection (Lewontin, 1974). There is no reason of which I am aware which would prevent organisms from adapting to a change from 40‰ (or so) down to 25‰ (or so) provided that the change occurred slowly (relative to an individual's life span). Many species whose individuals have life spans of 0.1 to 100 years have lived through salinity changes of 1 to 3.5‰ over time scales of 10^3 to 10^6 or 10^7 years, without any observed adverse effect.

The salinity and temperature explanations of animal extinction essentially lost their scientific basis with the discovery of immense amounts of genetic variability in organisms of all habitats (Lewontin, 1974). In sum, no density-independent mechanisms of extinction seem to me to be plausible as a major contribution to the broad patterns of the rise and fall of life over geologic time.

Biogeography

Biogeography will be treated as the synthesis of three points of view: (1) predicting a faunal equilibrium, (2) predicting the number and size of faunal provinces, and (3) relating these predictions to observed changes in faunal diversity.

(1) The essence of the notion that changes in habitable area control overall faunal diversity is that there is an equilibrium number of taxa per unit area. Total diversity can be altered either by adding or deleting the number of areas (i.e., number of faunal provinces) or by changing the size of each area. On the one hand, as a region opens for colonization, the rate of immigration into the area is high; but this rate must subsequently decrease as faunal diversity rises and space is occupied. On the other hand, in a virgin territory, the initial rate of extinction will be very low, but as the fauna increases, the rate of extinction then must increase. There is a point at which the fauna of a region is sufficiently numerous that the rate of addition of individuals (or taxa) should equal the rate of extinction of individuals (or taxa). Achievement of this point forms the basis for the notion of a faunal and floral equilibrium over ecologic time and, summed over eons, the basis for a biotic equilibrium over geologic time.

Faunal diversity as a function of area has been described for many taxa. This relationship applies to such disparate examples as decapod crustacean species on coral heads of different sizes (Abele, 1976), mite species on rodents of different sizes of range (Dritschilo et al., 1975), insect species on host trees of different sizes of range (Strong and Levin, 1975), as well as to such widely publicized examples as amphibians plus reptiles on islands of different sizes (MacArthur and Wilson, 1967). The taxon/area relationship has been determined for more than a hundred sets of data (Flessa and Sepkoski, 1978; Connor and McCoy, 1979). These data include species, genera, and families of both particular taxa and whole biotas. Area per se is a good predictor of faunal diversity.

There are three major explanations of why the taxon/area relationship should work. There is evidence for each of them, and they are not mutually exclusive (Connor and McCoy, 1979). The first explanation is that an increase in habitable area allows the population size of all of the actual or potentially inhabiting species to increase. Since population size for each species has increased, the chance of extinction for each has decreased, and total diversity therefore increases. The second explanation is that an increase in habitable area happens to include an increase in the variety of habitat types, thus allowing for more niches, and therefore an increase in taxonomic diversity. Third is that a larger area simply samples by chance a larger amount of the world's diversity.

In the examples cited previously, habitable area varies over as much as five orders of magnitude—a range far greater than is found in variation in taxonomic diversity. Thus on the customary taxon/area plot, the slope of the taxon/area line must be much less than 1.0. Although a wide range of values (most between 0.2 and 0.5) has been observed, the median value is very close to 0.30. The basic relationship is shown in Figure 7-9. $S = kA^z$, or $\log S = \log$

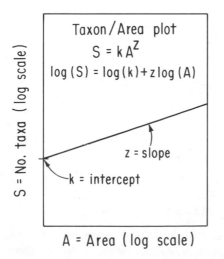

Figure 7-9   Plot of number of taxa (S) versus area (A) to show typical form and fitted constants (k = intercept; z = slope).

$k + z \log A$, where S = number of taxa, A is area, and k (the intercept) and z (the slope) are fitted constants.

In dealing with both recent and fossil materials, there is no necessary reason why one should use a log/log plot in preference to a log/arithmetic, arithmetic/log, or even arithmetic/arithmetic plot (Connor and McCoy, 1979). But a log/log convention seems to apply to more sets of data than any other type of plot.

The global faunal diversity of modern shallow marine faunas of the continental shelf is chiefly a reflection of high tropical diversity. This high tropical diversity exists for two reasons: (1) on the average, tropical faunal provinces are more than twice as large as temperate or boreal provinces (Schopf, Fisher, and Smith, 1978), and diversity is correlated with size; and (2) there are more provinces in the tropics than there are in any other latitudinal zone because of the high incidence of shallow seas at low latitudes, where the earth's circumference is greatest. Over geologic time, knowledge of the number and size of tropical provinces is probably much more important than comparable data on temperate or polar regions.

In modern seas, the number of marine genera per faunal province (for bivalves, foraminifera, and corals) is correlated significantly with the size of the faunal province, as shown in Figure 7-10. The scatter is considerable, but no greater than occurs in many other taxon/area plots. In addition, the fit is best for tropical and temperate provinces, and worst for boreal and polar provinces (E. D. McCoy, personal communication, 1978). This is an additional reason

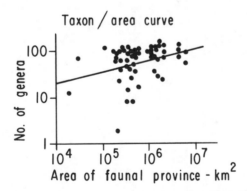

Figure 7-10   Scatter plot of number of genera (of bivalves, foraminiferans, and corals) versus area of faunal provinces of the continental shelf; $y = 1.19 \times ^{.25}$; $r = .37^*$; $n = 57$ (significant, 5 percent level). Data from Schopf, Fisher, and Smith, 1978.

why one should focus on tropical and temperate regions for ancient faunal provinces.

We can evaluate the relative importance of changing the number of faunal provinces versus changing the size of faunal provinces. If we let a given province increase by 5 times its original size, then the new diversity varies as 5^z (recalling that $S = kA^z$). If $z = 0.3$, then the new diversity increases by $5^{.3}$, or a factor of 1.6. If the size of the province were merely doubled, the new diversity would vary as $2^{.3}$, or a factor of 1.2. Similarly, a decrease in habitable area by a factor of 2 to 5 leads to a reduction of 20 to 40 percent in the original diversity. A change in the size of individual faunal provinces by a factor of 2 to 5 probably occurred as sea level changed during the Pleistocene as well as during other times of continental glaciation, and during some intervals of significant changes in the rate of sea-floor spreading. These changes in sea level should cause changes in the sizes of provinces that should be related to changes in diversity.

A significant change in diversity is more likely to be a result of the *number* of faunal provinces than the *size* of individual provinces. A single faunal province can be subdivided into two provinces, as when continental shelf regions move away from each other. If the sum of the area of the two new provinces is the same as the area of the old province, then diversity in each new province will vary as $(1/2)^z$. If, as before, we let $z = 0.3$, then the diversity in *each* new province will be 80 percent of the original diversity. If, as happens over geologic time, the faunal provinces evolve entirely different faunas, then the total diversity of these two provinces will be 1.6 times the original diversity. A subsequent increase in the area of each province to the original size would increase total diversity to about a factor of 2 over what it had been before subdividing the original province. The *number* of provinces thus seems to be the chief factor controlling diversity over periods of tens of millions of years, whereas the *size* of provinces become more critical over time scales of a few to many thousands of years.

(2) In using taxon/area relationships there needs to be a way to predict the number of faunal provinces over geologic time. At any given time, the marine fauna of the world is organized into a collection of faunal provinces, the sizes of which are determined by ocean currents and coastal geography. Typically, a modern province has 30 to 50 percent of its species endemic to that province (Schopf, 1979). The boundaries of faunal provinces are shifting if viewed on a local scale. However, since most provinces are on the order of 10^5 to 10^6 km², precise changes in boundaries have little effect on the total size.

The best guide to predicting the boundaries of modern faunal provinces is knowledge of the ocean currents, because currents of the open ocean and of continental shelves maintain a region of cohesive temperature, salinity, and other physical and chemical conditions, and distribute the larvae within that specific region (Reid et al., 1978). The prevalence of currents in controlling faunal provinces has been known for decades to oceanographers (e.g., Ekman, 1953:2, 22). For example, Bé (1977:32) wrote that "numerous plankton biogeographers . . . have noted a close relationship between plankton communities and the major hydrographic regions of the world ocean." More specifically, changes in the phytoplankton, zooplankton, and fish communities occur in the open Atlantic, passing northward from the equator toward Bermuda and then toward Greenland (Backus et al., 1965; Jahn and Backus, 1976; Maynard, 1976). Obviously no physical barriers keep the faunal provinces distinct. In addition, Gulf Stream rings have floras and faunas which are quite distinct from those in non–Gulf Stream water (Wiebe et al., 1976). Sedentary organisms (sipunculids and pogonophora) on the continental slope also are separated into regions by "the effects of bottom currents on larval dispersion" (Cutler, 1975). Indeed, precisely the same type of effect can be seen in the distribution of clay minerals, which are also transported by current systems (Hathaway, 1972).

In summary, species are more or less confined to separate current systems and seem to maintain discrete populations, not because a specific physical factor presents a fundamental physiological barrier to their distribution, but because the currents confine species to given regions. (An alternative view is that the steepness of the temperature gradient per se is of fundamental importance in regulating taxonomic diversity; see Valentine, 1973, and earlier). The view favored here is that currents control and predict the faunal provinces and that paleontological effort at predicting past faunal provinces both in the open ocean and on the sea bottom must focus on delineating ancient current systems.

The most conservative oceanographic guidelines for predicting ancient current systems are shown in Figure 7-11. These include the existence of (1) a westward-flowing equatorial current; (2) an intensification of flow along the western margin of an ocean, which spreads back eastward at 30° to 40° latitude, analogous to the Gulf Stream; (3) a more boreal westward-moving return gyre, which may merge into a polar clockwise gyre driven by the polar high, as exists today in the Arctic (Ostenso, 1966). Local geography will of course influence any patterns which obtain, but the basic current systems themselves are the result of atmospheric circulation on a sphere

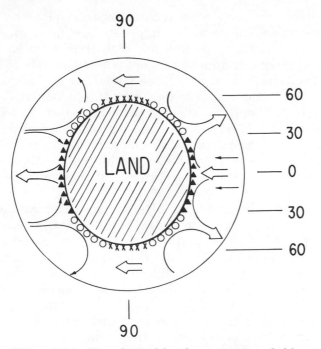

Figure 7-11 Hypothetical land mass surrounded by world ocean to illustrate current patterns and faunal provinces. Eight faunal provinces are indicated: two tropical (triangles), four temperate (circles), and two polar (X's). After Schopf, 1979.

combined with the laws of mass balance, and those two factors are not likely to have changed significantly over the Phanerozoic. As a start, I therefore advocate using this system to predict ancient ocean currents, and the concomitant faunal provinces.

In the simplest of all geographic worlds, we can conceive of a single continent, as existed in Pangaea, and can predict the ocean currents surrounding it (Chapter 3) and its biogeographic provinces (as in Figure 7-11). Such a world with current systems outlined as in the previous paragraph would be consistent with eight major faunal provinces: northern and southern polar, eastern and western tropical, and four intermediate provinces (one on each side of the continent, in both Northern and Southern hemispheres). These eight provinces (and their water temperature and chemistry) result from the patterns of ocean circulation.

There are two additional guidelines for determining faunal provinces: (1) by analogy with the modern world, a body of water as large as the Mediterranean constitutes a separate province; (2) marine species are not able to maintain genetic continuity over an infinitely large range. The tropical region of the Indian Ocean and Western Pacific, extending from Somali to the Solomons (about 15,- 000 km), seems to be divided into three faunal regions despite a "continuous" water connection. The reason for this biotic division is that dispersal is sufficiently weak over those large distances that gene flow is simply insufficient to maintain reproductive continuity. Thus although larval transport is *possible* for great distances (Scheltema, 1971), local and regional biogeographic differentiation is more significant when distances are greater than approximately 4,000 km.

This way of considering faunal diversity says nothing about *which* species exist in any given province, for that clearly depends on individual physiological adaptations. However, the model does allow one to do the following: (1) it allows prediction of the number of faunal provinces; (2) since species diversity is a function of faunal province size, the taxa/area curve allows one to predict how many species will exist per province; (3) it is then a simple matter (in principle) to sum up the number of provinces and their sizes and to estimate total diversity; (4) thus one can conclude that changes in number of and size of provinces "control" total diversity through time.

If one applies this general method of determining faunal provinces to the present world (Figure 7-12), 18 major faunal provinces are predicted. These provinces are very similar to the major provinces of the present world, and are the minimum number likely to exist in a world of similar geographic separation. For comparison, I give in Figure 7-13 the empirically derived modern faunal provinces for bryozoans, with 21 provinces. Recognition of minor differences and of oceanic islands (like Hawaii) as separate provinces permits finer and finer subdivision. Valentine (1973:356) cited 31 marine molluscan provinces of the world's continental shelves, but many of these are minor subdivisions of the major provinces shown for the bryozoans, and would not be recognizable in the fossil record. The major biogeographic provinces are indicated by considerations of the general oceanic circulation. Campbell and Valentine (1977) suggested they should be evident also from examining past faunas.

This model of predicting faunal provinces is designed for the large portion of Phanerozoic time when the extent of marine transgressions onto continental regions was less than, say, 40 percent. Certain aspects of the model relating to continents as barriers for ocean currents may not apply as well to those times when epicontinental seas are said to have covered as much as 75 percent of conti-

Predicted faunal provinces

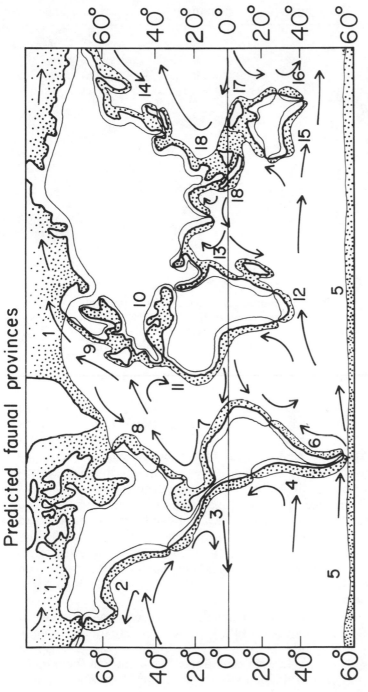

Figure 7-12 Predicted faunal provinces of modern shelf regions according to the deductive model discussed in text. Note that in comparison to the "real" faunal provinces (Figure 7-13), the predicted pattern omits the smaller oceanic islands (Hawaii, etc.) and such geographically controlled provinces as the Virginian (between Cape Cod and Cape Hatteras). Outlines of continents are generalized so as to be approximately commensurate with what might be expected for paleogeographic reconstructions. After Schopf, 1979.

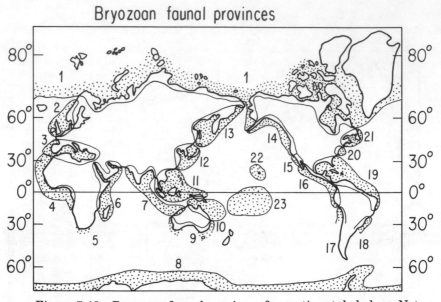

Figure 7-13   Bryozoan faunal provinces for continental shelves. Note the greater detail in comparison with Figure 7-12. After Schopf, 1979.

nental regions, especially for narrow continents (say, less than 5,000 km).

The differences between anticipated and real patterns of faunal provinces are due chiefly to (1) lack of recognition of *small* provinces of temperate latitudes and of oceanic islands, and (2) lack of knowledge about the exact configuration of tropical provinces where current systems are influenced strongly by local geography (e.g., Southeast Asia). The small "extra", temperate-latitude provinces are controlled chiefly geographically (e.g., the Virginian between Cape Hatteras and Cape Cod) and are characterized by steep temperature gradients at either border. These provinces *may* represent the degree of provincial addition which Valentine believed is characteristic of a world with steep thermal gradient. Such small provinces do not contribute nearly as much to total diversity as do the major tropical provinces.

(3) Let us now combine biogeographic theory and changes in number of faunal provinces into a general view of changing biotic diversity through time. I believe that the effect of both changes in endemism and changes in area can be seen separately in the famous late Paleozoic Permo-Triassic extinctions. Figure 7-14 shows a log-

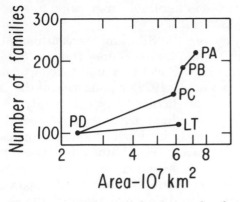

Figure 7-14 Plot on a log-log scale of number of families of Permian shallow marine invertebrates versus area of Permian shallow marine seas. PA is early Permian, PB and BC are middle Permian, and PD is late Permian; LT is early Triassic. After Schopf, 1979.

log plot of the diversity of Permian marine invertebrate families versus the area of Permian shallow marine seas. Note the steep slope going from lowermost Permian (PA) to middle Permian (PB, then PC). These periods of time correspond to a markedly changing paleogeography, with the addition of the Siberian Plate to Europe (and the formation of the Ural Mountains) and the addition of the Angaran Plate to southern Asia (paleogeographic maps of these intervals are given by Schopf, 1974; Figures 3, 4). For the early Permian, when the Siberian Plate and the Angaran Plate were set off as distinct continental regions, I estimated 14 faunal provinces, and by the end of the Permian and Pangaea, I estimated 8.

Empirical summaries of Pangaean brachiopod faunas of the late Permian also reveal only a few provinces: tropical, temperate, and polar faunas in both the Northern and Southern hemispheres (Waterhouse and Bonham-Carter, 1975). Thus the main reduction in diversity from the lower to upper Permian is attributed to a decreasing number of endemic regions, and this is consistent with the steep slope in the taxon/area curve.

Subsequent changes in diversity in the latest Permian and lower Triassic have a much lower slope on the log-log plot (PC to PD and PD to LT, Figure 7-14). There are no discernible major paleogeographic shifts during these times. The major change in diversity is attributed to the effect of changes in the size of the provinces which did exist (via lowering and then raising of sea level). No significant increase in endemism has been documented for the lower

Triassic; there is only a change in habitable area owing to a rise in sea level of approximately 200 m.

Further back in time, during the Silurian, the continental regions of Gondwana, North China, South China, Baltica, Siberia, Malaysia, Tibetia, Kazakhstania, and Laurentia appear to have been distinct (Ziegler, Hansen et al., 1977). Application of the principles outlined above yields a prediction of approximately 13 Silurian faunal provinces, some of which, however, have very little rock remaining today. This estimate is twice as high as is indicated in the empirical summaries (Boucot, 1974), although the empirical data are not plotted on drift reconstructions, a fact that makes it more difficult to evaluate provincialism. It is of critical importance to have a firm understanding of mid-Paleozoic provinciality, set in mid-Paleozoic geography, as a test for what we can call the biogeographic theory for faunal diversity.

It is not surprising that strictly empirical summaries of faunal regions through time (e.g., Valentine, Foin, and Peart, 1978) might grossly underestimate the number of *provinces*. As Ziegler, Scotese et al. (1977:16) commented, "Most of the paleobiogeographic units so far defined in the Paleozoic have the scale of realms or regions in the sense used by biogeographers working in the Recent. In the sense used for the Recent biota, provinces have yet to be consistently recognized in the fossil record. This is probably because an adequate paleogeographic base has not been available." Some of the ancient faunal provinces are today almost totally lacking in rock of the appropriate age. For this reason, strict empirical counting of provinces seen in the rock record, although previously strongly and well favored by some, may not reflect original patterns of diversity.

I conclude that, judged over Phanerozoic time, variation in diversity should be a reflection of changes in number of faunal provinces. Not nearly enough paleogeographic information presently is available to test fully this idea. However, it appears that predicted faunal provinces and total diversity change in parallel, as summarized in Figure 7-6.

Summary

This chapter largely has dealt with (1) production of organic carbon over geologic time; (2) patterns of change in taxonomic composition, taxonomic diversity, and faunal extinctions; and (3) biogeography. Productivity over geologic time has seemingly not varied by more than a factor of 2 for the past 3.3 b.y., as though methods for turning solar energy into organic carbon have been largely equivalent over geologic time even though the taxonomic groups have changed vastly. With regard to taxonomic composition, the oldest organisms are possibly 3.4 b.y., and certainly as old as 2.0 b.y.; procaryotes (blue-green algae and bacteria) had evolved into eucaryotes by 1.4 b.y; and metazoa and metaphyta are known 0.7 b.y. ago. Most of the major invertebrate taxa alive in the Cambrian are still represented today. With respect to patterns of taxonomic diversity, our knowledge of biologic diversity over geologic time is a function of (1) sampling of available rock exposures, (2) adaptations into new habitats, and (3) changes in the degree of endemism owing to different paleogeographic settings. With regard to biogeography, a method for determining number and placement of biogeographic provinces is outlined. This may be of help in evaluating traditional empirical summaries, which have grouped faunas together largely independently of consideration of paleogeographic setting.

The view adopted in this book is (1) that diversity is a function of habitable area; (2) that the amount of habitable area is a function of the number of biogeographic provinces and of the size of each province, as mediated by sea-level changes; (3) that the geologic record is capable of being deciphered with respect to ancient paleogeography and sea-level changes; (4) that therefore, biogeographical predictions in combination with empirical documentation should yield an accurate "signal" of changes in diversity (on the order of ± 25 percent) over geologic time; (5) that superimposed on the main geologic control of biological diversity are "steps" as new adaptive zones are entered, and as diversity increases to local saturation.

APPENDIX

Grain-Size Nomenclature

Sediments classified according to grain size following the Udden-Wentworth-Krumbein (UWK) scheme are designated by the phi notation: $\emptyset = -\log_2 d/d_0$, in which d is diameter in mm, and d_0 is a standard (1 mm). For sediment size decreasing to the right on plots of sediment versus frequency, $\emptyset$ has the property of increasing to the right. More than one size designation has been applied to the same sediments depending upon the classification scheme used (see Table A-1).

Table A-1 Three common size grade scales. Mesh size refers to U.S. standard sieves.

Udden-Wentworth-Krumbein	Ø value	U.S. Dept. of Agriculture and Soil Science Society of America	U.S. Corps of Engineers, Dept. of the Army, and Bureau of Reclamation
Cobbles		Cobbles	Boulders
—— 64 mm ——	−6		10 in
	−5	—— 80 mm ——	Cobbles
Pebbles	−4	Gravel	—— 3 in ——
	−3		Gravel
—— 4 mm ——	−2		
Granules			—— 4.76 mm (=4 mesh) ——
—— 2 mm ——	−1	—— 2 mm ——	Coarse sand
Very coarse sand		Very coarse sand	—— 2 mm (=10 mesh) ——
—— 1 mm ——	0	—— 1 mm ——	Medium sand
Coarse sand		Coarse sand	
—— 0.5 mm ——	+1	—— 0.5 mm ——	
Medium sand		Medium sand	
—— 0.25 mm ——	+2	—— 0.25 mm ——	—— 0.42 mm (=40 mesh) ——
Fine sand		Fine sand	Fine sand
—— 0.125 mm ——	+3	—— 0.10 mm ——	
Very fine sand		Very fine sand	
—— 0.0625 mm ——	+4	—— 0.05 mm ——	—— 0.074 mm (=200 mesh) ——
	+5		Fines
Silt	+6	Silt	
	+7		
—— 0.0039 mm ——	+8	—— 0.002 mm ——	
Clay		Clay	

REFERENCES

Abbott, B. M. 1973. Terminology of stromatoporoid shapes. *Journal of Paleontology 47*:805–806.

Abele, L. G. 1976. Comparative species richness in fluctuating and constant environments: coral-associated decapod crustaceans. *Science 192*:461–463.

Abelson, P. H. 1966. Chemical events on the primitive earth. *Proceedings of the National Academy of Sciences 55*:1365–1372.

Adam, D. P. 1975. The tropical cyclone as a global climatic stabilizing mechanism. *Geology 3*:625–626.

Adey, W. H., and Macintyre, I. G. 1973. Crustose coralline algae: a re-evaluation in the geological sciences. *Geological Society of America Bulletin 84*:883–904.

Alexander, R. R. 1974. Morphologic adaptations of the bivalve *Anadara* from the Pliocene of the Kettleman Hills, California. *Journal of Paleontology 48*:633–651.

Alexandersson, E. T. 1978. Distribution of submarine cements in a modern Caribbean fringing reef, Galata Point, Panama. A discussion. *Journal of Sedimentary Petrology 48*:665–688.

Allen, J. R. L. 1963. Asymmetrical ripple marks and the origin of water-laid cosets of cross-strata. *Liverpool and Manchester Geological Journal 3*:187–236.

———— 1965. A review of the origin and characteristics of recent alluvial sediments. *Sedimentology 5*:89–191.

———— 1967. Depth indicators of clastic sequences. *Marine Geology 5*:429–446.

Alvarez-Borrego, S., Guthrie, D., Culberson, C. H., and Park, P. K. 1975. Test of Redfield's model for oxygen-nutrient relationships using regression analysis. *Limnology and Oceanography 20*:795–805.

Anderson, A. T. 1974. Chlorine, sulfur, and water in magmas and oceans. *Geological Society of America Bulletin 85*:1485–1492.

———— 1975. Some basaltic and andesitic gases. *Reviews of Geophysics and Space Physics 13*:37–55.

Anderson, E. J. 1971. Environmental models for Paleozoic communities. *Lethaia 4*:287–302.

Arking, A. 1964. Latitudinal distribution of cloud cover from Tiros III photographs. *Science 143*:569–572.

Armstrong, R. L., and Hein, S. M. 1973. Computer simulation of Pb and Sr isotope evolution in the Earth's crust and upper mantle. *Geochimica et Cosmochimica Acta 37*:1–18.

Arrhenius, G. 1966. Sedimentary record of long-period phenomena. Pp.

155–174 in *Advances in Earth Science,* ed. P. M. Hurley. M.I.T. Press, Cambridge.

Audley-Charles, M. G., Curray, J. R., and Evans, G. 1977. Location of major deltas. *Geology 5*:341–344.

Aumento, F. 1971. Vesicularity of mid-ocean pillow lavas. *Canadian Journal of Earth Sciences 8*:1315–1319.

Backus, R. H., Mead, G. W., Haedrich, R. L. and Ebeling, A. W. 1965. The mesopelagic fishes collected during cruise 17 of the R/V Chain, with a method for analyzing faunal transects. *Bulletin of the Museum of Comparative Zoology 134*:139–158.

Bada, J. L., and Miller, S. L. 1968. Ammonium ion concentration in the primitive ocean. *Science 159*:423–425.

Bada, J. L., and Schroeder, R. A. 1975. Amino acid racemization reactions and their geochemical implications. *Naturwissenschaften 62*:71–79.

Badham, J. P. N., and Stanworth, C. W. 1977. Evaporites from the lower Proterozoic of the East Arm, Great Slave Lake. *Nature 268*:516–517.

Badiozamani, K., Mackenzie, F. T., and Thorstenson, D. C. 1977. Experimental carbonate cementation: salinity, temperature and vadose-phreatic effects. *Journal of Sedimentary Petrology 47*:529–542.

Baes, C. F., Jr., Goeller, H. E., Olson, J. S., and Rotty, R. M. 1977. Carbon dioxide and climate: the uncontrolled experiment. *American Scientist 65*:310–320.

Baker, D. J., Jr. 1970. Models of oceanic circulation. *Scientific American 222*:114–121.

Baker, V. R. 1977. Stream-channel response to floods, with examples from central Texas. *Geological Society of America Bulletin 88*:1057–1071.

Ball, M. M. 1967. Carbonate sand bodies of Florida and the Bahamas. *Journal of Sedimentary Petrology 37*:556–591.

Ball, M. M., Shinn, E. A., and Stockman, K. W. 1967. The geologic effects of Hurricane Donna in south Florida. *Journal of Geology 75*:583–597.

Bambach, R. K. 1977. Species richness in marine benthic habitats through the Phanerozoic. *Paleobiology 3*:152–167.

Bandy, O. L. 1960a. General correlation of foraminiferal structure with environment. Pp. 7–19 in *International Geological Congress, 21st Session, Part no. 22 (Norway),*

———— 1960b. Planktonic foraminiferal criteria for paleoclimatic zonation. Pp. 1–8 in *Science Reports, Tohoku University, Sendai, Japan, 2nd Series (Geology), Special Volume no. 4* (Professor Shoshiro Hanzawa Memorial Volume).

———— 1963. Larger living foraminifera of the continental borderland of southern California. *Contributions from the Cushman Foundation for Foraminiferal Research 14*:121–126.

———— 1972. A review of the calibration of deep-sea cores based upon species variation, productivity, and $^{16}O/^{18}O$ ratios of planktonic foraminifera—including sedimentation rates and climatic inferences. Pp. 37–61 in *Calibration of Hominoid Evolution,* ed. W. W. Bishop and J. A. Miller. University of Toronto Press, Toronto.

Bandy, O. L., and Arnal, R. E. 1960. Concepts of foraminiferal paleoecology. *American Association of Petroleum Geologists Bulletin 44*:1921–1932.

———— 1969. Middle Tertiary basin development, San Joaquin Valley, California. *Geological Society of America Bulletin 80*:783–820.

Banin, A. B., and Navrot, J. 1975. Origin of life: clues from relations between chemical compositions of living organisms and natural environments. *Science 189*:550–551.

Barnes, S. S. 1967. Minor element composition of ferromanganese nodules. *Science 157*:63–65.

Barth, T. F. W. 1968. The geochemical evolution of continental rocks. A model. Pp. 587–597 in *Origin and Distribution of the Elements,* ed. L. H. Ahrens. Pergamon Press, Oxford.

Bates, D. R., and Nicolet, M. 1950. The photochemistry of atmospheric water vapor. *Journal of Geophysical Research 55*:301–327.

Bathurst, R. G. C. 1971. *Carbonate Sediments and Their Diagenesis.* Developments in Sedimentology, no. 12. Elsevier, Amsterdam, 620 pp.

Bé, A. W. H. 1968. Shell porosity of Recent planktonic foraminifera as a climatic index. *Science 161*:881–884.

———— 1977. An ecological, zoogeographic and taxonomic review of recent planktonic foraminifera. Pp. 1–100 in *Oceanic Micropaleontology,* ed. A. T. S. Ramsay, vol. 1. Academic Press, New York.

Bé, A. W. H., and Duplessy, J. C. 1976. Subtropical convergence fluctuations and Quaternary climates in the middle latitudes of the Indian Ocean. *Science 194*:419–422.

Becker, R. H., and Clayton, R. N. 1972. Carbon isotopic evidence for the origin of a banded iron-formation in Western Australia. *Geochimica et Cosmochimica Acta 36*:557–595.

Behrens, E. W., and Frishman, S. A. 1971. Stable carbon isotopes in blue-green algal mats. *Journal of Geology 79*:94–100.

Bentor, Y. K., and Kastner, M. 1965. Notes on the mineralogy and origin of glauconite. *Journal of Sedimentary Petrology 35*:155–166.

Berger, W. H. 1969. Ecologic patterns of living planktonic foraminifera *Deep-Sea Research 16*:1–24.

———— 1970. Biogenous deep-sea sediments: fractionation by deep-sea circulation. *Geological Society of America Bulletin 81*:1385–1402.

———— 1972. Deep sea carbonates: dissolution facies and age-depth constancy. *Nature 236*:392–395.

———— 1973. Deep-sea carbonates: Pleistocene dissolution cycles. *Journal of Foraminiferal Research 3*:187–195.

———— 1974. Deep-sea sedimentation. Pp. 213–241 in *The Geology of Continental Margins,* ed. C. A. Burk and C. L. Drake. Springer-Verlag, New York.

———— 1976. Biogenous deep sea sediments: production, preservation and interpretation. Pp. 265–388 in *Chemical Oceanography,* ed. J. P. Riley and R. Chester, vol. 5, 2nd ed. Academic Press, London.

———— 1978a. Deep-sea carbonate: pteropod distribution and the aragonite compensation depth. *Deep-Sea Research 25*:447–452.

———— 1978b. Oxygen-18 stratigraphy in deep-sea sediments: additional evidence for the deglacial meltwater effect. *Deep-Sea Research 25*:473–480.

———— 1978c. Historical oceanography (review of *Oceanic Micropalaeontology,* ed. A. T. S. Ramsay, 1977). *Science 200*:1475.

Berger, W. H., and Gardner, J. V. 1975. On the determination of Pleistocene temperatures from planktonic foraminifera. *Journal of Foraminiferal Research 5*:102–113.

Berger, W. H., Johnson, R. F., and Killingley, J. S. 1977. "Unmixing" of the deep-sea record and the deglacial meltwater spike. *Nature 269*:661–663.

Berger, W. H., and Parker, F. L. 1970. Diversity of planktonic foraminifera in deep-sea sediments. *Science 168*:1345–1347.

Berger, W. H., and von Rad, U. 1972. Cretaceous and Cenozoic sediments from the Atlantic Ocean. *Initial Reports of the Deep Sea Drilling Project 14*:787–954.

Berger, W. H., and Winterer, E. L. 1974. Plate stratigraphy and the fluctuating carbonate line. Pp. 11–48 in *Pelagic Sediments: On Land and under the Sea,* ed. K. J. Hsü and H. C. Jenkyns. Special Publications of the International Association of Sedimentologists, no. 1.

Berggren, W. A., and Hollister, C. D. 1974. Paleogeography, paleobiogeography and the history of circulation in the Atlantic Ocean. Pp. 126–186 in *Studies in Paleo-oceanography,* ed. W. W. Hay. Society of Economic Paleontologists and Mineralogists, Special Publication no. 20.

Berkner, L. V., and Marshall, L. C. 1964. The history of growth of oxygen in the earth's atmosphere. Pp. 102–126 in *The Origin and Evolution of Atmospheres and Oceans,* ed. P. J. Brancazio and A. G. W. Cameron. John Wiley and Sons, New York.

Berner, R. A. 1971. *Principles of Chemical Sedimentology.* McGraw-Hill Book Co., New York. 240 pp.

———— 1972. Sulfate reduction, pyrite formation, and the oceanic sulfur budget. Pp. 347–361 in *The Changing Chemistry of the Oceans,* ed. D. Dyrssen and D. Jagner. Nobel Symposium Proceedings, no. 20. John Wiley and Sons, New York.

Bewers, J. M., and Yeats, P. A. 1977. Oceanic residence times of trace metals. *Nature 268*:595–598.

Birot, P. 1968. *The Cycle of Erosion in Different Climates.* University of California Press, Berkeley. 144 pp.

Biscaye, P. E., Kolla, V. R., and Hanley, A. 1977. Quartz abundances and accumulation rates in the sediments of the Atlantic Ocean in relation to climatology. *Geological Society of America Abstracts with Programs 9*:899.

Bluck, B. J. 1969. Particle rounding in beach gravels. *Geological Magazine 106*:1–14.

Blumenstock, D. I. 1958. Typhoon effects at Jaluit Atoll in the Marshall Islands. *Nature 182*:1267–1269.

Blumenstock, D. I., Fosberg, F. R., and Johnson, C. G. 1961. The re-survey of typhoon effects on Jaluit Atoll in the Marshall Islands. *Nature 189*:618–620.

Bodine, B. R. 1969. *Hurricane Surge Frequency Estimated for the Gulf Coast of Texas.* U.S. Army Corps of Engineers, Coastal Engineering Research Center, Technical Memorandum no. 26. 32pp.

Bogdanov, D. V. 1963. Map of the natural zones of the ocean. *Deep-Sea Research 10*:520–523. (From *Okeanologiya,* 1961, *1*:941–943.)

Bogorov, V. G., Vinogradov, M. Y., Voronina, N. M., Kanayeva, I. P., and Suyetova, I. A. 1969. Zooplankton biomass distribution in the ocean surface layer. *Doklady of the Academy of Sciences of the U.S.S.R. Earth Sciences Sections 182*:235–237. (Russian original dated 1968.)

Boltovskoy, E. 1973. Note on the determination of absolute surface water paleotemperature by means of the foraminifer *Globigerina bulloides* d'Orbigny. *Paläontologische Zeitschrift 47*:152–155.

Bond, G. 1976. Evidence for continental subsidence in North America during the Late Cretaceous global submergence. *Geology 4*:557–560.

—— 1978a. Evidence for late Tertiary uplift of Africa relative to North America, South America, Australia and Europe. *Journal of Geology 86*:47–65.

—— 1978b. Speculations on real sea-level changes and vertical motions of continents at selected times in the Cretaceous and Tertiary Periods. *Geology 6*:247–250.

Boon, J. D. III, and MacIntyre, W. G. 1968. The boron-salinity relationship in estuarine sediments of the Rappahannock River, Virginia. *Chesapeake Science 9*:21–26.

Bordovskiy, O. K. 1965. Accumulation of organic matter in bottom sediments. *Marine Geology 3*:33–82.

Bosellini, A., and Winterer, E. L. 1975. Pelagic limestone and radiolarite of the Tethyan Mesozoic: a genetic model. *Geology 3*:279–282.

Boucot, A. J. 1974. Silurian and Devonian biogeography. Pp. 165–176 in *Paleogeographic Provinces and Provinciality,* ed. C. A. Ross. Society of Economic Paleontologists and Mineralogists, Special Publication 21.

Boulos, M., and Manuel, O. K. 1971. The xenon record of extinct radioactivities in the earth. *Science 174*:1334–1336.

Bouma, A. H. 1972. Recent and ancient turbidites and contourites. *Transactions of the Gulf Coast Association of Geological Societies 22*:205–221.

Bowen, H. J. M. 1966. *Trace Elements in Biochemistry.* Academic Press, London. 241 pp.

Bradshaw, J. S. 1959. Ecology of living planktonic foraminifera in the North and equatorial Pacific. *Contributions of the Cushman Foundation for Foraminiferal Research 10*:25–64.

Bramlette, M. N. 1961. Pelagic sediments. Pp. 345–366 in *Oceanography,* ed. M. Sears. American Association for the Advancement of Science, Publication no. 67.

Bray, J. R. 1977. Pleistocene volcanism and glacial initiation. *Science 197*:251–254.

Brenner, R. L., and Davies, D. K. 1973. Storm-generated coquinoid sandstone: genesis of high-energy marine sediments from the Upper Jurassic of Wyoming and Montana. *Geological Society of America Bulletin* 84:1685–1698.

――― 1974. Storm-generated coquinoid sandstone: genesis of high-energy marine sediments from the Upper Jurassic of Wyoming and Montana: reply. *Geological Society of America Bulletin 85*:838.

Brewer, P. G. 1975. Minor elements in sea water. Pp. 415–496 in *Chemical Oceanography,* ed. J. P. Riley and G. Skirrow, vol. 1, 2nd ed. Academic Press, London.

Briden, J. C. 1970. Palaeolatitude distribution of precipitated sediments. Pp. 437–444 in *Palaeogeophysics,* ed. S. K. Runcorn. Academic Press, London.

Brinkman, R. T. 1969. Dissociation of water vapor and evolution of oxygen in the terrestrial atmosphere. *Journal of Geophysical Research* 74:5355–5368.

Broadhurst, F. M., and Simpson, I. M. 1973. Bathymetry on a carboniferous reef. *Lethaia 6*:367–381.

Brock, T. D. 1978. *Thermophilic Microorganisms and Life at High Temperatures.* Springer-Verlag, New York. 465 pp.

Broecker, W. S. 1964. Radiocarbon dating: a case against the proposed link between river mollusks and soil humus. *Science 143*:596–597.

――― 1970. A boundary condition on the evolution of atmospheric oxygen. *Journal of Geophysical Research 75*:3553–3557.

――― 1971. A kinetic model for the chemical composition of sea water. *Quaternary Research 1*:188–207.

――― 1973. Factors controlling CO_2 content in the oceans and atmosphere. Pp. 32–50 in *Carbon and the Biosphere,* ed. G. M. Woodwell and E. V. Pecan. Technical Information Center, Office of Information Services, U.S. Atomic Energy Commission.

Broecker, W. S., and Takahashi, T. 1978. The relationship between lysocline depth and *in situ* carbonate ion concentration. *Deep-Sea Research 25*:65–95.

Broecker, W. S., Takahashi, T., and Li, Y.-H. 1976. Hydrography of the central Atlantic—I. The two-degree discontinuity. *Deep-Sea Research 23*:1083–1104.

Brooks, R. R., Presley, B. J., and Kaplan, I. R. 1968. Trace elements in the interstitial waters of marine sediments. *Geochimica et Cosmochimica Acta 32*:397–414.

Brosche, P. and Sündermann, J. 1977. Tides around Pangaea. *Naturwissenschaften 64*:89–90.

Brown, C. W. 1939. Hurricanes and shore-line changes in Rhode Island. *Geographical Journal 29*:416–430.

Brown, H. 1949. Rare gases and the formation of the earth's atmosphere. Pp. 200–208 in *The Atmospheres of the Earth and Planets,* ed. G. P. Kuiper. University of Chicago Press, Chicago.

Bryson, R. A. 1974. A perspective on climatic change. *Science 184*:753–760.

Buchardt, B. 1978. Oxygen isotope palaeotemperatures from the Tertiary Period in the North Sea area. *Nature 275*:121–123.

Buchardt, B., and Fritz, P. 1978. Strontium uptake in shell aragonite from the freshwater gastropod *Limnaea stagnalis*. *Science 199*:291–292.

Budyko, M. I. 1969. The effect of solar radiation variations on the climate of the earth. *Tellus 21*:611–619.

—— 1972. The future climate. *EOS* (American Geophysical Union Transactions) 53:868–874.

Burke, K. 1975. Atlantic evaporites formed by evaporation of water spilled from Pacific, Tethyan, and southern oceans. *Geology 3*:613–616.

Burnett, W. C. 1977. Geochemistry and origin of phosphorite deposits from off Peru and Chile. *Geological Society of America Bulletin 88*:813–823.

Butler, W. A., Jeffery, P. M., Reynolds, J. H., and Wasserburg, G. J. 1963. Isotopic variations in terrestrial Xenon. *Journal of Geophysical Research 68*:3283–3291.

Byrnes, J. G. 1968. Notes on the nature and environmental significance of the Receptaculitaceae. *Lethaia 1*:368–381.

Cadogan, P. H. 1977. Palaeoatmospheric argon in Rhynie chert. *Nature 268*:38–41.

Calder, J. A., and Parker, P. L. 1973. Geochemical implications of induced changes in C^{13} fractionation by blue-green algae. *Geochimica et Cosmochimica Acta 37*:133–140.

Calvert, S. E. 1976. The mineralogy and geochemistry of near-shore sediments. Pp. 188–280 in *Chemical Oceanography,* ed. J. P. Riley and R. Chester. vol. 6, 2nd ed. Academic Press, London.

—— 1977. Mineralogy of silica phases in deep-sea cherts and porcelanites. *Philosophical Transactions of the Royal Society of London, Part A 286*:239–252.

Cameron, E. M., and Baumann, A. 1972. Carbonate sedimentation during the Archean. *Chemical Geology 10*:17–30.

Campbell, C. A., and Valentine, J. W. 1977. Comparability of modern and ancient marine faunal provinces. *Paleobiology 3*:49–57.

Carden, J. R., Connelly, W., Forbes, R. B., and Turner, D. L. 1977. Blue schists of the Kodiak Islands, Alaska: an extension of the Seldovia schist terrane. *Geology 5*:529–533.

Carey, S. N., and Sigurdsson, H. 1978. Deep-sea evidence for distribution of tephra from the mixed magma eruption of the Soufrière on St. Vincent, 1902: ash turbidites and air fall. *Geology 6*:271–274.

Carey, S. W. 1976. *The Expanding Earth.* Elsevier, New York. 488 pp.

Carlson, T. N., and Prospero, J. M. 1972. The large-scale movement of Saharan air outbreaks over the northern equatorial Atlantic. *Journal of Applied Meteorology 11*:283–297.

Carpenter, T. H., Holle, R. L., and Fernandez-Partagas, J. J. 1972. Observed relationships between lunar tidal cycles and formation of hurricanes and tropical storms. *Monthly Weather Review 100*:451–460.

Cawley, J. L., Burruss, R. C., and Holland, H. D. 1969. Chemical weathering in central Iceland: an analogy of pre-Silurian weathering. *Science* 165:391–392.

Chan, K. M., and Manheim, F. T. 1970. Interstitial water studies on small core samples, deep sea drilling project, leg 2. *Initial Reports of the Deep Sea Drilling Project* 2:367–371.

Charney, J., Stone, P. H., and Quirk, W. J. 1976. Drought in the Sahara: insufficient biogeographical feedback? *Science* 191:100–102.

Chaudhuri, A. 1977. Influence of eolian processes on Precambrian sandstones of the Godavari Valley, South India. *Precambrian Research* 4:339–360.

Cherry, R. D., Higgo, J. J. W., and Fowler, S. W. 1978. Zooplankton fecal pellets and element residence times in the ocean. *Nature* 274:246–248.

Chester, R. 1972. Geological, geochemical and environmental implications of the marine dust veil. Pp. 291–305 in *The Changing Chemistry of the Oceans,* ed. D. Dyrssen and D. Jagner. Nobel Symposium Proceedings, no. 20. John Wiley and Sons, New York.

Chester, R., Elderfield, H., Griffin, J. J., Johnson, L. R., and Padgham, R. C. 1972. Eolian dust along the eastern margins of the Atlantic Ocean. *Marine Geology* 13:91–105.

Chilingar, G. V. 1963. Ca/Mg and Sr/Ca ratios of calcareous sediments as a function of depth and distance from shore. *Journal of Sedimentary Petrology* 33:236.

Churchman, G. J., Clayton, R. N., Sridhar, K., and Jackson, M. L. 1976. Oxygen isotopic composition of aerosol size quartz in shales. *Journal of Geophysical Research* 81:381–386.

Churkin, M., Jr., Carter, C., and Johnson, B. R. 1977. Subdivision of Ordovician and Silurian time scale using accumulation rates of graptolitic shale. *Geology* 5:452—456.

Ciesielski, P. F., and Weaver, F. M. 1974. Early Pliocene temperature changes in the Antarctic seas. *Geology* 2:511–515.

Cifelli, R. 1971. On the temperature relationships of planktonic foraminifera. *Journal of Foraminiferal Research* 1:170–177.

Cifelli, R., and Bernier, C. S. 1976. Planktonic foraminifera from near the West African coast and a consideration of faunal parcelling in the North Atlantic. *Journal of Foraminiferal Research* 6:258–273.

Cifelli, R., and Smith, R. K. 1969. Problems in the distribution of recent planktonic foraminifera and their relationships with water mass boundaries in the North Atlantic. Pp. 68–81 in *Proceedings of the First International Conference on Planktonic Microfossils. Geneva,* ed. P. Brönnimann and H. H. Renz, vol. 2. E. J. Brill, Leiden.

Cin, R. D. 1968. Climatic significance of roundness and percentage of quartz in conglomerates. *Journal of Sedimentary Petrology* 38:1094–1099.

Clarke, F. W., and Wheeler, W. C. 1922. *The Inorganic Constituents of Marine Invertebrates.* U.S. Geological Survey Professional Paper 124. 62 pp.

Clarke, W. B., Beg, M. A., and Craig, H. 1969. Excess ³He in the sea: evidence for terrestrial primordial helium. *Earth and Planetary Science Letters 6*:213–220.

Clarkson, E. N. K. 1967. Environmental significance of eye-reduction in trilobites and recent arthropods. *Marine Geology 5*:367–375.

CLIMAP Project Members. 1976. The surface of the ice-age earth. *Science 191*:1131–1137.

Cline, R. M., and Hays, J. D., eds. 1976. *Investigation of Late Quarternary Paleoceanography and Paleoclimatology.* Geological Society of America, Memoir 145. 464 pp.

Cloud, P. E., Jr. 1968. Atmospheric and hydrospheric evolution on the primitive earth. *Science 160*:729–736.

—— 1971a. Precambrian of North America. *Geotimes 16*:13–18.

—— 1971b. The Precambrian. *Science 173*:851–854.

—— 1973. Paleoecological significance of the banded iron-formation. *Economic Geology 68*:1135–1143.

—— 1974a. Evolution of ecosystems. *American Scientist 62*:54–66.

—— 1974b. Rubey conference on crustal evolution. *Science 183*:878–881.

—— 1976a. Beginnings of biospheric evolution and their biogeochemical consequences. *Paleobiology 2*:351–387.

—— 1976b. Major features of crustal evolution. *Geological Society of South Africa,* Annexure to vol. 79, pp. 1–33.

Cohen, Y. 1978. Consumption of dissolved nitrous oxide in an anoxic basin, Saanich Inlet, British Columbia. *Nature 272*:235–237.

Collins, D. 1978. A coiled nautiloid preserved in life orientation from the Middle Ordovician of Ontario. *Canadian Journal of Earth Sciences 15*:1661–1664.

Condie, K. C. 1972. A plate tectonics evolutionary model of the South Pass Archean greenstone belt, southwest Wyoming. Pp. 104–112 in *International Geological Congress, 24th Session, Section I (Montreal).*

—— 1973. Archean magmatism and crustal thickening. *Geological Society of America Bulletin 84*:2981–2992.

—— 1976. *Plate Tectonics and Crustal Evolution.* Pergamon Press, New York. 288 pp.

Condie, K. C., and Potts, M. J. 1969. Calc-alkaline volcanism and the thickness of the early Precambrian crust in North America. *Canadian Journal of Earth Sciences 6*:1179–1184.

Connolly, J. R., and Ewing, M. 1965a. Ice-rafted detritus as a climatic indicator in Antarctic deep-sea cores. *Science 150*:1822–1824.

—— 1965b. Pleistocene glacial-marine zones in North Atlantic deep-sea sediments. *Nature 208*:135–138.

Connor, E. F., and McCoy, E. D. 1979. The statistics and biology of the species-area relationship. *American Naturalist 113*:791–833.

Cook, H. E., and Enos, P., eds. 1977. *Deep-Water Carbonate Environments.* Society of Economic Paleontologists and Mineralogists, Special Publication no. 25. 336 pp.

Cook, P. J. 1977. Loss of boron from shells during weathering and possible implications for the determination of palaeosalinity. *Nature 268*:426–427.

Couch, E. L. 1971. Calculation of paleosalinities from boron and clay mineral data. *American Association of Petroleum Geologists Bulletin 55*:1829–1837.

Craig, H. 1961. Standard for reporting concentrations of deuterium and oxygen-18 in natural waters. *Science 133*:1833–1834.

———— 1965. The measurement of oxygen isotope paleotemperatures. Pp. 161–182 in *Stable Isotopes in Oceanographic Studies and Paleotemperatures (Spoleto, 1965)*, ed. E. Tongiorgi. Consiglio Nazionale Delle Ricerche Laboratorio Di Geologia Nucleare, Pisa.

Craig, H., and Gordon, L. I. 1965. Isotopic oceanography: deuterium and oxygen 18 variations in the ocean and the marine atmosphere. Pp. 277–374 in *Symposium on Marine Geochemistry,* ed. D. R. Schink and J. T. Corless. Narragansett Marine Laboratory, University of Rhode Island, Occasional Publication no. 3.

Craig, H., and Lupton, J. E. 1976. Primordial neon, helium, and hydrogen in oceanic basalts. *Earth and Planetary Science Letters 31*:369–385.

Cram, J. M. 1979. The influence of continental shelf width on tidal range: paleoceanographic implications. *Journal of Geology 87*:in press.

Creager, J. S., and Sternberg, R. W. 1972. Some specific problems in understanding bottom sediment distribution and dispersal on the continental shelf. Pp. 347–362 *Shelf Sediment Transport,* ed. D. J. P. Swift, D. B. Duane, and O. H. Pilkey. Dowden, Hutchenson and Ross, Stroudsburg, Pa.

Crimes, T. P., and Harper, J. C., eds. 1970. *Trace Fossils.* Seel House Press, Liverpool. 547 pp.

———— 1977. *Trace Fossils 2.* Seel House Press, Liverpool. 351 pp.

Cronin, J. F. 1971. Recent volcanism and the stratosphere. *Science 172*:847–849.

Crook, K. A. W. 1968. Weathering and roundness of quartz sand grains. *Sedimentology 11*:171–182.

Cutler, E. B. 1975. Zoogeographical barrier on the continental slope off Cape Lookout, North Carolina. *Deep-Sea Research 22*:893–902.

Dales, R. P., Mangum, C. P., and Tichy, J. C. 1970. Effects of changes in oxygen and carbon dioxide concentrations on ventilation rhythms in Onuphid polychaetes. *Journal of the Marine Biological Association of the United Kingdom 50*:365–380.

Daly, R. A. 1909. First calcareous fossils and the evolution of the limestones. *Geological Society of America Bulletin 20*:153–170.

Damon, P. E. 1968. The relationship between terrestrial factors and climate. *Meteorological Monographs 8*:106–111.

Damuth, J. E. 1977. Late Quaternary sedimentation in the western equatorial Atlantic. *Geological Society of America Bulletin 88*:695–710.

Dansgaard, W. 1964. Stable isotopes in precipitation. *Tellus 16*:436–468.

Davies, R. D., and Allsopp, H. L. 1976. Strontium isotopic evidence relating to the evolution of the lower Precambrian granitic crust in Swaziland. *Geology* 4:553–556.

Davies, T. A., Hay, W. W., Southam, J. R., and Worsley, T. R. 1977. Estimates of Cenozoic oceanic sedimentation rates. *Science* 197:53–55.

Dearborn, D. S. P., and Newman, M. J. 1978. Efficiency of convection and time variation of the solar constant. *Science* 201:150–151.

Deelman, J. C. 1972. On mechanisms causing birdseye structures. *Neues Jahrbuch für Geologie und Paläontologie,* Monatshefte no. 10. pp. 582–595.

Defant, A. 1958. *Ebb and Flow.* University of Michigan Press, Ann Arbor. 121 pp.

Degens, E. T. 1966. Stable isotope distribution in carbonates. Pp. 193–208 in *Carbonate Rocks: Physical and Chemical Aspects,* ed. G. V. Chilingar, H. J. Bissell, and R. W. Fairbridge. Elsevier, Amsterdam.

———— 1969. Biogeochemistry of stable carbon isotopes. Pp. 304–329 in *Organic Chemistry,* ed. G. Eglinton and M. T. J. Murphy. Springer-Verlag, Berlin.

Degens, E. T., and Epstein, S. 1962. Relationship between O^{18}/O^{16} ratios in coexisting carbonates, cherts, and diatomites. *American Association of Petroleum Geologists Bulletin* 46:534–542.

Degens, E. T., Hunt, J. M., Reuter, J. H., and Reed, W. E. 1964. Data on the distribution of amino acids and oxygen isotopes in petroleum brine waters of various geologic ages. *Sedimentology* 3:199–225.

Degens, E. T., Williams, E. G., and Keith, M. L. 1958. Environmental studies of carboniferous sediments. Part II. Application of geochemical criteria. *American Association of Petroleum Geologists Bulletin* 42:981–997.

Delaca, T. E., and Lipps, J. H. 1972. The mechanism and adaptive significance of attachment and substrate pitting in the foraminiferan *Rosalina globularis* d'Orbigny. *Journal of Foraminiferal Research* 2:68–72.

Deuser, W. G. 1971. Organic-carbon budget of the Black Sea. *Deep-Sea Research* 18:995–1004.

———— 1975. Reducing environments. Pp. 1–37 in *Chemical Oceanography,* ed. J. P. Riley and G. Skirrow, vol. 3, 2nd ed. Academic Press, London.

Deuser, W. G., and Degens, E. T. 1967. Carbon isotope fractionation in the system CO_2 (gas)–CO_2 (aqueous)–HCO_3^- (aqueous). *Nature* 215:1033–1035.

Deuser, W. G., Degens, E. T., and Guillard, R. R. L. 1968. Carbon isotope relationships between plankton and sea water. *Geochimica et Cosmochimica Acta* 32:657–660.

Deuser, W. G., Ross, E. H., and Mlodzinska, Z. J. 1978. Evidence for and rate of denitrification in the Arabian Sea. *Deep-Sea Research* 25:431–445.

Dewis, F. J., Levinson, A. A., and Bayliss, P. 1972. Hydrogeochemistry of the surface waters of the Mackenzie River drainage basin, Canada. IV. Boron-salinity-clay mineralogy relationships in modern deltas. *Geochimica et Cosmochimica Acta* 36:1359–1375.

Dietz, R. S. 1964a. A re-cycled hydrosphere? *Nature 201*:279–281.

——— 1964b. Wave-base, marine profile of equilibrium, and wave-built terraces: reply. *Geological Society of America Bulletin. 75*:1275–1282.

——— 1965. Earth's original crust—lost quest? *Tectonophysics 6*:515–520.

Dietz, R. S., and Holden, J. C. 1965. Earth and moon: tectonically contrasting realms. *Annals of the New York Academy of Science 123*:631–640.

Dingle, R. V. 1965. Sand waves in the North Sea mapped by continuous reflection profiling. *Marine Geology 3*:391–400.

Distanov, U. G., and Sorokin, V. I. 1975. Two genetic types of glauconite in paleogene sediments of the Volga region. *Doklady Earth Science Sections 213*:181–183. (English translation of Russian original published in 1973.)

Dodd, J. R. 1967. Magnesium and strontium in calcareous skeletons: a review. *Journal of Paleontology 41*:1313–1329.

Dodd, J. R., and Schopf, T. J. M. 1972. Approches to biogeochemistry. Pp. 46–60 in *Models in Paleobiology,* ed. T. J. M. Schopf. Freeman, Cooper and Co., San Francisco.

Dodd, J. R., and Stanton, R. J., Jr. 1975. Paleosalinities within a Pliocene bay, Kettleman Hills, California: a study of the resolving power of isotopic and faunal techniques. *Geological Society of America Bulletin 86*:51–64.

Doemel, W. N., and Brock, T. D. 1974. Bacterial stromatolites: origin of laminations. *Science 184*:1083–1085.

Dole, M. 1949. The history of oxygen. *Science 109*:77–81, 96.

Donn, W. L., Donn, B. D., and Valentine, W. G. 1965. On the early history of the earth. *Geological Society of America Bulletin 76*:287–306.

Donn, W. L., and Shaw, D. M. 1977. Model of climate evolution based on continental drift and polar wandering. *Geological Society of America Bulletin 88*:390–396.

Donnelly, T. W. 1977. Chemistry of Cenozoic sedimentation in the world ocean. *Geological Society of America Abstracts with Programs 9*:953.

Dott, R. H., Jr. 1974. Cambrian tropical storm waves in Wisconsin: reply. *Geology 2*:613.

Douglas, I. 1967. Erosion of granite terrains under tropical rain forest in Australia, Malaysia, and Singapore. Pp. 31–40 in *Symposium on River Morphology, General Assembly of Bern, 25 Sept.–7 Oct. 1967.* Association Internationale d'Hydrologie Scientifique, Publication no. 75.

Douglas, R. G., and Savin, S. M. 1973. Oxygen and carbon isotope analyses of Cretaceous and Tertiary foraminifera from the central North Pacific. *Initial Reports of the Deep Sea Drilling Project 17*:591–605.

——— 1975. Oxygen and carbon isotope analyses of Tertiary and Cretaceous microfossils from Shatsky Rise and other sites in the North Pacific Ocean. *Initial Reports of the Deep Sea Drilling Project 32*:509–520.

Dowding, L. G. 1977. Sediment dispersal within the Cocos Gap, Panama Basin. *Journal of Sedimentary Petrology 47*:1132–1156.

Drake, C. L., and Burk, C. A. 1974. Geological significance of continental margins. Pp. 3–10 in *The Geology of Continental Margins,* ed. C. A. Burk and C. L. Drake. Springer-Verlag, New York.

Draper, L. 1967. Wave activity at the sea bed around northwestern Europe. *Marine Geology 5*:133–140.

Drever, J. I. 1974. Geochemical model for the origin of Precambrian banded iron formations. *Geological Society of America Bulletin 85*:1099–1106.

Drewry, G. E., Ramsay, A. T. S., and Smith, A. G. 1974. Climatically controlled sediments, the geomagnetic field, and trade wind belts in Phanerozoic time. *Journal of Geology 82*:531–553.

Dritschilo, W., Cornell, H., Nafus, D., and O'Connor, B. 1975. Insular biogeography: of mice and mites. *Science 190*:467–469.

Drury, S. A. 1978. Were Archaean continental geothermal gradients much steeper than today? *Nature 274*:720–721.

Duffield, W. A. 1978. Vesicularity of basalt erupted at Reykjanes Ridge crest. *Nature 274*:217–220.

Dunne, T. 1978. Rates of chemical denudation of silicate rocks in tropical catchments. *Nature 274*:244–246.

Duplessy, J. C., Chenouard. L., and Vila, F. 1975. Weyl's theory of glaciation supported by isotopic study of Norwegian core K 11. *Science 188*:1208–1209.

Duplessy, J. C., Lalou, C., and Vinot, A. C. 1970. Differential isotopic fractionation in benthic foraminifera and paleotemperatures reassessed. *Science 168*:250–251.

Durazzi, J. T., and Stehli, F. G. 1972. Average generic age, the planetary temperature gradient, and pole location. *Systematic Zoology 21*:384–389.

Durum, W. H., and Haffty, J. 1963. Implications of the minor element content of some major streams of the world. *Geochimica et Cosmochimica Acta 27*:1–11.

Dvořák, J. 1972. Shallow-water character of the nodula limestones and their paleogeographic interpretation. *Neues Jahrbuch für Geologie und Paläontologie,* Monatshefte 1972, no. 9, pp. 509–511.

Dymond, J., and Hogan, L. 1978. Factors controlling the noble gas abundance patterns of deep-sea basalts. *Earth and Planetary Science Letters 38*: 117–128.

Eaton, G. P. 1963. Volcanic ash deposits as a guide to atmospheric circulation in the geologic past. *Journal of Geophysical Research 68*:521–528.

Eddy, J. A. 1976. The Maunder minimum. *Science 192*:1189–1202.

——— 1977. Climate and the changing sun. *Climatic Change 1*:173–190.

Eddy, J. A., Gilman, P. A., and Trotter, D. E. 1977. Anomalous solar rotation in the early seventeenth century. *Science 198*:824–829.

Eichelberger, J. C. 1978. Andesitic volcanism and crustal evolution. *Nature 275*:21–27.

Einsele, G., and Wiedmann, J. 1975. Faunal and sedimentological evidence for upwelling in the Upper Cretaceous coastal basin of Tarfaya, Morocco. Pp. 67–74 in *IXth International Congress of Sedimentology* (Nice, 1975).

Ekman, S. 1953. *Zoogeography of the Sea.* Sidgwick and Jackson, London. 417 pp. (1967 paperback edition by the same publisher.)

El-Hinnawi, E. E., and Loukina, S. M. 1972. On the distribution of strontium in some Egyptian carbonate rocks. *Neues Jahrbuch für Geologie und Paläontologie,* Monatshefte no. 2, pp. 72–77.

Elias, M. K. 1937. Depth of deposition of the Big Blue (Late Paleozoic) sediments in Kansas. *Geological Society of America Bulletin 48*:403–432.

——— 1964. Depth of Late Paleozoic sea in Kansas and its megacyclic sedimentation. Pp. 87–106 in *Symposium on Cyclic Sedimentation,* ed. D. F. Merriam, vol. 1. Geological Survey of Kansas Bulletin 169.

Ellis, H. T., and Pueschel, R. F. 1971. Solar radiation: absence of air pollution trends at Mauna Loa. *Science 172*:845–846.

Emery, K. O. 1955. Grain size of marine beach gravels. *Journal of Geology 63*:39–49.

——— 1965. Characteristics of continental shelves and slopes. *Bulletin of the American Association of Petroleum Geologists 49*:1379–1384.

——— 1966. *Atlantic Continental Shelf and Slope of the United States. Geologic Background.* Pp. A1–A23 in U.S. Geological Survey, Professional Paper 529-A.

——— 1968. Relict sediments on continental shelves of world. *American Association of Petroleum Geologists Bulletin 52*:445–464.

——— 1970. Continental margins of the world. Pp. 7–29 in *The Geology of the East Atlantic Continental Margin, 1.* General and Economic Papers, ICSU/SCOR, Working Party 31, Symposium, Cambridge, Rept. 70/13.

Emery, K. O., and Csanady, G. T. 1973. Surface circulation of lakes and nearly land-locked seas. *Proceedings of the National Academy of Sciences 70*:93–97.

Emery, K. O., and Milliman, J. D. 1978. Suspended matter in surface waters: influence of river discharge and of upwelling. *Sedimentology 25*:125–140.

Emery, K. O., Orr, W. L., and Rittenberg, S. C. 1955. Nutrient budgets in the ocean. Pp. 147–157 in *Essays in the Natural Sciences in Honor of Captain Allan Hancock.* University of Southern California Press, Los Angeles.

Emery, K. O., and Uchupi, E. 1972. *Western North Atlantic Ocean: Topography, Rocks, Structure, Water, Life, and Sediments.* American Association of Petroleum Geologists, Memoir 17. 532 pp.

Emiliani, C. 1954. Temperatures of Pacific bottom waters and polar superficial waters during the Tertiary. *Science 119*:853–855.

——— 1955. Pleistocene temperatures. *Journal of Geology 63*:538–578.

——— 1966. Isotopic paleotemperatures. *Science 154*:851–857.

——— 1971. The amplitude of Pleistocene climatic cycles at low latitudes and isotopic composition of glacial ice. Pp. 183–197 in *The Late Cenozoic Glacial Ages,* ed. K. K. Turekian. Yale University Press, New Haven.

——— 1977. Oxygen isotopic analysis of the size fraction between 62 and 250 micrometers in Caribbean cores P6304-8 and P6304-9. *Science 198*:1255–1256.

Emiliani, C., and Shackleton, N. J. 1974. The Brunhes Epoch: isotopic paleo-temperatures and geochronology. *Science 183*:511–514.

Engel, A. E. J. 1969. The Barberton Mountain Land: clues to the differentiation of the Earth. *Geological Society of South Africa,* Annexure to vol. 71 (Symposium on Rhodesian Basement Complex), pp. 255–270. (Reprinted in *Adventures in Earth History,* ed. P. E. Cloud, Jr. [W. H. Freeman and Co., San Francisco, 1970], pp. 431–445.)

Epstein, S. 1968. Distribution of carbon isotopes and their biochemical and geochemical significance. Pp. 5–14 in *CO₂: Chemical, Biochemical and Physiological Aspects,* ed. R. E. Forster, J. Edsall, A. B. Otis, and F. J. W. Roughton. NASA Office of Technology Utilization. Symposium at Haverford, Pa., 1968.

Epstein, S., Buchsbaum, R., Lowenstam, H. A., and Urey, H. C. 1953. Revised carbonate-water isotopic temperature scale. *Geological Society of America Bulletin 64*:1315–1326.

Epstein, S., and Mayeda, T. 1953. Variation of O^{18} content of waters from natural sources. *Geochimica et Cosmochimica Acta 4*:213–224.

Epstein, S., Sharp, R. P., and Gow, A. J. 1970. Antarctic ice sheet: stable isotope analyses of Byrd station cores and interhemispheric climatic implications. *Science 168*:1570–1572.

Epstein, S., Thompson, P., and Yapp, C. J. 1977. Oxygen and hydrogen isotopic ratios in plant cellulose. *Science 198*:1209–1215.

Erez, J. 1978. Vital effect on stable-isotope composition seen in foraminifera and coral skeletons. *Nature 273*:199–202.

Ericsson, B. 1972. The chlorinity of clays as a criterion of the palaeosalinity. *Geologiska Föreningens i Stockholm Förhandlingar 94*:5–21.

Eriksson, K. A., Hobday, D, K., and Klein, G. deV. 1976. Tidal sedimentation. *Geotimes 21*:(10):19–20.

Eriksson, K. A., McCarthy, T. S., and Truswell, J. F. 1975. Limestone formation and dolomitization in a Lower Proterozoic succession from South Africa. *Journal of Sedimentary Petrology 45*:604–614.

Eugster, H. P., and Chou, I-M. 1973. The depositional environments of Precambrian banded iron-formations. *Economic Geology 68*:1144–1168.

Eugster, H. P., and Munoz, J. 1966. Ammonium micas: possible sources of atmospheric ammonia and nitrogen. *Science 151*:683–686.

Evans, R. 1978. Origin and significance of evaporites in basins around Atlantic margin. *American Association of Petroleum Geologists Bulletin 62*:223–234.

Ewing, J. A. 1973. Wave-induced bottom currents on the outer shelf. *Marine Geology 15*:M31–M35.

Fairbridge, R. W. 1964. The importance of limestone and its Ca/Mg content to paleoclimatology. Pp. 431–478, 521–530 in *Problems in Paleoclimatology,* ed. A. E. M. Nairn. Interscience Publishers, London.

Fanale, F. P. 1971. A case for catastrophic early degassing of the earth. *Chemical Geology 8*:79–105.

Fanale, F. P., and Cannon, W. A. 1971. Physical adsorption of rare gas on terrigenous sediments. *Earth and Plentary Science Letters 11*:362–368.

Faure, G., Assereto, R., and Tremba, E. L. 1978. Strontium-isotope composition of marine carbonates of Middle Triassic to Early Jurassic age, Lombardic Alps, Italy. *Sedimentology, 25*:523–543.

Faure, G., and Powell, J. L. 1972. *Strontium Isotope Geology.* Springer-Verlag, Heidelberg. 188 pp.

Ferris, J. P., and Joshi, P. C. 1978. Chemical evolution from hydrogen cyanide: photochemical decarboxylation of orotic acid and orotate derivatives. *Science 201*:361–362.

Ferris, J. P., and Nicodem, D. E. 1972. Ammonia photolysis and the role of ammonia in chemical revolution. *Nature 238*:268–269.

Fischer, A. G. 1960. Latitudinal variations in organic diversity. *Evolution 14*:50–74.

Fisher, J. S., and Stauble, D. K. 1977. Impact of Hurricane Belle on Assateague Island washover. *Geology 5*:765–768.

Fleagle, R. G., and Businger, J. A. 1975. The "Greenhouse effect." *Science 190*:1042–1043.

Fleming, R. H., and Elliott, F. E. 1952. Some physical aspects of the inshore environment of the coastal waters of the United States and Mexico. Pp. 409–416 in *Proceedings Eighth General Assembly and Seventeenth International Congress of the International Geographical Union, Washington, D.C., August 8–15, 1952* (no editor).

Flessa, K. W., and Sepkoski, J. J., Jr. 1978. On the relationship between Phanerozoic diversity and changes in habitable area. *Paleobiology 4*:359–366.

Flohn, H. 1969. *Climate and Weather,* trans. B. V. de G. Walden. McGraw-Hill Book Co., New York. 253 pp.

Floran, R. J., and Papike, J. J. 1975. Petrology of the low-grade rocks of the Gunflint Iron-Formation, Ontario, Minnesota. *Geological Society of America Bulletin 86*:1169–1190.

Flörke, O. W., Hollmann, R., von Rad, U., and Rösch, H. 1976. Intergrowth and twinning in opal-CT lepispheres. *Contributions to Mineralogy and Petrology 58*:235–242.

Foerste, A. F. 1930. The color patterns of fossil cephalopods and brachiopods, with notes on gastropods and pelecypods. *Contributions from the Museum of Paleontology.* (University of Michigan) *3*:109–150.

Folk, R. L. 1951. Stages of textural maturity in sedimentary rocks. *Journal of Sedimentary Petrology 21*:127–130.

——— 1966. A review of grain-size parameters. *Sedimentology 6*:73–93.

——— 1974. The natural history of crystalline calcium carbonate: effect of magnesium content and salinity. *Journal of Sedimentary Petrology 44*:40–53.

Folk, R. L., and Land, L. S. 1975. Mg/Ca ratio and salinity: two controls over crystallization of dolomite. *American Association of Petroleum Geologists Bulletin 59*:60–68.

Folk, R. L., and Robles, R. 1964. Carbonate sands of Isla Perez, Alacran Reef Complex, Yucatan. *Journal of Geology 72*:255–292.

Folk, R. L., and Siedlecka, A. 1974. The "Schizohaline" environment: its sedimentary and diagenetic fabrics as exemplified by Late Paleozoic rocks of Bear Island, Svalbard. *Sedimentary Geology 11*:1–15.

Fontugne, M., and Duplessy, J. C. 1978. Carbon isotope ratio of marine plankton related to surface water masses. *Earth and Planetary Science Letters 41*:365–371.

Forchhammer, G. 1865. On the composition of sea-water in the different parts of the ocean. *Philosophical Transactions of the Royal Society of London 155*:203–262.

Forney, G. G. 1975. Permo-Triassic sea-level change. *Journal of Geology 83*:773–779.

Fortey, R. A. 1975. Early Ordovician trilobite communities. *Fossils and Strata 4*:339–360.

Fowler, W. B., and Helvey, J. D. 1975. Irrigation increases rainfall? *Science 188*:281.

Fox, W. T., and Davis, R. A., Jr. 1976. Weather patterns and coastal processes. Pp. 1–23 in *Beach and Nearshore Sedimentation,* ed. R. A. Davis, Jr., and R. L. Ethington. Society of Economic Paleontologists and Mineralogists Special Publication no. 24.

Frank, P. W. 1975. Latitudinal variation in the life history features of the black turban snail *Tegula funebralis* (Prosobranchia: Trochidae). *Marine Biology 31*:181–192.

French, B. M. 1966. Some geological implications of equilibrium between graphite and a C-H-O gas phase at high temperatures and pressures. *Reviews of Geophysics 4*:223–253.

Frerichs, W. E., Heiman, M. E., Borgman, L. E., and Bé, A. W. H. 1972. Latitudinal variations in planktonic foraminiferal test porosity: Part I. Optical studies. *Journal of Foraminiferal Research 2*:6–13.

Frey, R. W., ed. 1975. *The Study of Trace Fossils.* Springer-Verlag, New York. 562 pp.

Friedman, G. M. 1967. Dynamic processes and statistical parameters compared for size frequency distribution of beach and river sands. *Journal of Sedimentary Petrology 37*:327–354.

Frölich. C. 1977. Contemporary measures of the solar constant. Pp. 93–109 in *The Solar Output and Its Variation,* ed. O. R. White. Colorado Associated University Press, Boulder.

Fryer, B. J. 1977. Rare earth evidence in iron-formations for changing Precambrian oxidation states. *Geochimica et Cosmochimica Acta 41*:361–367.

Füchtbauer, H. 1963. Zum Einfluss des Ablagerungsmilieus auf die Farbe von Biotiten und Turmalinen. *Fortschritte der Geologie Rheinland und Westfalen 10*:331–336.

Fullard, H., ed. 1971. *World Patterns: The Aldine College Atlas.* Aldine Publishing Co. and Scott, Foresman and Co., Chicago. 128 pp.

Furnes, H. 1973. Variolitic structure in Ordovician pillow lava and its possible significance as an environmental indicator. *Geology 1*:27–30.

Furst, M., Lowenstam, H. A., and Burnett, D. S. 1976. Radiographic study of the distribution of boron in recent mollusc shells. *Geochimica et Cosmochimica Acta 40*:1381–1386.

Galimov, E. M., Kuznetsova, N. G., and Prokhorov, V. S. 1968. The composition of the former atmosphere of the earth as indicated by isotopic analysis of Precambrian carbonates. *Geochemistry International 5*:1126–1131. (Translation from *Geokhimiya,* 1968, no. 11, pp. 1376–1381.)

Galloway, W. E. 1975. Process framework for describing the morphologic and stratigraphic evolution of deltaic depositional systems. Pp. 87–98 in *Deltas,* ed. M. L. Broussard. Houston Geological Society, Houston.

Garlick, S., Oren, A., and Padan, E. 1977. Occurrence of facultative anoxygenic photosynthesis among filamentous and unicellular Cyanobacteria. *Journal of Bacteriology 129*:623–629.

Garrels, R. M., and Mackenzie, F. T. 1969. Sedimentary rock types: relative proportions as a function of geological time. *Science 163*:570–571.

———— 1971a. *Evolution of Sedimentary Rocks.* W. W. Norton, New York. 397 pp.

———— 1971b. Gregor's denudation of continents. *Nature 231*:382–383.

Garrels, R. M., and Perry, E. A., Jr. 1974. Cycling of carbon, sulfur, and oxygen through geologic time. Pp. 303–336 in *The Sea,* ed. E. D. Goldberg, vol. 5. John Wiley and Sons, New York.

Garrison, R. E. 1974. Radiolarian cherts, pelagic limestones, and igneous rocks in eugeosynclinal assemblages. Pp. 367–399 in *Pelagic Sediments: On Land and under the Sea,* ed. K. J. Hsü and H. C. Jenkyns. Special Publications of the International Association of Sedimentologists, no. 1.

Gartner, S., and Keany, J. 1978. The terminal Cretaceous event: a geologic problem with an oceanographic solution. *Geology, 6*:708–712.

Gates, E. L. 1976. Modeling the ice-age climate. *Science 191*:1138–1144.

Gibbs, R. J. 1973. Mechanisms of trace metal transport in rivers. *Science 180*:71–73.

Gibbs, R. J., Matthews, M. D., and Link, D. A. 1971. The relationship between sphere size and settling velocity. *Journal of Sedimentary Petrology 41*:7–18.

Gibson, L. B. 1966. Some unifying characteristics of species diversity. *Contributions from the Cushman Foundation for Foraminiferal Research 17*:117–124.

Gibson, T. G. 1967. Stratigraphy and paleoenvironment of the phosphatic Miocene strata of North Carolina. *Geological Society of America Bulletin 78*:631–650.

———— 1968. Stratigraphy and paleoenvironment of the phosphatic Miocene strata of North Carolina: reply. *Geological Society of America Bulletin 79*:1437–1448.

Ginsburg, R. N., ed. 1975. *Tidal Deposits.* Springer-Verlag, New York. 428 pp.

Ginsburg, R. N., and Hardie, L. A. 1975. Tidal and storm deposits, northwestern Andros Island, Bahamas. Pp. 201–208 in *Tidal Deposits,* ed. R. N. Ginsburg. Springer-Verlag, New York.

Glaesner, M. F., and Wade, M. 1966. The late Precambrian fossils from Ediacara, South Australia. *Palaeontology 9*:599–628.

Glasby, G. P. 1972a. Effect of pressure on deposition of manganese oxides in the marine environment. *Nature Physical Science 237*:85–86.

—— 1972b. The mineralogy of manganese nodules from a range of marine environments. *Marine Geology 13*:57–72.

Glennie, K. W. 1970. *Desert Sedimentary Environments.* Developments in Sedimentology, no. 14. Elsevier, Amsterdam. 222 pp.

Glennie, K. W., and Evans, G. 1976. A reconnaissance of the Recent sediments of the Ranns of Kutch, India. *Sedimentology 23*:625–647.

Goldberg, E. D. 1965. Minor elements in sea water. Pp. 163–196 in *Chemical Oceanography,* ed. J. P. Riley and G. Skirrow, vol. 1. Academic Press, London.

Goldich, S. S. 1938. A study in rock weathering. *Journal of Geology 46*:17–58.

Goldich, S. S., Hedge, C. E., Stern, T. W. 1970. Age of the Morton and Montevideo gneisses and related rocks, southwestern Minnesota. *Geological Society of America Bulletin 81*:3671–3696.

Goldring, R., and Bridges, P. 1973. Sublittoral sheet sandstones. *Journal of Sedimentary Petrology 43*:736–747.

Gonfiantini, R. 1978. Standards for stable isotope measurements in natural compounds. *Nature 271*:534–536.

Gooch, J. L., and Schopf, T. J. M. 1972. Genetic variability in the deep sea: relation to environmental variability. *Evolution 26*:545–552.

Goodwin, A. M. 1973. Archean iron-formations and tectonic basins of the Canadian Shield. *Economic Geology 68*:915–933.

Goodwin, P. W., and Anderson, E. J. 1974. Associated physical and biogenic structures in environmental subdivision of a Cambrian tidal sand body. *Journal of Geology 82*:779–794.

Gramberg, I. S., and Spiro, N. S. 1964. Evolution of the composition of water of the Arctic Ocean basin during the Upper Paleozoic and Mesozoic. Pp. 511–521 in *Chemistry of the Earth,* ed. A. P. Vinogradov, vol. 2. (English translation 1967: Israel program for scientific translations.)

Grandstaff, D. E. 1974. Uraninite oxidation and the Precambrian atmosphere. *Transactions of the American Geophysical Union, 55*:457.

Gray, W. M. 1968. Global view of the origin of tropical disturbances and storms. *Monthly Weather Review 96*:669–700.

Grazzini, C. V. 1975. ^{18}O changes in foraminifera carbonates during the last 10^5 years in the Mediterranean Sea. *Science 190*:272–274.

Green, D. H. 1975. Genesis of Archean peridotitic magmas and constraints on Archean geothermal gradients and tectonics. *Geology 3*:15–18.

Green, D. H., Nicholls, I. A., Viljoen, M. J., and Viljoen, R. P. 1975. Experimental demonstration of the existence of peridotitic liquids in earliest Archean magmatism. *Geology 3*:11–14.

Greene, H. G. 1970. Microrelief on an arctic beach. *Journal of Sedimentary Petrology 40*:419–427.

Gregor, C. B. 1967. The geochemical behavior of sodium with special reference to post-Algonkian sedimentation. *Koninklijk Nederlandse Akademie van Wetenschappen, Afdeeling Naturkunde, Verhandelingen 24*:1–67.

—— 1968. The rate of denudation in Post-Algonkian time. *Proceedings of* the Koninkijke Nederlandse Akademie van Wetenschappen; Series B, Physical Sciences 71:22–30.

—— 1970. Denudation of the continents. Nature 228:273–275.

Griffin, J. J., Windom, H., and Goldberg, E. D. 1968. The distribution of clay minerals in the world ocean. Deep-Sea Research 15:433–459.

Gross, M. G. 1972. Oceanography. Prentice-Hall, Englewood Cliffs, N.J. 581 pp.

Gulbrandsen, R. A. 1969. Physical and chemical factors in the formation of marine apatite. Economic Geology 64:365–382.

Hagan, G. M., and Logan, B. W. 1974. Development of carbonate banks and hypersaline basins, Shark Bay, Western Australia. American Association of Petroleum Geologists Memoir 22:61–139.

Håkansson, E., Bromley, R., and Perch-Nielsen, K. 1974. Maastrictian chalk of northwest Europe—a pelagic shelf sediment. Pp. 211–233 in Pelagic Sediments: On Land and under the Sea, ed. K. J. Hsü and H. C. Jenkyns. Special Publications of the International Association of Sedimentologists, no. 1.

Hallam, A. 1975. Jurassic Environments. Cambridge University Press, Cambridge. 269 pp.

—— 1977a. Axonic events in the Cretaceous ocean. Nature 268:15–16.

—— ed. 1977b. Patterns of Evolution As Illustrated by the Fossil Record. Elsevier, Amsterdam. 591 pp.

Halley, R. B. 1977. Ooid fabric and fracture in the Great Salt Lake and the geologic record. Journal of Sedimentary Petrology 47:1099–1120.

Hanks, T. C., and Anderson, D. L. 1969. The early thermal history of the earth. Physics of the Earth and Planetary Interiors 2:19–29.

Hare, P. E. 1977. Amino acid dating—limitations and potential. Geological Society of America Abstracts with Programs 9:1004–1005.

Hargraves, R. B. 1970. Sedimentologic evidence of strong tidal currents in the Early Proterozoic. Pp. 471–478 in Palaeogeophysics, ed. S. K. Runcorn. Academic Press, London.

Harland, W. B., and Herod, K. N. 1975. Glaciations through time. Pp. 189–216 in Ice Ages: Ancient and Modern, ed. A. E. Wright and F. Moseley. Seel House Press, Liverpool.

Harris, L. D. 1973. Dolomotization model for Upper Cambrian and Lower Ordovician carbonate rocks in the eastern United States. Journal of Research 1:63–78.

Harriss, R. C. 1969. Boron regulation in the oceans. Nature 223:290–291.

Hart, R. A. 1973. Geochemical and geophysical implications of the reaction between seawater and the oceanic crust. Nature 243:76–78.

Hathaway, J. C. 1972. Regional clay mineral facies in estuaries and continental margin of the United States east coast. Pp. 293–316 in Environmental Framework of Coastal Plain Estuaries, ed. B. W. Nelson. Geological Society of America Memoir 133.

Hayden, B. P. 1975. Storm wave climates at Cape Hatteras, North Carolina: recent secular variations. Science 190:981–983.

Hayes, M. O. 1964. Lognormal distribution of inner continental shelf widths and slopes. *Deep-Sea Research 11*:53–78.

—— 1967a. *Hurricanes as Geological Agents: Case Studies of Hurricanes Carla, 1961, and Cindy, 1963*. Bureau of Economic Geology, Report of Investigations, no. 61, University of Texas. 54 pp.

—— 1967b. Relationship between coastal climate and bottom sediment type on the inner continental shelf. *Marine Geology 5*:111–132.

Haynes, J. 1965. Symbiosis, wall structure and habitat in foraminifera. *Contributions from the Cushman Foundation for Foraminiferal Research 16*:40–43.

Heath, G. R. 1974. Dissolved silica and deep-sea sediments. Pp. 77–93 in *Paleo-Oceanography*, ed. W. W. Hay. Society of Economic Paleontologists and Mineralogists, Special Publication no. 20.

Hecht, A. D. 1973a. A model for determining Pleistocene paleotemperatures from planktonic foraminiferal assemblages. *Micropaleontology 19*:68–77.

—— 1973b. Faunal and oxygen isotopic paleotemperatures and the amplitude of glacial/interglacial temperature changes in the equatorial Atlantic, Caribbean Sea and Gulf of Mexico. *Journal of Quaternary Research 3*:671–690.

—— 1974a. Intraspecific variation in recent populations of *Globigerinoides ruber* and *Globigerinoides trilobus* and their application to paleoenvironmental analysis. *Journal of Paleontology 48*:1217–1234.

—— 1974b. Quantitative micropaleontology and the amplitude of glacial/interglacial temperature changes in the Caribbean Sea, Gulf of Mexico and Equatorial Atlantic. *Colloques Internationaux du Centre National de la Recherche Scientifique, 219*:211–220.

Hedges, J. I., and Parker, P. L. 1976. Land-derived organic matter in surface sediments from the Gulf of Mexico. *Geochimica et Cosmochimica Acta 40*:1019–1029.

Heezen, B. C. 1963. Turbidity currents. Pp. 742–775 in *The Sea*, ed. M. N. Hill, vol. 3. Wiley Interscience, New York.

Heezen, B. C., and Hollister, C. D. 1964. Deep-sea current evidence from abyssal sediments. *Marine Geology 1*:141–174.

Heezen, B. C., Hollister, C. D., and Ruddiman, W. F. 1966. Shaping of the continental rise by deep geostrophic currents. *Science 152*:502–508.

Hein, J. R., Scholl, D. W., Miller, J. 1978. Episodes of Aleutian ridge explosive volcanism. *Science 199*:137–141.

Heinrichs, T. K., and Reimer, T. O. 1977. A sedimentary barite deposit from the Archean Fig Tree Group of the Barberton Mountain Land (South Africa). *Economic Geology 72*:1426–1441.

Heling, D. 1969. Relationships between initial porosity of Tertiary argillaceous sediments and paleosalinity in the Rheintalgraben (SW-Germany). *Journal of Sedimentary Petrology 39*:246–254.

Henderson, L. J. 1913. *The Fitness of the Environment*. Beacon Press, Boston. 317 pp. (Paperback edition published 1958.)

Hertweck, G. 1971. Der Golf von Gaeta (Tyrrhenisches Meer). V. Abfolge der Biofaziesbereiche in den Vorstand- und Shelfsedimenten. *Senckenbergiana Maritima 3*:247–276.

Hess, H. H. 1962. History of ocean basins. Pp. 599–620 in *Petrologic Studies: A Volume to Honor A. F. Buddington,* ed. A. E. J. Engel, H. L. James, and B. F. Leonard. Geological Society of America.

Hirst, D. M. 1968. Relationships between minor elements, mineralogy and depositional environment in carboniferous sedimentary rocks from a borehole at Rookhope (northern Pennines). *Sedimentary Geology 2*:5–12.

Hjulström, P. F. 1939. Transportation of detritus by moving water. Pp. 5–31 in *Recent Marine Sediments,* ed. P. D. Trask. (Dover reprint, 1955.)

Hobbs, P. V., and Harrison, H. 1974. Solar energy absorption. *Science 185*:101.

Hobday, D. K., and Reading, H. G. 1972. Fair weather versus storm processes in shallow marine sand bar sequences in the Late Precambrian of Finmark, north Norway. *Journal of Sedimentary Petrology 42*: 318–324.

Hoffman, P. 1975. Shoaling-upward shale-to-dolomite cycles in the Rocknest Formation (Lower Proterozoic), Northwest Territories, Canada. Pp. 257–265 in *Tidal Deposits,* ed. R. N. Ginsburg. Springer-Verlag, New York.

Holeman, J. N. 1968. The sediment yield of major rivers of the world. *Water Resources Research 4*:737–747.

Holland, H. D. 1968. The abundance of CO_2 in the Earth's atmosphere through geologic time. Pp. 949–954 in *Origin and Distribution of the Elements,* ed. L. H. Ahrens. Pergamon Press, London.

—— 1972. The geologic history of sea water—an attempt to solve the problem. *Geochimica et Cosmochimica Acta 36*:637–651.

—— 1973. Systematics of the isotopic composition of sulfur in the oceans during the Phanerozoic and its implications for atmospheric oxygen. *Geochimica et Cosmochimica Acta 37*:2605–2616.

—— 1976. The evolution of seawater. Pp. 559–567 in *The Early History of the Earth,* ed. B. F. Windley. John Wiley and Sons, New York.

—— 1978. *The Chemistry of the Atmosphere and Oceans.* Wiley Interscience, New York. 351 pp.

Holser, W. T. 1977. Catastrophic chemical events in the history of the ocean. *Nature 267*:403–408.

Howard, J. D. 1972. Trace fossils as criteria for recognizing shorelines in stratigraphic record. Pp. 215–225 in *Recognition of Ancient Sedimentary Environments,* ed. J. K. Rigby and W. K. Hamblin. Society of Economic Paleontologists and Mineralogists, Special Publication no. 16.

Hoyle, F. 1972. The history of the earth. *Quarterly Journal of the Royal Astronomical Society 13*:328–345.

Hsü, K. J. 1976. *Paleoceanography of the Mesozoic Alpine Tethys.* Geological Society of America, Special Paper 170. 44 pp.

Hsü, K. J., Montadert, L., Bernoulli, D., Cita, M. B., Erickson, A., Garrison,

R. E., Kidd, R. B., Mèlierés, F., Müller, C., Wright, R. 1977. History of the Mediterranean salinity crisis. *Nature* 267:399–403.

Huang, T. C., Watkins, N. D., Shaw, D. M., and Kennett, J. P. 1973. Atmospherically transported volcanic dust in South Pacific deep sea sedimentary cores at distances over 3000 km from the eruptive source. *Earth and Planetary Science Letters* 20:119–124.

Hunten, D. M. 1973. The escape of light gases from planetary atmospheres. *Journal of the Atmospheric Sciences* 30:1481–1494.

Hutchins, L. W. 1947. The bases for temperature zonation in geographical distribution. *Ecological Monographs* 17:325–335.

Idso, S. G., and Brazel, A. J. 1978. Climatological effects of atmospheric particulate pollution. *Nature* 274:781–782.

Il'in, A. V., and Lisitzin, A. P. 1968. Origin of submarine canyons as related to their actual extent in the Atlantic Ocean. *Doklady of the Academy of Sciences of the USSR, Earth Science Section* 183:221–224. (American Geological Institute translation.)

Imbrie, J., and Buchanan, H. 1965. Sedimentary structures in modern carbonate sands of the Bahamas. Pp. 149–172 in *Primary Sedimentary Structures and Their Hydrodynamic Interpretation,* ed. G. V. Middleton. Society of Economic Paleontologists and Mineralogists, Special Publication no. 12.

Imbrie, J., and Kipp, N. G. 1971. A new micropaleontological method for quantitative paleoclimatology: application to a Late Pleistocene Caribbean core. Pp. 72–181 in *The Late Cenozoic Glacial Ages,* ed. K. K. Turekian. Yale University Press, New Haven.

Imbrie, J., van Donk, J., and Kipp, N. G. 1973. Paleoclimatic investigation of a Late Pleistocene Caribbean deep-sea core: comparisons of isotopic and faunal methods. *Quaternary Research* 3:10–38.

Inman, D. L., and Brush, B. M. 1973. The coastal challenge. *Science* 181:20–32.

Inman, D. L., and Nordstrom, C. E. 1971. On the tectonic and morphologic classification of coasts. *Journal of Geology* 79:1–22.

Irwin, M. L. 1965. General theory of epeiric clear water sedimentation. *American Association of Petroleum Geologists Bulletin* 49:445–459.

Jackson, T. A., and Keller, W. D. 1970. A comparative study of the role of lichens and "inorganic" processes in the chemical weathering of recent Hawaiian lava flows. *American Journal of Science* 269:446–466.

Jahn, A. E., and Backus, R. H. 1976. On the mesopelagic fish faunas of slope water, Gulf Stream, and northern Sargasso Sea. *Deep-Sea Research* 23:223–234.

Jakeš, P., and White, A. J. R. 1971. Composition of island arcs and continental growth. *Earth and Planetary Science Letters* 12:224–230.

Jenkins, W. J., Beg, M. A., Clarke, W. B., Wangersky, P. J., and Craig, H. 1972. Excess ^{3}He in the Atlantic Ocean. *Earth and Planetary Science Letters* 16:122–126.

Jenkins, W. J., Edmond, J. M., and Corliss, J. B. 1978. Excess ^{3}He and ^{4}He in Galapagos submarine hydrothermal waters. *Nature* 272:156–158.

Jenkyns, H. C. 1977. Fossil nodules. Pp. 87–108 in *Marine Manganese Deposits*, ed. G. P. Glasby. Elsevier, Amsterdam.

Johannes, R. E., et al. 1972. The metabolism of some coral reef communities: a team study of nutrient and energy flux at Eniwetok. *BioScience* 22:541–543.

Johnson, T. C., Hamilton, E. L., and Berger, W. H. 1977. Physical properties of calcareous ooze: control by dissolution at depth. *Marine Geology* 24:259–277.

JOIDES Executive Committee. 1977. *The Future of Scientific Drilling.* Department of Oceanography, University of Washington, Seattle, Wash. 92 pp.

Jones, J. B., and Segnit, E. R. 1971. The nature of opal. I. Nomenclature and constituent phases. *Journal of the Geological Society of Australia 18*:57–68.

Jones, J. G. 1969. Pillow lavas as depth indicators. *American Journal of Science 267*:181–195.

Jones, J. G., and Nelson, P. H. H. 1970. The flow of basalt lava from air into water—its structural expression and stratigraphic significance. *Geological Magazine 107*:13–19.

Jones, M. L., and Dennison, J. M. 1970. Oriented fossils as paleocurrent indicators in Paleozoic lutites of southern Appalachians. *Journal of Sedimentary Petrology 40*:642–649.

Joyner, W. B. 1967. Basalt-eclogite transition as a cause for subsidence and uplift. *Journal of Geophysical Research 72*:4977–4998.

Junge, C. E. 1972. The cycle of atmospheric gases—natural and man made. *Quarterly Journal of the Royal Meteorological Society 98*:711–729.

Kahle, C. F., and Floyd, J. C. 1971. Stratigraphic and environmental significance of sedimentary structures in Cayugan (Silurian) tidal flat carbonates, northwestern Ohio. *Geological Society of America Bulletin 82*:2071–2098.

Kastner, M., Keene, J. B., and Gieskes, J. M. 1977. Diagenesis of siliceous oozes. I. Chemical controls of siliceous oozes. I. Chemical controls on the rate of opal-A to opal-CT transformation—an experimental study. *Geochemica et Cosmochimica Acta 41*:1041–1059.

Kazakov, A. V. 1937. The phosphorite facies and the genesis of phosphorites. Pp. 95–113 in *Scientific Institute Fertilizers and Insecto Fungicides: Transactions 17th International Geological Congress.* Publication 142. Leningrad, 1937.

Keary, R. 1970. Coastal climate and shelf-bottom sediments: a comment. *Marine Geology 8*:363–365.

Keene, J. B. 1975. Cherts and porcellanites from the North Pacific, DSDP Leg 32. *Initial Reports of the Deep Sea Drilling Project 32*:429–507.

Keith, M. L., and Degens, E. T. 1959. Geochemical indications of marine and fresh-water sediments. Pp. 38–61 in *Researches in Geochemistry*, ed. P. H. Abelson. John Wiley and Sons, New York.

Keller, W. D. 1956. Clay minerals as influenced by environments of their formation. *American Association of Petroleum Geologists Bulletin 40*:2689–2710.

Kellogg, W. W., Cadle, R. D., Allen, E. R., Lazrus, A. L., and Martell, E. A. 1972. The sulfur cycle. *Science 175*:587–596.

Kellogg, W. W., and Schneider, S. H. 1974. Climate stabilization: for better or for worse? *Science 186*:1163–1171.

Kemp, P. H. 1956. *The Chemistry of Borates. Part I.* Borax Consolidated, London. 90 pp.

Kendall, C. G. St. C. 1969. An environmental reinterpretation of the Permian evaporite/carbonate shelf sediments of the Guadalupe Mountains. *Geological Society of America Bulletin 80*:2503–2526.

Kennett, J. P. 1966. Foraminiferal evidence of a shallow calcium carbonate solution boundary, Ross Sea, Antarctica. *Science 153*:191–193.

—— 1968a. *Globorotalia truncatulinoides* as a paleo-oceanographic index. *Science 159*:1461–1463.

—— 1968b. Latitudinal variation in *Globigerina pachyderma* (Ehrenberg) in surface sediments of the southwest Pacific Ocean. *Micropaleontology 14*:305–318.

—— 1976. Phenotypic variation in some Recent and late Cenozoic planktonic foraminifera. Pp. 1–60 in *Foraminifera,* ed. R. H. Hedley and C. G. Adams. Vol. 2. Academic Press, London.

Kennett, J. P., Burns, R. E., Andrews, J. E., Churkin, M., Jr., Davies, T. A., Dumitrica, P., Edwards, A. R., Galehouse, J. S., Packham, G. H., and van der Lingen, G. J. 1972. Australian-Antarctic continental drift, palaecirculation changes and Oligocene deep-sea erosion. *Nature Physical Sciences 239*:51–55.

Kennett, J. P., Huddlestun, P., and Clark, H. C. 1974. Paleoclimatology, paleomagnetism and tephrochronology of Late Pleistocene sedimentary cores, Gulf of Mexico. *Colloques Internationaux du Centre National de la Recherche Scientifique. 219*:239–250.

Kennett, J. P., McBirney, A. R., and Thunell, R. C. 1977. Episodes of Cenozoic volcanism in the circum-Pacific region. *Journal of Volcanology and Geothermal Research 2*:145–163.

Kennett, J. P., and Shackleton, N. J. 1975. Laurentide ice sheet meltwater recorded in Gulf of Mexico deep-sea cores. *Science 188*:147–150.

Kennett, J. P., and Thunell, R. C. 1977. On explosive Cenozoic volcanism and climatic implications. *Science 196*:1231–1234.

Kennett, J. P., and Watkins, N. D. 1975. Deep-sea erosion and manganese nodule development in the southeast Indian Ocean. *Science 188*:1011–1013.

Keulegan, G. H., and Krumbein, W. C. 1949. Stable configuration of bottom slope in a shallow sea and its bearing on geological processes. *Transactions American Geophysical Union 30*:855–861.

King, C. A. M. 1959. *Beaches and Coasts.* Edward Arnold, London. 403 pp.

King, K., Jr. 1978. γ-carboxyglutamic acid in fossil bones and its significance for amino acid dating. *Nature 273*:41–43.

Kinsman, D. J. J. 1966. Gypsum and anhydrite of recent age, Trucial Coast, Persian Gulf. Pp. 302–326 in *Second Symposium on Salt,* ed. J. L. Rau, vol. 1. The Northern Ohio Geological Society.

—— 1969. Modes of formation, sedimentary associations, and diagnostic features of shallow-water and supratidal evaporites. *American Association of Petroleum Geologists Bulletin 53*:830–840.

——— 1975. Salt floors to geosynclines. *Nature 255*:375–378.

——— 1976. Evaporites: relative humidity control of primary mineral facies. *Journal of Sedimentary Petrology. 46*:273–279.

Kipp, N. G. 1976. New transfer function for estimating past sea-surface conditions from sea-bed distribution of planktonic foraminiferal assemblages in the North Atlantic. Pp. 3–41 in *Investigations of Late Quaternary Paleoceanography and Paleoclimatology,* ed. R. M. Cline and J. D. Hays. Geological Society of America Memoir 145.

Klein, C., and Bricker, O. P. 1977. Some aspects of the sedimentary and diagenetic environment of Proterozoic banded iron-formation. *Economic Geology 72*:1457–1470.

Klein, G. deV. 1970. Tidal origin of a Precambrian Quartzite—the lower fine-grained quartzite (Middle Dalradian) of Islay, Scotland. *Journal of Sedimentary Petrology 40*:973–985.

——— 1971. A sedimentary model for determining paleotidal range. *Geological Society of America Bulletin 82*:2585–2592.

——— 1972. Sedimentary model for determining paleotidal range: reply. *Geological Society of America Bulletin 83*:539–546.

——— 1974. Estimating water depths from analysis of barrier island and deltaic sedimentary sequences. *Geology 2*:409–412.

——— 1977. *Clastic Tidal Facies.* Continuing Education Publication Co., Champaign, Ill. 149 pp.

Klein, G. deV., and Ryer, T. A. 1978. Tidal circulation patterns in Precambrian, Paleozoic, and Cretaceous epeiric and mioclinal shelf seas. *Geological Society of America Bulletin 89*:1050–1058.

Knauth, L. P., and Epstein, S. 1976. Hydrogen and oxygen isotope ratios in nodular and bedded cherts. *Geochimica et Cosmochimica Acta 40*:1095–1108.

Knauth, L. P., and Lowe, D. R. 1978. Oxygen isotopic composition of cherts from the Onverwacht Group (3.4 b. yrs.), South Africa, with implications for secular variations in the isotopic composition of cherts. *Geological Society of America Abstracts with Programs 10*:436.

Knoll, A. H., and Barghoorn, E. S. 1977. Archean microfossils showing cell division from the Swaziland System of South Africa. *Science 198*:396–398.

Kobayashi, K., and Nomura, M. 1972. Iron Sulfides in the sediment cores from the sea of Japan and their geophysical implications. *Earth and Planetary Science Letters 16*:200–208.

Kolehamainen, S. E., and Morgan, T. O. 1972. Mangrove root communities in a thermally altered bay in Puerto Rico. In *Abstracts of Papers for 35th Annual Meeting of American Society of Limnology and Oceanography, Tallahassee, Florida* (unpaged). 1972.

Kolodny, Y., and Epstein, S. 1976. Stable isotope geochemistry of deep sea cherts. *Geochimica et Cosmochimica Acta 40*:1195–1209.

Komar, P. D., Neudeck, R. H., and Kulm, L. D. 1972. Observations and significance of deep-water oscillatory ripple marks on the Oregon continental shelf. Pp. 601–619 in *Shelf Sediment Transport: Process and Pattern,* ed. D. J. P. Swift, D. B. Duane, and O. H. Pilkey. Dowden, Hutchinson and Ross, Stroudsburg, Pa.

Kozary, M. T., Dunlap, J. C., and Humphrey, W. E. 1968. Incidence of saline deposits in geologic time. *Geological Society of America Special Paper 88*:43–57.

Kramer, J. R. 1965. History of sea water: constant temperature-pressure equilibrium models compared to liquid inclusion analyses. *Geochimica et Cosmochimica Acta 29*:921–945.

Kriausakul, N., and Mitterer, R. M. 1978. Isoleucine epimerization in peptides and proteins: kinetic factors and application to fossil proteins. *Science 201*:1011–1014.

Krinsley, D. H., and Donahue, J. 1968. Environmental interpretation of sand grain surface textures by electron microscopy. *Geological Society of America Bulletin 79*:743–748.

Krinsley, D. H., Friend, P. F., and Klimentidis, R. 1976. Eolian transport textures on the surfaces of sand grains of Early Triassic age. *Geological Society of America Bulletin 87*:130–132.

Ku, T. L. 1977. Rates of accretion. Pp. 249–267 in *Marine Manganese Deposits,* ed. G. P. Glasby. Elsevier, Amsterdam.

Kuenen, P. H. 1950. *Marine Geology.* John Wiley and Sons, New York. 568 pp.

Kukal, Z. 1971. *Geology of Recent Sediments.* Academic Press, London. 490 pp.

LaBarbera, M. 1978. Precambrian geological history and the origin of the Metazoa. *Nature 273*:22–25.

Labeyric, L., Jr. 1974. New approach to surface seawater palaeotemperatures using $^{18}O/^{16}O$ ratios in silica of diatom frustules. *Nature 248*:40–42.

Lafon, G. M., and Mackenzie, F. T. 1974. Early evolution of the oceans—a weathering model. Pp. 205–218 in *Paleo-Oceanography,* ed. W. W. Hay. Society of Economic Paleontologists and Mineralogists, Special Publication no. 20.

Landergren, S. 1945. Contribution to the geochemistry of boron. II. the distribution of boron in some Swedish sediments, rocks, and iron ores. The boron cycle in the upper lithosphere. *Arkiv för Kemi, Mineralogi och Geologi 19A* (26):1–31.

Landergren, S., and Manheim, F. T. 1963. Über die Abhängigkeit der Verteilung von Schwermetallen von der Fazies. *Fortschritte in der Geologie von Rheinland und Westfalen 10*:173–192.

Lantzy, R. J., Dacey, M. F., and Mackenzie, F. T. 1977. Catastrophe theory: application to the Permian mass extinction. *Geology 5*:724–728.

Lavelle, J. W., Keller, G. H., Clarke, T. L. 1975. Possible bottom current response to surface winds in the Hudson Shelf Channel. *Journal of Geophysical Research 80*:1953–1956.

Ledbetter, M. T., and Johnson, D. A. 1976. Increased transport of Antarctic bottom water in the Vema Channel during the last ice age. *Science 194*:837–839.

Leeder, M. R., and Zeidan, R. 1977. Giant late Jurassic sabkhas of Arabian Tethys. *Nature 268*:42–44.

Lees, A., and Buller, A. T. 1972. Modern temperature-water and warm-water shelf carbonate sediments contrasted. *Marine Geology 13*:M67–M73.

Leetmaa, A. 1977. Effects of the winter 1976–1977 on the northwestern Sargasso Sea. *Science 198*:188–189.

Le Pichon, X., Melguen, M., and Sibuet, J. C. 1978. A schematic model of the evolution of the South Atlantic. Pp. 1–48 in *Advances in Oceanography*, ed. H. Charnock and G. Deacon. Plenum Press, New York.

Lepp, H., and Goldich, S. S. 1959. Chemistry and origin of iron formations. *Geological Society of America Bulletin 70*:1637.

——— 1964. Origin of Precambrian iron formations. *Economic Geology 59*:1025–1060.

Lerman, J. C. 1974. Isotope "paleothermometers" on continental matter: assessment. *Colloques Internationaux du Centre National de la Recherche Scientifique 219*:163–181.

Leutze, W. P. 1968. Stratigraphy and paleoenvironments of the phosphatic Miocene strata of North Carolina: discussion. *Geological Society of America Bulletin 79*:1433–1436.

Lewis, K. B. 1968. Size of fossil animals as an indicator of paleotemperatures. *Tuatara 16*:62–68.

Lewis, K. B. 1971. Slumping on a continental slope inclined at 1°-4°. *Sedimentology 16*:97–110.

Lewis, W. M., Jr. 1976. Surface/volume ratio: implications for phytoplankton morphology. *Science 192*:885–887.

Lewontin, R. C. 1974. *The Genetic Basis of Evolutionary Change.* Columbia University Press, New York. 346 pp.

Linacre, E. T. 1967. Further notes on a feature of leaf and air temperatures. *Archiv für Meteorologie, Geophysik und Bioklimatologie, Serie B, Allgemeine und Biologische Klimatologie 15*:422–436.

Lisitzin, A. P. 1972. *Sedimentation in the World Ocean.* Society of Economic Paleontologists and Mineralogists Special Publication no. 17. 218 pp.

Livingston, W. C. 1978. Cooling of the sun's photosphere coincident with increased sunspot activity. *Nature 272*:340–341.

Lloyd, R. M. 1964. Variations in the oxygen and carbon isotope ratios of Florida Bay mollusks and their environmental significance. *Journal of Geology 72*:84–111.

——— 1966. Oxygen isotope enrichment of sea water by evaporation. *Geochimica et Cosmochimica Acta 30*:801–814.

——— 1967. Oxygen-18 composition of oceanic sulfate. *Science 156*:1228–1231.

Lochman-Balk, C. 1971. The Cambrian of the Craton of the United States. Pp. 79–167 in *Cambrian of the New World*, ed. C. H. Holland. Wiley-Interscience, London.

Lockwood, G. W. 1975. Planetary brightness changes: evidence for solar variability. *Science 190*:560–562.

Lohmann, K. C. 1976. Lower Dresbachian (Upper Cambrian) platform-to-basin transition in eastern Nevada and western Utah: an evaluation through lithologic cycle correlation. *Brigham Young University Geology Studies 23*:111–122.

Longinelli, A., and Craig, H. 1967. Oxygen-18 variations in sulfate ions in sea water and saline lakes. *Science 156*:56–59.

Longinelli, A., and Nuti, S. 1973. Revised phosphate-water isotopic temperature scale. *Earth and Planetary Science Letters 19*:373–376.

Lovelock, J. E., and Lodge, J. P., Jr. 1972. Oxygen in the contemporary atmosphere. *Atmospheric Environment 6*:575–578.

Lowenstam, H. A. 1972. Biogeochemistry of hard tissues, their depth and possible pressure relationships. Pp. 19–32 in *Barobiology and the Experimental Biology of the Deep Sea*, ed. R. W. Brauer. University of North Carolina Press, Chapel Hill.

Lucia, F. J. 1972. Recognition of evaporite-carbonate shoreline sedimentation. Pp. 160–191 in *Recognition of Ancient Sedimentary Environments*, ed. J. K. Rigby and W. K. Hamblin. Society of Economic Paleontologists and Mineralogists, Special Publication no. 16.

Ludwick, J. C. 1970. Sand Waves in the tidal entrance to Chesapeake Bay: preliminary observations. *Chesapeake Science 11*:98–110.

Lupton, J. E., Weiss, R. F., and Craig, H. 1977a. Mantle helium in hydrothermal plumes in the Galapagos Rift. *Nature 267*:603–604.

——— 1977b. Mantle helium in the Red Sea brines. *Nature 267*:244–246.

Luyendyk, B. P., Forsyth, D., and Phillips, J. D. 1972. Experimental approach to the paleocirculation of the oceanic surface waters. *Geological Society of America Bulletin 83*:2649–2664.

MacArthur, R. H., and Wilson, E. O. 1967. *The Theory of Island Biogeography*. Princeton University Press, Princeton, N.J. 203 pp.

Macintyre, I. G., and Pilkey, O. H. 1969. Preliminary comments on linear sand-surface features, Onslow Bay, North Carolina continental shelf: problems in making detailed sea-floor observations. *Maritime Sediments, 5*:26–29.

Mack, G. H., and Suttner, L. J. 1977. Paleoclimate interpretation from a petrographic comparison of Holocene sands and the Fountain Formation (Pennsylvanian) in the Colorado Front Range. *Journal of Sedimentary Petrology 47*:89–100.

Mackenzie, F. T. 1969. Chemistry of seawater. Pp. 106–112 in *The Encyclopedia of Marine Resources*, ed. F. E. Firth. Van Nostrand Reinhold Co. New York.

——— 1974. Oceans, development of. *Encyclopedia Britannica*, 476–482.

——— 1975. Sedimentary cycling and the evolution of sea water. Pp. 309–364 in *Chemical Oceanography*, ed. J. P. Riley and G. Skirrow, vol. 1, 2nd ed. Academic Press, London.

Mackenzie, F. T., Stoffyn, M., and Wollast, R. 1978. Aluminum in seawater: control by biologic activity. *Science 199*:680–682.

Malmgren, B. A., and Kennett, J. P. 1978. Test size variation in *Globigerina bulloides* in response to Quaternary palaeoceanographic changes. *Nature 275*:123–124.

Manabe, S., Bryan, K., and Spelman, M. J. 1975. A global ocean-atmosphere climate model. Part I. The atmospheric circulation. *Journal of Physical Oceanography 5*:3–29.

Mangum, C. P. 1972. Temperature sensitivity of metabolism in offshore and intertidal Onuphid polychaetes. *Marine Biology 17*:108–114.

Manheim, F. T. 1965. Manganese-iron accumulations in the shallow marine environment. Pp. 217–276 in *Symposium on Marine Geochemistry,* ed. D. R. Schink and J. T. Corless. Narragansett Marine Laboratory, University of Rhode Island, Occasional Publication no. 3-1965.

Manheim, F. T., and Chan, K. M. 1974. Interstitial waters of Black Sea sediments: new data and review. *American Association of Petroleum Geologists Memoir 20*:155–180.

Manheim, F. T., Hathaway, J. C., and Uchupi, E. 1972. Suspended matter in surface waters of the northern Gulf of Mexico. *Limnology and Oceanography 17*:17–27.

Manheim, F. T., and Horn, M. K. 1968. Composition of deeper subsurface waters along the Atlantic continental margin. *Southeastern Geology 9*:215–236.

Manheim, F. T., Rowe, G. T., and Jipa, D. 1975. Marine phosphorite formation off Peru. *Journal of Sedimentary Petrology 45*:243–251.

Manheim, F. T., and Sayles, F. L. 1970. Brines and interstitial brackish water in drill cores from the deep Gulf of Mexico. *Science 170*:57–61.

Margolis, S. V., and Kennett, J. P. 1971. Cenozoic paleoglacial history of Antarctica recorded in subantarctic deep-sea cores. *American Journal of Science 271*:1–36.

Margolis, S. V., Kroopnick, P. M., Goodney, D. E., Dudley, W. C., and Mahoney, M. E. 1975. Oxygen and carbon isotopes from calcareous nanofossils as palaeoceanographic indicators. *Science 189*:555–557.

Mason, C. C., and Folk, R. L. 1958. Differentiation of beach, dune, and aeolian flat environments by size analysis, Mustang Island, Texas. *Journal of Sedimentary Petrology 28*:211–226.

Mason, T. R., and von Brunn, V. 1977. Three-Gyr-old stromatolites from South Africa. *Nature 266*:47–49.

Maynard, N. G. 1976. Relationship between diatoms in surface sediments of the Atlantic Ocean and the biological and physical oceanography of overlying waters. *Paleobiology 2*:99–121.

Mazor, E. 1972. Paleotemperatures and other hydrological parameters deduced from noble gases dissolved in groundwaters; Jordan Rift Valley, Israel. *Geochimica et Cosmochimica Acta 36*:1321–1336.

McAlester, A. L., and Rhoads, D. C. 1967. Bivalves as bathymetric indicators. *Marine Geology 5*:383–388.

McCave, I. N. 1970. Deposition of fine-grained suspended sediment from tidal currents. *Journal of Geophysical Research 75*:4151–4159.

——— 1971. Sand waves in the North Sea off the coast of Holland. *Marine Geology 10*:199–225.

McClennen, C. E. 1973. Sands on continental shelf off New Jersey move in response to waves and currents. *Maritimes 17*:14–16.

McElhinny, M. W., Taylor, S. R., and Stevenson, D. J. 1978. Limits to the expansion of Earth, Moon, Mars, and Mercury and to changes in the gravitational constant. *Nature 271*:316–321.

McGovern, W. E. 1969. The primitive earth: thermal models of the upper atmosphere for a methane-dominated environment. *Journal of the Atmospheric Sciences 26*:623–635.

McIntyre, A., Ruddiman, W. F., and Jantzen, R. 1972. Southward penetrations of the North Atlantic polar front: faunal and floral evidence of large-scale surface water mass movements over the last 225,000 years. *Deep-Sea Research 19*:61–77.

McKee, E. D. 1959. Storm sediments on a Pacific atoll. *Journal of Sedimentary Petrology 29*:354–364.

McKee, E. D., Chronic, J., and Leopold, E. B. 1959. Sedimentary belts in lagoon of Kapingamarangi atoll. *American Association of Petroleum Geologists Bulletin 43*:501–562.

McKerrow, W. S., ed. 1978. *The Ecology of Fossils.* M.I.T. Press, Cambridge. 383 pp.

McKerrow, W. S. 1979. Ordovician and Silurian changes in sea level. *Journal of the Geological Society of London 136*:137–146.

McKirdy, D. M., and Powell, T. G. 1974. Metamorphic alteration of carbon isotopic composition in ancient sedimentary organic matter: new evidence from Australia and South Africa. *Geology 2*:591–595.

McLellan, H. J. 1965. *Elements of Physical Oceanography.* Pergamon Press, Oxford. 150 pp.

McManus, D. A. 1970. Criteria of climatic change in the inorganic components of marine sediments. *Quarternary Research 1*:72–102.

Meade, R. H. 1964. *Removal of Water and Rearrangement of Particles during the Compaction of Clayey Sediments—Review.* Pp. B1–B23 in U.S. Geological Survey, Professional Paper 497-B.

———— 1969. Errors in using modern stream-load data to estimate natural rates of denudation. *Geological Society of America Bulletin 80*:1265–1274.

———— 1972. Transport and deposition of sediments in estuaries. Pp. 91–120. Environmental Framework of Coastal Plain Estuaries, ed. B. W. Nelson. Geological Society of America Memoir 133.

———— 1974. Net transport of sediment through the mouths of estuaries: seaward or landward? International symposium on interrelationships of estuaries and continental shelf sedimentation. *Mémoires de l'Institut de Géologie du Bassin d'Aquitaine 7*:207–213.

———— 1976. Sediment problems in the Savannah River basin. Pp. 105–129 in *The Future of the Savannah River,* (proceedings of a symposium held at Hickory Knob State Park), ed. B. L. Dillman and J. M. Stepp. Water Resources Research Institute, Clemson University.

Meade, R. H., Sachs, P. L., Manheim, F. T., Hathaway, J. C., and Spencer, D. W. 1975. Sources of suspended matter in waters of the middle Atlantic Bight. *Journal of Sedimentary Petrology 45*:171–188.

Meade, R. H., and Trimble, S. W. 1974. Changes in sediment loads in rivers of the Atlantic drainage of the United States since 1900. Pp. 99–104 in *Effects of Man on the Interface of the Hydrological Cycle with the Physical Environment—Symposium* (proceedings of the Paris Symposium, September 1974). IAHS-AISH Publication no. 113.

Meigs, P. 1952. Arid and semiarid climatic types of the world. Pp. 135–138 in *Proceedings Eighth General Assembly and Seventeenth International Congress of the International Geographical Union, Washington D.C., August 8–15, 1952* (no editor).

Menard, H. W. 1961. Some rates of regional erosion. *Journal of Geology* 69:154–161.

——— 1973. Epeirogeny and plate tectonics. *EOS* (American Geophysical Union Transactions) 54:1244–1255.

Menard, H. W., and Smith, S. M. 1966. Hysometry of ocean basin provinces. *Journal of Geophysical Research* 71:4305–4325.

Menzel, D. W. 1974. Primary productivity, dissolved and particulate organic matter, and the sites of oxidation of organic matter. Pp. 659–678 in *The Sea*, ed. E. D. Goldberg, vol. 5. John Wiley and Sons, New York.

Merifield, P. M., and Lamar, D. L. 1968. Sand waves and early earth-moon history. *Journal of Geophysical Research* 73:4767–4774.

——— 1970. Paleotides and the geologic record. Pp. 31–40 in *Palaeogeophysics*, ed. S. K. Runcorn. Academic Press, London.

Merriam, M. F. 1974. Solar energy absorption. *Science* 185:101.

Meyerson, A. L. 1972. Pollen and paleosalinity analyses from a Holocene tidal marsh sequence, Cape May County, New Jersey. *Marine Geology* 12:335–357.

Michel, R., and Williams, P. M. 1973. Bomb-produced tritium in the Antarctic Ocean. *Earth and Planetary Science Letters* 20:381–384.

Middleton, G. V. 1973. Johannes Walther's law of the correlation of facies. *Geological Society of America Bulletin* 84:979–988.

——— 1976. Hydraulic interpretation of sand size distributions. *Journal of Geology* 84:405–426.

Mikkelsen, N. 1978. Preservation of diatoms in glacial to Holocene deep-sea sediments of the equatorial Pacific. *Geology* 6:553–555.

Mikkelsen, N., Labeyrie, L., Jr., and Berger, W. H. 1978. Silica oxygen isotopes in diatoms: a 20,000 year record in deep-sea sediments. *Nature* 271:536–538.

Milliman, J. D. 1972. *Atlantic Continental Shelf and Slope of the United States. Petrology of the Sand Fraction of Sediments, Northern New Jersey to Southern Florida.* Pp. J1–J40 in U.S. Geological Survey Professional Paper 529-J.

Milliman, J. D., and Müller, J. 1977. Characteristics and genesis of shallow-water and deep-sea limestones. Pp. 655–672 in *The Fate of Fossil Fuel CO_2 in the Oceans,* ed. N. R. Andersen and A. Malahoff. Plenum Press, New York.

Mitchell, R. 1974. The evolution of thermophily in hot springs. *The Quarterly Review of Biology* 49:229–242.

Mohr, P. A., and Wood, C. A. 1976. Volcano spacings and lithospheric attenuation in the eastern rift of Africa. *Earth and Planetary Science Letters* 33:126–144.

Möller, F. 1963. On the influence of changes in the CO_2 concentration in air on the radiation balance of the earth's surface and on the climate. *Journal of Geophysical Research* 68:3877–3886.

Mook, W. G. 1971. Paleotemperatures and chlorinities from stable carbon and oxygen isotopes in shell carbonate. *Palaeogeography, Palaeoclimatology, Palaeoecology* 9:245–263.

Mook, W. G., Bommerson, J. C., and Staverman, W. H. 1974. Carbon isotope fractionation between dissolved bicarbonate and gaseous carbon dioxide. *Earth and Planetary Science Letters* 22:169–176.

Mook, W. G., and Vogel, J. C. 1968. Isotopic equilibrium between shells and their environment. *Science* 159:874–875.

Moorbath, S. 1976. Age and isotope constraints for the evolution of Archean crust. Pp. 351–360 in *The Early History of the Earth,* ed. B. F. Windley. John Wiley and Sons, London.

———— 1977. Ages, isotopes and evolution of Precambrian continental crust. *Chemical Geology* 20:151–187.

Moorbath, S., Wilson, J. F., Goodwin, R., and Humm, M. 1977. Further Rb-Sr age and isotope data on early and late Archaean rocks from the Rhodesian craton. *Precambrian Research* 5:229–239.

Moore, A. E. 1976. Controls of post-Gondwanaland alkaline volcanism in southern Africa. *Earth and Planetary Science Letters* 31:291–296.

Moore, J. G. 1970. Submarine basalt from the Revillagigedo Islands region, Mexico. *Marine Geology* 9:331–345.

Mörner, N-A. 1976. Eustasy and geoid changes. *Journal of Geology* 84:123–151.

Muehlenbachs, K., and Clayton, R. N. 1976. Oxygen isotope composition of the oceanic crust and its bearing on seawater. *Journal of Geophysical Research* 81:4365–4369.

Müller, G. 1969. Sedimentary phosphate method for estimating paleosalinities: limited applicability. *Science* 163:812–813.

Müller, P. J. 1977. C/N ratios in Pacific deep-sea sediments: effect of inorganic ammonium and organic nitrogen compounds sorbed by clays. *Geochimica et Cosmochimica Acta* 41:765–776.

Müller, P. J., and Suess, E. 1979. Productivity, sedimentation rate, and sedimentary organic carbon content in the oceans. *Deep-sea Research:* in press.

Murray, J. W. 1971. Living foraminiferids of tidal marshes: a review. *Journal of Foraminiferal Research* 1:153–161.

———— 1973. *Distribution and Ecology of Living Benthic Foraminiferids.* Crane, Russak and Co., New York. 274 pp.

———— 1976. A method of determining proximity of marginal seas to an ocean. *Marine Geology* 22:103–119.

Murray, S. P. 1970. Bottom currents near the coast during Hurricane Camille. *Journal of Geophysical Research* 75:4579–4582.

Myers, J. S. 1976. Granitoid sheets, thrusting, and Archean crustal thickening in West Greenland. *Geology* 4:265–268.

Nagle, J. S. 1967. Wave and current orientation of shells. *Journal of Sedimentary Petrology* 37:1124–1138.

Nance, W. B., and Taylor, S. R. 1977. Rare earth element patterns and crustal evolution—II. Archean sedimentary rocks from Kalgoorlie, Australia. *Geochimica et Cosmochimica Acta* 41:225–231.

Nanz, R. H., Jr. 1953. Chemical composition of pre-Cambrian slates with notes of the geochemical evolution of lutites. *Journal of Geology* 61:51–64.

Naqvi, S. M., Rao, V. D., and Narain, H. 1978. The primitive crust: evidence from the Indian Shield. *Precambrian Research* 6:323–345.

Nelson, B. W. 1967. Sedimentary phosphate method for estimating paleosalinities. *Science* 158:917–920.

Newell, N. D., Imbrie, J., Purdy, E. G., and Thurber, D. L. 1959. Organism communities and bottom facies, Great Bahama Bank. *American Museum of Natural History Bulletin* 117:177–228.

Newell, R. E. 1971a. Modification of stratospheric properties by trace constituent changes. Pp. 697–699 in *Man's Impact on the Climate,* ed. W. H. Matthews, W. W. Kellogg, and G. D. Robinson. M.I.T. Press, Cambridge.

——— 1971b. The global circulation of atmospheric pollutants. *Scientific American* 224:32–42.

Newell, R. E., and Weare, B. C. 1976. Factors governing troposperic mean temperature. *Science* 194:1413–1414.

Newman, M. J., and Rood, R. T. 1977. Implications of solar evolution for the earth's early atmosphere. *Science* 198:1035–1037.

Nicholls, G. D. 1967. Trace elements in sediments: an assessment of their possible utility as depth indicators. *Marine Geology* 5:539–555.

Nichols, M. M. 1977. Response and recovery of an estuary following a river flood. *Journal of Sedimentary Petrology* 47:1171–1186.

Nicol, D. 1967. Some characteristics of coldwater marine pelecypods. *Journal of Paleontology* 41:1330–1340.

Niino, H., and Emery, K. O. 1966. Continental shelf off northeastern Asia. *Journal of Sedimentary Petrology* 36:152–161.

Nikolaeva, I. V. 1972. Glauconite and its paleogeographic reconstructions. *Academy of Science of the USSR, Siberian Branch, Geology and Geophysics* 6:51–57. (In Russian.)

Ninkovich, D., and Donn, W. L. 1976. Explosive Cenozoic volcanism and climatic implications. *Science* 194:899–906.

Odum, E. P. 1976. Earth as a productive system (review of *Primary Productivity of the Biosphere,* ed. H. Leith and R. H. Whittaker, 1975). *Science* 193:138.

Oehler, D. Z., Schopf, J. W., and Kvenvolden, K. A. 1972. Carbon isotopic studies of organic matter in Precambrian rocks. *Science* 175:1246–1248.

Oehler, D. Z., and Smith, J. W. 1977. Isotopic composition of reduced and oxidized carbon in Early Archean rocks from Isua, Greenland. *Precambrian Research* 5:221–228.

Off, T. 1963. Rhythmic linear sand bodies caused by tidal currents. *American Association of Petroleum Geologists Bulletin* 47:324–341.

Olausson, E. 1969. Le climat au Pleistocène et la circulation des océans. *Revue de Géographie Physique et de Géologie Dynamique* 11:251–264.

Olson, W. S. 1972. Sedimentary model for determining paleotidal range: discussion. *Geological Society of America Bulletin* 83:537–538.

Olsson, R. K. 1974. Pleistocene paleoceanography and *Globigerina pachyderma* (Ehrenberg) in site 36, DSDP, northeastern Pacific. *Journal of Foraminiferal Research* 4:47–60.

Öpik, E. J. 1968. Climatic change and the onset of the ice ages. *Irish Astronomical Journal* 8:153–157.

Orr, W. L., and Gaines, A. G., Jr. 1973. Observations on rate of sulfate reduction and organic matter oxidation in the bottom waters of an estuarine basin: the upper basin of the Pettaquamscutt River (Rhode Island). Pp. 791–812 in *Advances in Organic Geochemistry, 1973* (proceedings of the 6th International Congress of Organic Geochemistry, France).

Orr, W. N. 1967. Secondary calcification in the foraminiferal genus *Globorotalia*. *Science* 157:1554–1555.

Ostenso, N. A. 1966. Arctic Ocean. Pp. 49–55 in *The Encyclopedia of Oceanography*, ed. R. W. Fairbridge. Van Nostrand Rheinhold Co., New York.

Östlund, H. G., Dorsey, H. G., and Rooth, C. G. 1974. GEOSECS North Atlantic radiocarbon and tritium results. *Earth and Planetary Science Letters* 23:69–86.

Oversby, V. M. 1978. Lead isotopes in Archaean plutonic rocks. *Earth and Planetary Science Letters* 38:237–248.

Owens, J. P., and Sohl, N. F. 1973. Glauconites from New Jersey–Maryland coastal plain: their K-Ar ages and application in stratigraphic studies. *Geological Society of America Bulletin* 84:2811–2838.

Ozima, M. 1975. Ar isotopes and earth-atmosphere evolution models. *Geochimica et Cosmochimica Acta* 39:1127–1134.

Pannella, G. 1972. Paleontological evidence on the earth's rotational history since early Precambrian. *Astrophysics and Space Science* 16:212–237.

———— 1976. Geophysical inferences from stromatolite lamination. Pp. 673–685 in *Stromatolites*, ed. M. R. Walter. Developments in Sedimentology, no. 20. Elsevier, Amsterdam.

Parker, R. J., and Siesser, W. G. 1972. Petrology and origin of some phosphorites from the South Aftican continental margin. *Journal of Sedimentary Petrology* 42:434–440.

Parmenter, C., and Folger, D. W. 1974. Eolian biogenic detritus in deep sea sediments: a possible index of equatorial ice age aridity. *Science* 185:695–697.

Parsons, B., and Sclater, J. G. 1977. An analysis of the variation of ocean floor bathymetry and heat flow with age. *Journal of Geophysical Research* 82:803–827.

Passega, R. 1977. Significance of CM diagrams of sediments deposited by suspensions. *Sedimentology 24*:723–733.

Patrick, W. H., Jr., Gotoh, S., and Williams, B. G. 1973. Strengite dissolution in flooded soils and sediments. *Science 179*:564–565.

Patterson, C., and Tatsumoto, M. 1964. The significance of lead isotopes in detrital feldspar with respect to chemical differentiation within the earth's mantle. *Geochimica et Cosmochimica Acta 28*:1–22.

Perry, E. A., Jr. 1972. Diagenesis and the validity of the boron paleosalinity technique. *American Journal of Science 272*:150–160.

Perry, E. C., Jr. 1967. The oxygen isotope chemistry of ancient cherts. *Earth and Planetary Science Letters 3*:62–66.

Perry, E. C., Jr., Ahmad, S. N., and Swulius, T. M. 1978. The oxygen isotope composition of 3,800 m.y. old metamorphosed chert and iron formation from Isukasia, West Greenland. *Journal of Geology 86*:223–239.

Perry, E. C., Jr., Monster, J., and Reimer, T. O. 1971. Sulfur isotopes in Swaziland system barites and the evolution of the earth's atmosphere. *Science 171*:1015–1016.

Perry, E. C., Jr., and Tan, F. C. 1972. Significance of oxygen and carbon isotope variations in early Precambrian cherts and carbonate rocks of southern Africa. *Geological Society of America Bulletin 83*:647–664.

Peters, K. E., Sweeney, R. E., and Kaplan, I. R. 1978. Correlation of carbon and nitrogen stable isotope ratios in sedimentary organic matter. *Limnology and Oceanography 23*:598–604.

Petterssen, S. 1969. *Introduction to Meteorology,* 3rd ed. McGraw-Hill Book Co, New York. 333 pp.

Phleger, F. B. 1964. Foraminiferal ecology and marine geology. *Marine Geology 1*:16–43.

Pilkey, O. H., and Blackwelder, B. W. 1968. Mineralogy of the sand size carbonate fraction of some recent marine terrigenous and carbonate sediments. *Journal of Sedimentary Petrology 38*:799–810.

Pilkey, O. H., Trumbull, J. V. A., and Bush, D. M. 1978. Equilibrium shelf sedimentation, Rio de la Plata shelf, Puerto Rico. *Journal of Sedimentary Petrology 48*:389–400.

Pimentel, D., Terhune, E. C., Dyson-Hudson, R., Rochereau, S., Samis, R., Smith, E. A., Denman, D., Reifschneider, D., and Shepard, M. 1976. Land degradation: effects on food and energy resources. *Science 194*: 149–155.

Piper, D. Z., and Codispoti, L. A. 1975. Marine phosphorite deposits and the nitrogen cycle. *Science 188*:15–18.

Playford, P. E., Cockbain, A. E., Druce, E. C., and Wray, J. L. 1976. Devonian stromatolites from the Canning Basin, Western Australia. Pp. 543–563 in Stromatolites, ed. M. R. Walter. Elsevier, Amsterdam. Developments in Sedimentology, no. 20.

Plumley, W. J., Risley, G. A., Graves, R. W., Jr., and Kaley, M. E. 1962. Energy index for limestone interpretation and classification. Pp. 85–107 in *classification of Carbonate Rocks,* ed. W. E. Ham. American Association of Petroleum Geologists Memoir 1.

Pocklington, R. 1978. Climatic trends in the North Atlantic. *Nature* *273*:407.

Pollack, J. B., Toon, O. B., Sagan, C., Summers, A., Baldwin, B., and Van Camp, W. 1976. Volcanic explosions and climatic change: a theoretical assessment. *Journal of Geophysical Research 81*:1071–1083.

Pople, W., and Mensah, M. A. 1971. Evaporation as the upwelling mechanism in Ghanaian coastal waters. *Nature Physical Science 234*:18–20.

Porrenga, D. H. 1967a. Glauconite and chamosite as depth indicators in the marine environment. *Marine Geology 5*: 495–501.

––––––– 1967b. Influence of grinding and heating of layer silicates on boron sorption. *Geochimica et Cosmochimica Acta 31*:309–312.

Postma, H. 1967. Sediment transport and sedimentation in the estuarine environment. Pp. 158–179 in *Estuaries,* ed. G. H. Lauff. Ed. American Association for the Advancement of Science, Publication no. 83, Washington, D.C.

Potter, P. E. 1978. Significance and origin of big rivers. *Journal of Geology 86*:13–33.

Potter, P. E., and Pettijohn, F. J. 1977. *Paleocurrents and Basin Analysis,* 2nd ed. Springer-Verlag, New York. 426 pp.

Potter, P. E., Shimp, N. F., and Witters, J. 1963. Trace elements in marine and fresh-water argillaceous sediments. *Geochimica et Cosmochimica Acta 27*:669–694.

Pratt, R. M. 1963. Bottom currents on the Blake Plateau. *Deep-Sea Research 10*:245–249.

Priestley, C. H. B. 1966. The limitation of temperature by evaporation in hot climates. *Agricultural Meterology 3*:241–246.

Prince, R. A., Resig, J. M., Kulm, L. D., and Moore, T. C., Jr. 1974. Uplifted turbidite basins on the seaward wall of the Peru Trench. *Geology 2*:607–611.

Prospero, J. M., and Nees, R. T. 1977. Dust concentration in the atmosphere of the equatorial North Atlantic: possible relationship to the Sahelian. *Science 196*:1196–1198.

Purser, B. H., ed. 1973. *The Persian Gulf* Springer-Verlag, New York. 471 pp.

Ramsay, A. T. S., ed. 1977a. *Oceanic Micropaleontology.* 2 vols. Academic Press, New York. 1,453 pp.

––––––– 1977b. Sedimentological clues to Palaeo-Oceanography. Pp. 1371–1453 in *Oceanic Micropaleontology,* ed. A. T. S. Ramsay, vol. 2. Academic Press, New York.

Rankama, K. 1955. Geologic evidence of chemical composition of the Precambrian atmosphere. Pp. 651–664 in *Crust of the Earth,* ed. A. Poldevaart. Geological Society of America Special Paper 62.

Rasool, S. I., and Schneider, S. H. 1971. Atmospheric carbon dioxide and aerosols: effects of large increases on global climate. *Science 173*:138–141.

––––––– 1972. Aerosol concentrations: effect on planetary temperatures. *Science 175*:95–96.

Ratner, M. I., and Walker, J. C. G. 1972. Atmospheric ozone and the history of life. *Journal of the Atmospheric Sciences 29*:803–808.

Raup, D. M. 1976a. Species diversity in the Phanerozoic: a tabulation. *Paleobiology 2*:279–288.

——— 1976b. Species diversity in the Phanerozoic: an interpretation. *Paleobiology 2*:289–297.

Redfield, A. C. 1942. The processes determining the concentrations of oxygen, phosphate, and other organic derivatives within the depths of the Atlantic Ocean. *Papers in Physical Oceanography and Meterology, Massachusetts Institute of Technology and Woods Hole Oceanographic Institution 9*:1–22.

——— 1958a. The biological control of chemical factors in the environment. *American Scientist 46*:205–221.

——— 1958b. The influence of the continental shelf on the tides of the Atlantic Coast of the United States. *Journal of Marine Research 17*:432–448.

Reid, J. L. 1965. *Intermediate Waters of the Pacific Ocean.* Johns Hopkins Oceanographic Studies, no. 2. Johns Hopkins University Press, Baltimore. 85 pp.

Reid, J. L., Brinton, E., Fleminger, A., Venrick, E. L., and McGowan, J. A. 1978. Ocean circulation and marine life. Pp. 65–130 in *Advances in Oceanography,* ed. H. Charnock and G. Deacon. Plenum Press, New York.

Reid, R. E. H. 1968. Bathymetric distribution of Calcarea and Hexactinellida in the present and the past. *Geological Magazine 105*:546–559.

Reineck, H.-E., Gutmann, W. F., and Hertweck, G. 1967. Das Schlickgebiet südlich Helgoland als Beispiel rezenter Schelfablagerungen, *Senckenbergiana Lethaea 48*:219–275.

Reineck, H.-E., and Singh, I. B. 1973. *Depositional Sedimentary Environments.* Springer-Verlag, Berlin. 439 pp.

Reineck, H.-E., and Wunderlich, F. 1968. Classification and origin of flaser and lenticular bedding. *Sedimentology 11*:99–104.

Revelle, R. 1955. On the history of the oceans. *Journal of Marine Research 14*:446–461.

Reynolds, R. C., Jr. 1963. Potassium-rubidium ratios and polymorphism in illites and microclines from the clay size fractions of proterozoic carbonate rocks. *Geochimica et Cosmochimica Acta 27*:1097–1112.

——— 1965. The concentration of boron in Precambrian seas. *Geochimica et Cosmochimica Acta 29*:1–16.

Richards, F. A. 1970. The enhanced preservation of organic matter in anoxic marine environments. Pp. 399–411 in *Organic Matter in Natural Waters,* ed. D. W. Hood. Institute of Marine Science, Occasional Publication no. 1.

——— 1971. Anoxic verus oxic environments. Pp. 201–217 in *Impingement of Man on the Oceans,* ed. D. W. Hood. John Wiley and Sons, New York.

Riech, V. and von Rad, U. 1979. Silica diagenesis in the Atlantic Ocean: diagenetic potential and transformations. *Results of Deep Drilling in Atlantic Ocean,* ed. M. Talwani and W. B. F. Ryan. *Proceedings of the Second Maurice Ewing Symposium* (1978). v. 2. American Geophysical Union. In press.

Robertson, A. H. F., and Hudson, J. D. 1974. Pelagic sediments in the Cretaceous and Tertiary history of the Troodos Massif, Cyprus. Pp. 403–436 in *Pelagic Sediments: On Land and Under the Sea,* ed. K. J. Hsü and H. C. Jenkyns. Special Publications of the International Association of Sedimentologists, no. 1.

Rogers, J. J. W. 1977. Three arguments for continual evolution of sial throughout geologic time. Pp. 27–39 in *Chemical Evolution of the Early Precambrian,* ed. C. Ponnamperuma. Academic Press, New York.

Rogers, M. A., and Koons, C. B. 1969. Organic carbon δC^{13} values from Quarternary marine sequences in the Gulf of Mexico: a reflection of paleotemperature changes. *Transactions of the Gulf Coast Association of Geological Societies 19*:531–534.

Ronov, A. B. 1968. Probable changes in the composition of sea water during the course of geological time. *Sedimentology 10*:25–43.

Ronov, A. B., Khain, V. Ye., Balukhovskiy, A. N., and Seslavinskiy, K. B. 1977. Changes in distribution, volumes, and rates of deposition of sedimentary and volcanogenic deposits during the Phanerozoic (with the present continents). *International Geology Review 19*:1297–1304.

Ronov, A. B., and Korzina, G. A. 1960. Phosphorus in sedimentary rocks. *Geochemistry,* no. 8, pp. 805–829.

Ronov, A. B., and Migdisov, A. A. 1971. Geochemical history of the crystalline basement and the sedimentary cover of the Russian and North American platforms. *Sedimentology 16*:137–185.

Roscoe, S. M. 1968. *Huronian Rocks and Uraniferous Conglomerates.* Geological Survey of Canada, Paper 68–40. 205 pp.

Rosenberg, G. D., and Runcorn, S. K., eds. 1975. *Growth Rhythms and the History of the Earth's Rotation.* Wiley-Interscience, New York. 560 pp.

Ross, C. A. 1968. Paleoecology of fusulinaceans. Pp. 301–318 in *1968 Proceedings International Páleontological Union, 23rd International Geological Congress.*

Ross, R. J., Jr. 1976. Ordovician sedimentation in the western United States. Pp. 73–105 in *The Ordovician System,* ed. M. G. Bassett. University of Wales Press, Cardiff.

Rossiter, J. R. 1954. The North Sea storm surge of 31 January and 1 February 1953. *Philosophical Transactions of the Royal Society of London, Series A 246*:371–400.

Rubey, W. W. 1951. Geologic history of sea water. *Geological Society of America Bulletin 62*:1111–1147. (Reprinted in *The Origin and Evolution of Atmospheres and Oceans,* ed. P. J. Brancazio and A. G. W. Cameron [John Wiley and Sons, New York, 1964], pp. 1–63.)

Ruddiman, W. F. 1977. Investigations of Quaternary climate based on planktonic foraminifera. Pp. 101–162 in *Oceanic Micropaleontology,* ed. A. T. S. Ramsay, vol. 1. Academic Press, New York.

Russell, K. L. 1968. Oceanic ridges and eustatic changes in sea level. *Nature 218*:861–862.

Sackett, W. M., Eckelmann, W. R., Bender, M. L., and Bé, A. W. H. 1965.

Temperature dependence of carbon isotope composition in marine plankton and sediments. *Science 148*:235–237.

Sackett, W. M., and Moore, W. S. 1966. Isotopic variations of dissolved inorganic carbon. *Chemical Geology 1*:323–328.

Sackett, W. M., Poag, C. W., and Eadie, B. J. 1974. Kerogen recycling in the Ross Sea, Antarctica. *Science 185*:1045–1047.

Sagan, C., and Mullen, G. 1972. Earth and Mars: evolution of atmospheres and surface temperatures. *Science 177*:52–56.

Sagoe, K. O., and Visher, G. S. 1977. Population breaks in grain-size distributions of sand—a theoretical model. *Journal of Sedimentary Petrology 47*:285–310.

Sandberg, P. A. 1975. New interpretations of Great Salt Lake ooids and ancient non-skeletal carbonate mineralogy. *Sedimentology 22*:497–537.

Sanders, H. L., Hessler, R. R., and Hampson, G. R. 1965. An introduction to the study of deep-sea benthic faunal assemblages along the Gay Head–Bermuda transect. *Deep-Sea Research 12*:845–867.

Sapper, K. 1927. *Vulkankunde*. Englehorn, Stuttgart. 358 pp.

Sarnthein, M., and Diester-Haass, L. 1977. Eolian-sand turbidites. *Journal of Sedimentary Petrology 47*:868–890.

Saunders, W. B., and Wehman, D. A. 1977. Shell strength of *Nautilus* as a depth limiting factor. *Paleobiology 3*:83–89.

Savin, S. M., Douglas, R. G., and Stehli, F. G. 1975. Tertiary marine paleotemperatures. *Geological Society of America Bulletin 86*:1499–1510.

Savin, S. M., and Stehli, F. G. 1974. Interpretation of oxygen isotope paleotemperature measurements: effect of the O^{18}/O^{16} ratio of sea water, depth stratification of foraminifera, and selective solution. *Colloques Internationaux du Centre National de la Recherche Scientifique 219*:183–191.

Saxena, S. K. 1977. The Charnockite geotherm. *Science 198*:614–617.

Sayles, F. L., and Manheim, F. T. 1975. Interstitial solutions and diagenesis in deeply buried marine sediments: results from the Deep Sea Drilling Project. *Geochimica et Cosmochimica Acta 39*:103–127.

Scheltema, R. S. 1971. Larval dispersal as a means of genetic exchange between geographically separated populations of shallow-water benthic marine gastropods. *Biological Bulletin 140*:284–322.

Schidlowski, M., Appel, P. W. U., Eichmann, R. and Junge, C. E. 1979. Carbon isotope geochemistry of the 3.7×10^9 yr old Isua sediments, West Greenland: implications for the Archean carbon and oxygen cycles. *Geochimica et Cosmochimica Acta 43*:1–11.

Schidlowski, M., Eichmann, R., and Junge, C. E. 1975. Precambrian sedimentary carbonates: carbon and oxygen isotope geochemistry and implications for the terrestrial oxygen budget. *Precambrian Research 2*:1–69.

Schidlowski, M., Junge, C. E., and Pietrek, H. 1977. Sulfur isotope variations in marine sulfate evaporites and the Phanerozoic oxygen budget. *Journal of Geophysical Research 82*:2557–2565.

Schiegl, W. E. 1972. Deuterium content of peat as a paleoclimatic recorder. *Science 175*:512–513.

Schilling, J.-G., Unni, C. K., and Bender, M. L. 1978. Origin of chlorine and bromine in the oceans. *Nature 273*:631–636.

Schindler, D. W. 1977. Evolution of phosphorus limitation in lakes. *Science 195*:260–262.

Schmalz, R. F. 1969. Deep-water evaporite deposition: a genetic model. *American Association of Petroleum Geologists Bulletin 53*:798–823.

Schneider, S. H. 1972a. Atmospheric particles and climate: can we evaluate the impact of man's activities? *Quaternary Research 2*:425–435.

—— 1972b. Cloudiness as a global climatic feedback mechanism: the effects on the radiation balance and surface temperature of variations in cloudiness. *Journal of the Atmospheric Sciences 29*:1413–1422.

Schnitker, D. 1969. Distribution of foraminifera on a portion of the continental shelf of the Golfe de Gascogne (Gulf of Biscay), *Bulletin Centre Recherche Pau-SNPA 3*:33–64.

—— 1974. West Atlantic abyssal circulation the past 120,000 years. *Nature 248*:385–387.

Scholle, P. A., and Arthur, M. A. 1976. Carbon-isotopic fluctuations in Upper Cretaceous sediments: an indicator of paleo-oceanic circulation. *Geological Society of America Abstracts with Programs 8*:1089.

Schopf, J. W. 1978. The evolution of the earliest cells. *Scientific American 239*:110–138.

Schopf, T. J. M. 1966. *Conodonts of the Trenton Group (Ordovician) in New York, Southern Ontario and Quebec.* New York State Museum and Science Service, Bulletin 405. 105 pp.

—— 1969. Paleoecology of ectoprocts (bryozoans). *Journal of Paleontology 43*:234–244.

—— 1970. Taxonomic diversity gradients of ectoprocts and bivalves and their geologic implications. *Geological Society of America Bulletin 81*:3765–3768.

—— 1974. Permo-Triassic extinctions: relation to sea-floor spreading. *Journal of Geology 82*:129–143.

—— 1975. Oceans of the geological past (Review of *Studies in Paleo-Oceanography*, ed. W. W. Hay, 1974). *Science 187*:1189–1190.

—— 1978. Fossilization potential of an intertidal fauna: Friday Harbor, Washington. *Paleobiology 4*:261–270.

—— 1979. The role of biogeographic provinces in regulating marine faunal diversity through geologic time. Pp. 449–457 in *Historical Biogeography, Plate Tectonics, and the Changing Environment,* ed. J. Gray and A. J. Boucot. Biology Colloquium, Oregon State University.

Schopf, T. J. M., Farmanfarmaian, A., and Gooch, J. L. 1971. Oxygen consumption rates and their paleontological significance. *Journal of Paleontology 45*:247–252.

Schopf, T. J. M., Fisher, J. B., Smith, C. A. F., III. 1978. Is the marine latitudinal diversity gradient merely another example of the species area curve? Pp. 365–386 in *Genetics, Ecology, and Evolution of Marine Organisms,* ed. J. Beardmore and B. Battaglia. Plenum Press, New York.

Schopf, T. J. M., and Gooch, J. L. 1972. A natural experiment to test the hypothesis that loss of genetic variability was responsible for mass extinctions of the fossil record. *Journal of Geology 80*:481–483.

Schreiber, B. C., and Schreiber, E. 1977. The salt that was. *Geology 5*:527–528.

Schroeder, R. A. and Bada, J. L. 1973. Glacial-postglacial temperature difference deduced from aspartic acid racemization in fossil bones. *Science 182*:479–482.

—— 1976. A review of the geochemical applications of the amino acid racemization reaction. *Earth-Science Reviews 12*:347–391.

Schubel, J. R. 1974. Effects of Tropical Storm Agnes on the suspended solids of the northern Chesapeake Bay. Pp. 113–132 in *Suspended Solids in Water,* ed. R. J. Gibbs. Plenum Press, New York. 320 pp.

Schumm, S. A. 1968. Speculations concerning paleohydrologic controls of terrestrial sedimentation. *Geological Society of America Bulletin 79*:1573–1588.

Schwab, F. L. 1976. Modern and ancient sedimentary basins: comparative accumulation rates. *Geology 4*:723–727.

—— 1978. Secular trends in the composition of sedimentary rock assemblages—Archean through Phanerozoic time. *Geology 6*:532–536.

Schwartzman, D. W. 1973. Argon degassing models of the earth. *Nature Physical Science 245*:20–21.

Sclater, J. G., Abbott, D., and Thiede, J. 1977. Paleobathymetry and sediments of the Indian Ocean. Pp. 25–59 in *Indian Ocean Geology and Biostratigraphy Studies Following Deep-Sea Legs 22 to 29,* ed. J. R. Heirtzler, H. M. Bolli, T. A. Davies, J. V. Saunders, and J. G. Sclater. American Geophysical Union, Washington, D.C.

Sclater, J. G., Hellinger, S., Tapscott, C. 1977. The paleobathymetry of the Atlantic Ocean from the Jurassic to the present. *Journal of Geology 85*:509–552.

Sellers, W. D. 1969. A global climatic model based on the energy balance of the earth-atmosphere system. *Journal of Applied Meterology 8*:392–400.

—— 1973. A new global climatic model. *Journal of Applied Meterology 12*:241–254.

Sellwood, B. W. 1972. Regional environmental changes across a Lower Jurassic stage-boundary in Britain. *Palaeontology 15*:125–157.

Sen Gupta, B. K., and Kilbourne, R. T. 1974. Diversity of benthic formaminifera on the Georgia continental shelf. *Geological Society of America Bulletin 85*:969–972.

Sepkoski, J. J., Jr. 1976. Species diversity in the Phanerozoic: species-area effects. *Paleobiology 2*:298–303.

—— 1978. A kinetic model of Phanerozoic taxonomic diversity. I. Analysis of marine orders. *Paleobiology 4*:223–251.

Shackleton, N. J. 1967. Oxygen isotope analyses and Pleistocene temperatures re-assessed. *Nature 215*:15–17.

—— 1974. Attainment of isotopic equilibrium between ocean water and the benthonic foraminifera genus *Uvigerina*: isotopic changes in the ocean during the last glacial. *Colloques Internationaux du Centre National de la Recherche Scientifique 219*:203–209.

———— 1977. Carbon-13 in *Uvigerina*: tropical rainforest history and the equatorial Pacific carbonate dissolution cycles. Pp. 401–427 in *The Fate of Fossil Fuel CO_2 in the Oceans,* ed. N. R. Andersen and A. Malahoff. Plenum Press, New York.

Shackleton, N. J., and Kennett, J. P. 1975. Paleotemperature history of the Cenozoic and the initiation of Antarctic glaciation: oxygen and carbon isotope analyses in DSDP Sites 277, 279, 281. *Initial Reports of the Deep Sea Drilling Project 29*:743–755.

Shackleton, N. J., and Opdyke, N. D. 1973. Oxygen isotope and palaeomagnetic stratigraphy of equatorial Pacific Core V28–238: oxygen isotope temperatures and ice volumes on a 10^5 year and 10^6 year scale. *Journal of Quaternary Research 3*:39–55.

Sharma, G. D., Naidu, A. S., and Hood, D. W. 1972. Bristol Bay: model contemporary graded shelf. *American Association of Petroleum Geologists Bulletin 56*:2000–2012.

Shaw, A. B. 1964. *Time in Stratigraphy*. McGraw-Hill Book Co., New York. 365 pp.

Shaw, D. M. 1976. Development of the early continental crust. Part 2, Prearchean, Protoarchean and later eras. Pp. 33–53 in *The Early History of the Earth,* ed. B. F. Windley. John Wiley and Sons, London.

Shea, J. H. 1974. Deficiencies of clastic particles of certain sizes. *Journal of Sedimentary Petrology 44*:985–1003.

Sheldon, R. P., Maughan, E. K., and Cressman, E. R. 1967. Environment of Wyoming and adjacent states: Interval B. Pp. 48–54 in *Paleotectonic Maps of the Permian System,* by E. D. McKee and S. S. Oriel et al. U.S. Geological Survey, Miscellaneous Geologic Investigations Map, I-450.

Shepard, F. P. 1963. *Submarine Geology,* 2nd ed. Harper & Row, New York. 557 pp.

Shepard, F. P., Emery, K. O., and Gould, H. R. 1949. Distribution of sediments on East Asiatic Continental Shelf. Pp. 1–64 in *Allan Hancock Foundation Publications, Occasional Paper no. 9.*

Shepard, F. P. and Marshall, N. F. 1973. Currents along floors of submarine canyons. *American Association of Petroleum Geologists Bulletin 57*:244–264.

Shepard, F. P., and Young, R. 1961. Distinguishing between beach and dune sands. *Journal of Sedimentary Petrology 31*:196–214.

Shimp, N. F., Witters, J., Potter, P. E., and Schleicher, J. A. 1969. Distinguishing marine and freshwater muds. *Journal of Geology 77*:566–580.

Shinn, E. A., Lloyd, R. M., and Ginsburg, R. N. 1969. Anatomy of a modern carbonate tidal-flat, Andros Island, Bahamas. *Journal of Sedimentary Petrology 39*:1202–1228.

Shreve, R. L. 1974. Variation of mainstream length with basin area in river networks. *Water Resources Research 10*:1167–1177.

Sibley, D. F., and Vogel, T. A. 1976. Chemical mass balance of the earth's crust: the calcium dilemma (?) and the role of pelagic sediments. *Science* *192*:551–553.

Siesser, W. G. 1974. Relict and Recent beachrock from southern Africa. *Geological Society of America Bulletin 85*:1849–1854.

Siever, R. 1968. Sedimentological consequences of a steady-state ocean-atmosphere. *Sedimentology 11*:5–29.

Sillén, L. G. 1966. Regulation of O_2, N_2, and CO_2 in the atmosphere: thoughts of a laboratory chemist. *Tellus 18*:198–206.

Silver, M. W., Shanks, A. L., and Trent, J. D. 1978. Marine snow: microplankton habitat and source of small-scale patchiness in pelagic populations. *Science 201*:371–373.

Silverman, M. P., and Munoz, E. F. 1970. Fungal attack on rock: solubilization and altered infrared spectra. *Science 169*:985–987.

Simpson, P. R., and Bowles, J. F. W. 1977. Uranium mineralization of the Witwatersrand and Dominion Reef systems. *Philosophical Transactions of the Royal Society of London, Series A 286*:527–548.

Sindowski, K. 1957. Die synoptische Methode des Kornkurven-Vergleiches zur Ausdeutung fossiler Sedimentationsräume. *Geologisches Jahrbuch herausgegeben von den Geologischen Landesanstalten der Bundesrepublik Deutschland 73*:235–278.

Singh, I. B. 1976. Depositional environment of the Upper Vindhyan sediments in the Satna-Maihar area, Madhya Pradesh, and its bearing on the evolution of Vindhyan sedimentation basin. *Journal of the Palaeontological Society of India 19*:48–70 (for 1974).

Sloss, L. L. 1969. Evaporite deposition from layered solutions. *American Association of Petroleum Geologists Bulletin 53*:776–789.

Smith, B. N. 1972. Natural abundance of the stable isotopes of carbon in biological systems. *BioScience 22*:226–231.

Smith, E. V. P., and Gottlieb, D. M. 1974. Solar flux and its variations. *Space Science Reviews 16*:771–802.

Smith, J. D., and Hopkins, T. S. 1972. Sediment transport on the continental shelf off of Washington and Oregon in light of Recent current measurements. Pp. 143–180 in *Shelf Sediment Transport: Process and Pattern,* ed. D. J. P. Swift, D. B. Duane, and O. H. Pilkey. Dowden, Hutchinson and Ross, Stroudsburg, Pa.

Soutar, A., and Crill, P. A. 1977. Sedimentation and climatic patterns in the Santa Barbara Basin during the nineteenth and twentieth centuries. *Geological Society of America Bulletin 88*:1161–1172.

Spencer, D. W. 1963. The interpretation of grain size distribution curves of clastic sediments. *Journal of Sedimentary Petrology 33*:180–190.

——— 1966. Factors affecting element distributions in a Silurian graptolite band. *Chemical Geology 1*:221–249.

Spencer, D. W., Degens, E. T., and Kulbicki, G. 1968. Factors affecting element distribution in sediments. Pp. 981–998 in *Origin and Distribution of the Elements,* ed. L. H. Ahrens. Pergamon Press, Oxford.

Srinivasan, M. S., and Kennett, J. P. 1974. Secondary calcification of the planktonic foraminifer *Neogloboquadrina pachyderma* as a climatic index. *Science 186*:630–632.

Stahl, W., and Jordan, R. 1969. General considerations on isotopic paleotemperature determinations and analyses on Jurassic ammonites. *Earth and Planetary Science Letters 6*:173–178.

Stehli, F. G. 1966. Some applications of foraminiferal ecology. Pp. 223–240 in *Proceedings of the 2nd West African Micropaleontological Colloquim (Ibadan, 1965)*, ed. J. E. van Hinte. E. H. Brill, Leiden.

Stehli, F. G., Douglas, R. G. Kafescioglu, I. A. 1972. Models for the evolution of planktonic foraminifera. Pp. 116–129 in *Models in Paleobiology*, ed. T. J. M. Schopf. Freeman, Cooper & Co. San Francisco.

Stehli, F. G., Douglas, R. G., and Newell, N. D. 1969. Generation and maintenance of gradients in taxonomic diversity. *Science 164*:947–949.

Stehli, F. G., and Wells, J. W. 1971. Diversity and age patterns in hermatypic corals. *Systematic Zoology 20*:115–126.

Steiner, J. 1977. An expanding earth on the basis of sea-floor spreading. *Geology 5*:313–318.

Stevens, C. H. 1969. Water depth control of fusulinid distribution. *Lethaia 2*:121–132.

——— 1977. Was development of brackish oceans a factor in Permian extinctions? *Geological Society of America Bulletin 88*:133–138.

Stevenson, F. J. 1959. On the presence of fixed ammonium in rocks. *Science 130*:221–222.

Stewart, A. D. 1978. Limits to palaeogravity since the late Precambrian. *Nature 271*:153–155.

Stewart, W. D. P. 1971. Nitrogen fixation in the sea. Pp. 537–564 in *Fertility of the Sea*, ed. J. D. Costlow, Jr., vol. 2. Gordon and Breach Scientific Publishers, New York.

Stidd, C. K. 1975. Irrigation increases rainfall? *Science 188*:279–280.

Strauch, F. 1971. Some remarks on *Hiatella* as an indicator of sea temperatures. *Palaeogeography, Palaeoclimatology, and Palaeoecology 9*:59–64.

Streeter, S. S. 1973. Bottom water and benthonic foraminifera in the North Atlantic—Glacial-interglacial contrasts. *Quaternary Research 3*:131–141.

Strong, D. F., and Stevens, R. K. 1974. Possible thermal explanation of contrasting Archean and Proterozoic geological regimes. *Nature 249*:545–546.

Strong, D. R., Jr., and Levin, D. A. 1975. Species richness of the parasitic fungi of British trees. *Proceedings of the National Academy of Sciences 72*:2116–2119.

Stuart, W. D. 1973. Wind-driven model for evaporite deposition in a layered sea. *Geological Society of America Bulletin 84*:2691–2704.

Suess, E. 1976. Nutrients near the depositional interface. Pp. 57–79 in *The Benthic Boundary Layer*, ed. I. N. McCave. Plenum Press, New York.

Summerhayes, C. P., Birch, G. F., Rogers, J., and Dingle, R. V. 1973. Phosphate in sediments off south-western Africa. *Nature 243*:509–511.

Summerhayes, C. P., de Melo, U., and Barretto, H. T. 1976. The influence of upwelling on suspended matter and shelf sediments off southeastern Brazil. *Journal of Sedimentary Petrology 46*:819–828.

Sutton, R. G., and Ramsayer, G. R. 1975. Association of lithologies and sedimentary structures in marine deltaic paleoenvironments. *Journal of Sedimentary Petrology 45*:799–807.

Sverdrup, H. U., Johnson, M. W., and Fleming, R. H. 1942. *The Oceans.* Prentice-Hall, Englewood Cliffs, N.J. 1060 pp.

Swann, D. H., Lineback, J. A., and Frund, E. 1965. *The Borden Silstone (Mississippian) Delta in Southwestern Illinois.* Illinois State Geological Survey, Circular 386. 20 pp.

Swinchatt, J. P. 1969. Algal boring: a possible depth indicator in carbonate rocks and sediments. *Geological Society of America Bulletin 80*:1391–1396.

Symons, D. T. A. 1975. Huronian glaciation and polar wander from Gowganda Formation, Ontario. *Geology 3*:303–306.

Tan, F. C., and Hudson, J. D. 1971. Isotopic composition of carbonates in a marginal marine formation. *Nature Physical Sciences 232*:87–88.

Tanner, W. F. 1959. Near-shore studies in sedimentology and morphology along the Florida Panhandle coast. *Journal of Sedimentary Petrology 29*:564–574.

——— 1960. Florida coastal classification. *Transactions: of the Gulf Coast Association of Geological Societies 10*:259–266.

——— 1961. Offshore shoals in area of energy deficit. *Journal of Sedimentary Petrology 31*:87–95.

——— 1971. Numerical estimates of ancient waves, water depth and fetch. *Sedimentology 16*:71–88.

Tappan, H., and Loeblich, A. R., Jr. 1973. Evolution of the oceanic plankton. *Earth-Science Reviews 9*:207–240.

Tarney, J., and Windley, B. F. 1977. Chemistry, thermal gradients, and evolution of the lower continental crust. *Journal of the Geological Society of London 134*:153–172.

Taylor, H. P., Jr. 1977. Water/rock interactions and the origin of H_2O in grantitic batholiths. *Journal of the Geological Society 133*:509–558.

Thiede, J., and Van Andel, T. H. 1977. The paleoenvironment of anaerobic sediments in the Late Mesozoic South Atlantic Ocean. *Earth and Planetary Science Letters 33*:301–309.

Thierstein, H. R., and Berger, W. H. 1978. Injection events in ocean history. *Nature 276*:461–466.

Thomas, K. 1978. The rise of the fork (review of *The Civilizing Process: The History of Manners,* by Norbert Elias, 1978; and of *Human Figurations: Essays for Norbert Elias,* ed. P. R. Gleichmann, J. Goudsblom, and H. Korte, 1977). *The New York Review of Books,* March 9, pp. 28–31.

Thompson, G. 1968. Analyses of B, Ga, Rb, and K in two deep-sea sediment cores; consideration of their use as paleoenvironmental indicators. *Marine Geology 6*:463–477.

Thompson, G., and Bowen, V. T. 1969. Analyses of coccolith ooze from the deep tropical Atlantic. *Journal of Marine Research 27*:32–38.

Thompson, G., and Melson, W. G. 1970. Boron contents of serpentinites and metabasalts in the oceanic crust: implications for the boron cycle in the oceans. *Earth and Planetary Science Letters 8*:61–65.

Till, R. 1970. The relationship between environment and sediment composition (geochemistry and petrology) in the Bimini Lagoon, Bahamas. *Journal of Sedimentary Petrology 40*:367–385.

Todd, T. W. 1968. Paleoclimatology and the relative stability of feldspar minerals under atmospheric conditions. *Journal of Sedimentary Petrology 38*:832–844.

Tont, S. A. 1976. Short-period climatic fluctuations: effects on diatom biomass. *Science 194*:942–944.

Tooms, J. S., Summerhayes, C. P., and Cronan, D. S. 1969. Geochemistry of marine phosphate and manganese deposits. *Oceanography and Marine Biology Annual Review 7*:49–100.

Tourtelot, H. A. 1964. Minor-element composition and organic carbon content of marine and non-marine shales of Late Cretaceous age in the western interior of the United States. *Geochimica et Cosmochimica Acta 28*:1579–1604.

Towe, K. M. 1978. Early Precambrian oxygen: a case against photosynthesis. *Nature 274*:657–661.

Tschudy, R. N., and Scott, R. A., eds. 1969. *Aspects of Palynology: An Introduction to Plant Microfossils in Time.* Wiley-Interscience, New York. 510 pp.

Turcotte, D. L., and Burke, K. 1978. Global sea-level changes and the thermal structure of the earth. *Earth and Planetary Science Letters 41*:341–346.

Uchupi, E. 1967. Slumping on the continental margin southeast of Long Island, New York. *Deep-Sea Research 14*:635–639.

Ulrich, R. K. 1975. Solar neutrinos and variations in the solar luminosity. *Science 190*:619–624.

Urey, H. C. 1947. The thermodynamic properties of isotopic substances. *Journal of the Chemical Society.* pp. 562–581.

Vail, P. R., Mitchum, R. M., Jr., and Thompson, S., III. 1978. Seismic stratigraphy and global changes in sea level, part 4: global cycles of relative changes in sea level. Pp. 83–97 in *Seismic Stratigraphy: Applications to Hydrocarbon Exploration* ed. C. E. Payton. American Association of Petroleum Geologists, Memoir 26.

Valentine, J. W. 1972. Conceptual models of ecosystem evolution. Pp. 192–215 in *Models in Paleobiology,* ed. T. J. M. Schopf. Freeman, Cooper & Co., San Francisco.

——— 1973. *Evolutionary Paleoecology of the Marine Biosphere.* Prentice-Hall, Englewood Cliffs, N.J. 511 pp.

Valentine, J. W., Foin, T. C., and Peart, D. 1978. A provincial model of Phanerozoic marine diversity. *Paleobiology 4*:55–66.

Valentine, J. W., and Moores, E. M. 1972. Global tectonics and the fossil record. *Journal of Geology 80*:167–184.

Van Andel, T. H., Thiede, J., Sclater, J. G., and Hay, W. W. 1977. Depositional history of the South Atlantic Ocean during the last 125 million years. *Journal of Geology 85*:651–698.

van Donk, J. 1976. O¹⁸ Record of the Atlantic Ocean for the entire Pleistocene Epoch. Pp. 147–163 in *Investigation of Late Quaternary Paleoceanography and Paleoclimatology,* ed. R. M. Cline and J. D. Hays. Geological Society of America Memoir 145.

van Straaten, L. M. J. U. 1954. Sedimentology of Recent tidal flat deposits and the Psammites du Condroz (Devonian). *Geologie en Mijnbouw 16*:25–47.

Van Valen, L. 1969. Climate and evolutionary rate. *Science 166*:1656–1658.

Veevers, J. J. 1977. Paleobathymetry of the crest of spreading ridges related to the age of ocean basins. *Earth and Planetary Science Letters 34*:100–106.

Veizer, J. 1971. Do palaeogeographic data support the expanding earth hypothesis? *Nature 229*:480–481.

—— 1973. Sedimentation in geologic history: recycling vs. evolution or recycling with evolution. *Contributions to Mineralogy and Petrology 38*:261–278.

—— 1974. Chemical diagenesis of belemnite shells and possible consequences for paleotemperature determinations. *Neues Jahrbuch für Geologie und Paläontologie, Abhandlungen 147*:91–111.

—— 1976a. Evolution of ores of sedimentary affiliation through geologic history; relations to the general tendencies in evolution of the crust, hydrosphere, atmosphere, and biosphere. Pp. 1–41 in *Handbook of Stratabound and Stratiform Ore Deposits III,* ed. K. H. Wolf. Elsevier, Amsterdam.

—— 1976b. ⁸⁷Sr/⁸⁶Sr evolution of seawater during geologic history and its significance as an index of crustal evolution. Pp. 569–578 in *The Early History of the Earth,* ed. B. F. Windley. John Wiley and Sons, New York.

—— 1977. Diagenesis of pre-Quaternary carbonates as indicated by tracer studies. *Journal of Sedimentary Petrology 47*:565–581.

—— 1978. Secular variations in the composition of sedimentary carbonate rocks, II. Fe, Mn, Ca, Mg, Si, and minor constituents. *Precambrian Research 6*:381–413.

Veizer, J., and Compston, W. 1974. ⁸⁷Sr/⁸⁶Sr composition of seawater during the Phanerozoic. *Geochimica et Cosmochimica Acta 38*:1461–1484.

—— 1976. ⁸⁷Sr/⁸⁶Sr in Precambrian carbonates as an index of crustal evolution. *Geochimica et Cosmochimica Acta 40*:905–914.

Veizer, J., and Demovič, R. 1974. Strontium as a tool in facies analysis. *Journal of Sedimentary Petrology 44*:93–115.

Veizer, J., and Garrett, D. E. 1978. Secular variations in the composition of sedimentary carbonate rocks. I. Alkali metals. *Precambrian Research 6*:367–380.

Veizer, J., and Hoefs, J. 1976. The nature of ¹⁸O/¹⁶O and ¹³C/¹²C secular trends in sedimentary carbonate rocks. *Geochimica et Cosmochimica Acta 40*:1387–1395.

Veizer, J., Lemieux, J., Jones, B., Gibling, M. R., and Savelle, J. 1977. Sodium: paleosalinity indicator in ancient carbonate rocks. *Geology* 5:177–179.

―――― 1978. Paleosalinity and dolomitization of a Lower Paleozoic carbonate sequence, Somerset and Prince of Wales Islands, Arctic Canada. *Canadian Journal of Earth Sciences* 15:1448–1461.

Vinogradov, A. P., Ronov, A. B., and Ratynskii, V. M. 1957. Variation in the chemical composition of carbonate rocks of Russian platform. *Geochimica et Cosmochimica Acta* 12:273–276.

Visher, G. S. 1961. A simplified method for determining relative energies of deposition of stratigraphic units. *Journal of Sedimentary Petrology* 31:291–293.

―――― 1969. Grain size distributions and depositional processes. *Journal of Sedimentary Petrology 39*:1074–1106.

Vogel, K. 1975. Endosymbiotic algae in rudists. *Palaeogeography, Palaeoclimatology, Palaeoecology 17*:327–332.

Vogt, P. 1974. Volcano height and plate thickness. *Earth and Planetary Science Letters 23*:337–348.

von Arx, W. S. 1957. An experimental approach to problems in physical oceanography. *Physics and Chemistry of the Earth 2*:1–29.

von Brunn, V., and Hobday, D. K. 1976. Early Precambrian tidal sedimentation in the Pongola supergroup of South Africa. *Journal of Sedimentary Petrology 46*:670–679.

Vonder Haar, T. H., and Suomi, V. E. 1969. Satellite observations of the earth's radiation budget. *Science 163*:667–669.

Vos, R. G. 1975. An alluvial plain and lacustrine model for the Precambrian Witwatersrand deposits of South Africa. *Journal of Sedimentary Petrology 45*:480–493.

Wakita, H., Fujii, N., Matsuo, S., Notsu, K., Nagao, K., and Takaoka, N. 1978. "Helium spots": caused by a diapiric magma from the upper mantle. *Science 200*:430–432.

Walker, C. T. 1963. Size fractionation applied to geochemical studies of boron in sedimentary rocks. *Journal of Sedimentary Petrology 33*:694–702.

―――― 1968. Evaluation of boron as a paleosalinity indicator and its application to offshore prospects. *American Association of Petroleum Geologists Bulletin 52*:751–766.

Walker, C. T., and Price, N. B. 1963. Departure curves for computing paleosalinity from boron in illites and shales. *American Association of Petroleum Geologists Bulletin 47*:833–841.

Walker, J. C. G. 1974. Stability of atmospheric oxygen. *American Journal of Science 274*:193–214.

―――― 1977. *Evolution of the Atmosphere*. Macmillan Publishing Co., New York. 318 pp.

―――― 1978. Oxygen and hydrogen in the primitive atmosphere. *Pure and Applied Geophysics 116*:222–231.

Walker, R. G., and Harms, J. C. 1975. Shorelines of weak tidal activity: Upper Devonian Catskill Formation, central Pennsylvania. Pp. 103–108 in *Tidal Deposits,* ed. R. N. Ginsburg. Springer-Verlag, New York.

Walker, R. N., Muir, M. D., Diver, W. L., Williams, N., and Wilkins, N. 1977. Evidence of major sulphate evaporite deposits in the Proterozoic McArthur Group, Northern Territory, Australia. *Nature 265*:526–529.

Walter, M. R. 1970. Stromatolites used to determine the time of nearest approach of earth and moon. *Science 170*:1331–1332.

Wang, W. C., Yung, Y. L., Lacis, A. A., Mo, T., and Hansen, J. E. 1976. Greenhouse effects due to man-made perturbations of trace gases. *Science 194*:685–690.

Wanless, H. R., Ziebell, W. G., Ziemba, E. A., and Carozzi, A. 1957. Limestone texture as a key to interpreting depth of deposition. Pp. 65–82 in *International Geological Congress, 20th Session, Section V, (Mexico, 1956).*

Waskom, J. D. 1958. Roundness as an indicator of environment along the coast of panhandle Florida. *Journal of Sedimentary Petrology 28*:351–360.

Wasserburg, G. J. 1964. Comments on the outgassing of the earth. Pp. 83–85 in *The Origin and Evolution of Atmospheres and Oceans,* ed. P. J. Brancazio and A. G. W. Cameron. John Wiley and Sons, New York.

Waterhouse, J. B., and Bonham-Carter, G. 1975. Global distribution and character of Permian biomes based on brachiopod assemblages. *Canadian Journal of Earth Science 12*:1085–1146.

Watkins, N. D., Keany, J., Leidbetter, M. T., and Huang, T. C. 1974. Antarctic glacial history from analyses of ice-rafted deposits in marine sediments: new model and initial tests. *Science 186*:533–536.

Watkins, N. D., and Kennett, J. P. 1972. Regional sedimentary disconformities and Upper Cenozoic changes in bottom water velocities between Australasia and Antarctica. Pp. 273–293 in *Antarctic Oceanology, II: The Australian–New Zealand Sector,* ed. D. E. Hayes. American Geophysical Union, Antarctic Research Series, vol. 19.

Weaver, F. M., and Wise, S. W., Jr. 1974. Opaline sediments of the southeastern coastal plain and horizon A: biogenic origin. *Science 184*:899–901.

Weber, J. N. 1964. Chloride ion concentration in liquid inclusions of carbonate rocks as a possible environmental indicator. *Journal of Sedimentary Petrology 34*:677–680.

——— 1973a. Deep-sea ahermatypic scleractinian corals: isotopic composition of the skeleton. *Deep-Sea Research 20*:901–909.

——— 1973b. Temperature dependence of magnesium in echinoid and asteroid skeletal calcite: a reinterpretation of its significance. *Journal of Geology 81*:543–556.

Weber, J. N., and Woodhead, P. M. J. 1971. Diurnal variations in the isotopic composition of dissolved inorganic carbon in seawater from coral reef environments. *Geochimica et Cosmochimica Acta 35*:891–902.

Wedepohl, K. H. 1959. The contribution of minor element data of clays from the Atlantic Ocean to the geochemistry of pelagic sediments. Pp. 987–989 in *International Oceanographic Congress, Preprints,* ed. M. Sears. American Association for the Advancement of Science, Washington, D.C.

―――― 1971. Environmental influences on the chemical composition of shales and clays. *Physics and Chemistry of the Earth 8*:305–333.

Weeks, W. B. 1938. South Arkansas stratigraphy with emphasis on the older coastal plain beds. *American Association of Petroleum Geologists Bulletin 22*:953–983.

Wegener, A. 1929. *The Origin of Continents and Oceans.* 246 pp. (Dover Publications reprint of 1966 of the translation of the 4th rev. ed. of *Die Entstehung der Kontinente und Ozeane.*)

Wehmiller, J. F., Hare, P. E., and Kujala, G. A. 1976. Amino acids in fossil corals: racemization (epimerization) reactions and their implications for diagenetic models and geochronological studies. *Geochimica et Cosmochimica Acta 40*:763–776.

Wells, J. W. 1967. Corals as bathometers. *Marine Geology 5*:349–365.

Wells, P. R. A. 1976. Late Archean metamorphism in the Buksefjorden region, southwest Greenland. *Contributions to Mineralogy and Petrology 56*:229–242.

Westermann, G. E. G. 1973. Strength of concave septa and depth limits of fossil cephalopods. *Lethaia 6*:383–403.

Wetherald, R. T., and Manabe, S. 1975. The effects of changing the solar constant on the climate of a general circulation model. *Journal of Atmospheric Sciences 32*:2044–2059.

Weyl, P. K. 1968. The role of the oceans in climatic change: a theory of the ice ages. *Meteorological Monographs 8*:37–62.

―――― 1972. The salinity of the North Atlantic Ocean and the next glaciation. *Quaternary Research 2*:399–400.

Whelan, J. F., and Rye, R. O. 1977. Sulfur isotopic variations of Precambrian anhydrite units from the Grenville Series, New York. *Geological Society of America Abstracts with Programs 9*:1222–1223.

Wickstrom, C. E., and Castenholz, R. W. 1973. Thermophilic ostracod: aquatic metazoan with the highest known temperature tolerance. *Science 181*:1063–1064.

Wiebe, P. H., Hulburt, E. M., Carpenter, E. J., Jahn, A. E., Knapp, G. P., III, Boyd, S. H., Ortner, P. B., and Cox, J. L. 1976. Gulf Stream cold core rings: large-scale interaction sites for open ocean plankton communities. *Deep-Sea Research 23*:695–710.

Wigley, R. L., and Stinton, F. C. 1973. Distribution of macroscopic remains of recent animals off Massachusetts. *Fishery Bulletin 71*:1–40.

Williams, D. F., Thunell, R. C., and Kennett, J. P. 1978. Periodic freshwater flooding and stagnation of the eastern Mediterranean Sea during the late Quaternary. *Science 201*:252–254.

Wilson, J. L. 1969. Microfacies and sedimentary structures in "deeper water" lime mudstones. Pp. 4–16 in *Depositional Environments in Carbonate Rocks,* ed. G. M. Friedman. Society of Economic Paleontologists and Mineralogists Special Publication no. 14.

Wilson, L. 1969. Les relations entre les processus géomorphologiques et le climat moderne comme méthode de paléoclimatologie. *Revue de Géographie Physique et de Géologie Dynamique 2*:303–314.

——— 1973. Variations in mean annual sediment yield as a function of mean annual precipitation. *American Journal of Science 273*:335–349.

Wilson, T. R. S. 1975. Salinity and the major elements of sea water. Pp. 365–413 in *Chemical Oceanography,* ed. J. P. Riley and G. Skirrow, 2nd ed. vol. 1. Academic Press, London.

——— 1978. Evidence for denitrification in aerobic pelagic sediments. *Nature 274*:354–356.

Windley, B. F., and Davies, F. B. 1978. Volcano spacings and lithospheric crustal thickness in the Archaean. *Earth and Planetary Science Letters 38*:291–297.

Windom, H. L. 1975. Eolian contributions to marine sediments. *Journal of Sedimentary Petrology 45*:520–529.

Wise, S. W., Jr. and Kelts, K. R. 1972. Inferred diagenetic history of a weekly silicified deep sea chalk. *Transactions of the Gulf Coast Association of Geological Societies 22*:177–203.

Wofsy, S. C., McConnell, J. C., and McElroy, M. B. 1972. Atmospheric CH_4, CO, and CO_2. *Journal of Geophysical Research 77*:4477–4493.

Wofsy, S. C., and McElroy, M. B. 1973. On vertical mixing in the upper stratosphere and lower mesosphere. *Journal of Geophysical Research 78*:2619–2624.

Wolman, M. G., and Miller, J. P. 1960. Magnitude and frequency of forces in geomorphic processes. *Journal of Geology 68*:54–74.

Wooster, W. S., Lee, A. J., and Dietrich, G. 1969. Redefinition of salinity. *Limnology and Oceanography 13*:437–438.

Worsley, T. R., and Davies, T. A. 1978. Eustatic sea level changes and Pacific Ocean sedimentation rates. *Geological Society of America Abstracts with Programs 10*:519.

Worthington, L. V. 1959. The 18° water in the Sargasso Sea. *Deep-Sea Research 5*:297–305.

——— 1968. Genesis and evolution of water masses. *Meteorological Monographs 8*:63–67.

——— 1976. *On the North Atlantic Circulation.* John Hopkins Oceanographic Studies, no. 6. John Hopkins University Press, Baltimore. 110 pp.

——— 1977. The case for near-zero production of Antarctic bottom water. *Geochimica et Cosmochimica Acta 41*:1001–1006.

Wright, A. E., and Moseley, F., eds. 1975. *Ice Ages: Ancient and Modern.* Seel House Press, Liverpool. 320 pp.

Wright, R. P. 1974. Storm-generated coquinoid sandstone: genesis of high-energy marine sediments from the Upper Jurassic of Wyoming and Montana: discussion. *Geological Society of America Bulletin 85*:837.

Wrucke, C. T., Churkin, M., Jr., and Heropoulos, C. 1978. Deep-sea origin of Ordovician pillow basalt and associated sedimentary rocks, northern Nevada. *Geological Society of America Bulletin 89*:1272–1280.

Wüst, G., Borgmus, W., and Noodt, E. 1954. Die zonale Verteilung von Salzehalt, Niederschlag, Verdunstung, Temperatur und Dichte an der Oberfläche der Ozeane. *Kieler Meeresforschungen 10*:137–161.

Wyllie, P. J. 1971. *The Dynamic Earth*. John Wiley and Sons, New York. 416 pp.

Young, S. W., Basu, A., Suttner, L. J., Mack, G. H., and Darnell, N. A. 1975. Use of size composition trends in Holocene soil and fluvial sand for paleoclimate interpretation. Pp. 201–209 in *Proceedings IXth International Congress of Sedimentology Theme 1* (Nice 1975).

Ziegler, A. M., Hansen, K. S., Johnson, M. E., Kelly, M. A., Scotese, C. R., and Van der Voo, R. 1977. Silurian continental distributions, paleogeography, climatology, and biogeography. *Tectonophysics 40*:13–51.

Ziegler, A. M., Scotese, C. R., McKerrow, W. S., Johnson, M. E., and Bambach, R. K. 1977. Paleozoic biogeography of continents bordering the Iapteus (Pre-Caledonian) and Rheic (Pre-Hercynian) oceans. Pp. 1–22 in *Paleontology and Plate Tectonics with Special Reference to the History of the Atlantic Ocean,* ed. R. M. West. Milwaukee Public Museum, Special Publications in Biology and Geology, no. 2.

INDEX OF NAMES

INDEX OF SUBJECTS